Patrick Moore's Practical Astronomy Series

Other Titles in this Series

(Continued after Index)

The Night Sky Companion

Tammy Plotner

Consulting Editor–Ken Vogt

With a Foreword by

Terry Mann

 Springer

Tammy Plotner
Warren Rupp Observatory, USA
8215 Center Street
Caledonia, Ohio 43314
theastronomer2@gmail.com

Library of Congress Control Number: 2007933079

Patrick Moore's Practical Astronomy Series ISSN 1617-7185

ISBN-13: 978-0-387-71608-4   e-ISBN-13: 978-0-387-71609-1

Printed on acid-free paper.

9 8 7 6 5 4 3 2 1

springer.com

"The best thing that we're put here for's to see;
The strongest thing that's given us to see with's
A telescope"

...

I recollect a night of broken clouds
And underfoot snow melted down to ice,
And melting further in the wind to mud.
Bradford and I had out the telescope.
We spread our two legs as we spread its three,
Pointed our thoughts the way we pointed it,
And standing at our leisure till the day broke,
Said some of the best things we ever said.
That telescope was christened the Star-Splitter

—Robert Frost

Star-Splitter (Credit—Joe Orman).

Contents

Contents

Foreword

The book you are holding in your hands is one of the best learning and outreach tools you could want for your collection. *The Night Sky Companion* is full of information about astronomy. Think of how many times you have walked outside, looked at the stars and wondered what new things you could explore. Maybe you are a beginner and you want to learn the basics of stargazing, or perhaps you're a seasoned sky veteran. Either way, I bet you will see and learn something new each night.

I have used Tammy Plotner's previous books to help plan my outreach programs. Not only do they tell you what you can see in the night sky, just as this book does, but they also explain some of the history and science which has brought us to this point. You will quickly realize astronomy is what you make it. It can be extremely challenging or very relaxing: it's up to you to decide.

With this book, you can use telescope or binoculars of any size—or just kick back under the Milky Way and enjoy the view. It will be your "companion" to help you discover our galaxy. Picture yourself enjoying a warm August night watching the Perseid meteor shower, or seeing the Harvest Moon rise. While the spring holds the promise of bright planets and distant galaxies, there is something to be said about those long, cold winter evenings too! You can spend hours observing and still get to bed before midnight. With each changing season, with every constellation wheeling overhead, Tammy will show you the best of the night sky.

The wealth of information here about the Moon will amaze you. Wait until you point a telescope its way and find a specific crater. Suddenly, the lunar terrain takes on a new look. It has a stark beauty which will keep you coming back.

Ohio Aurora in May (Credit—Terry Mann).

Both beginning and advanced amateur astronomers will enjoy this book. There is something for every level of knowledge and any type of astronomical interest—for each night of the coming year. The vivid descriptions and beautiful images presented here will help you *see* objects more clearly, and Tammy's insights into what you're looking at will guide you through your observations. Her intimate knowledge of the sky shines throughout her book.

Use this book as a learning tool for yourself and your family. It is a great primer for science students.

Don't let another night fade away, let's go out and see what it has to offer!

Terry Mann

Terry L. Mann
President, Astronomical League

Acknowledgments

When one has weighed the Sun in balance, and measured the steps of the Moon, and mapped out the
seven heavens, there still remains oneself. Who can calculate the orbit of his own soul?
...
We are all on our backs in the gutter, only some of us are looking at the stars.

—Oscar Wilde

Throughout the years, it has been my great pleasure and privilege to know many professional and amateur astronomers. I have been inspired by countless books, magazines, and programs, and have spent years of starry nights at the eyepiece in study. It is a combination of all of these things that makes this book what it is.

And makes me who I am.

For all of you out there? I thank you. Your support and kind words have made a dream come true. Now let's thank the people who made it come about...

Ricardo Borba

Ricardo Borba is an amateur astronomer living in Ottawa, Ontario, and a member of the Royal Astronomical Society of Canada. Between observing sessions he is an application software engineer at Natural Convergence and his photographic skills are becoming highly acclaimed as they are featured in various publications. **www.borba.com**

NGC 1579—(Credit—R. Jay GaBany).

Alan Chu

As a resident of Hong Kong, Alan might joke that the Moon is all he has to observe due to light pollution, but Alan's work is no laughing matter. He has produced one of the finest amateur Photographic Atlases of the Moon available and it is a great honor to include his fine photography in these pages. **www.alanchuhk.com**

John Chumack

With over 500 of his images published in such notable periodicals as *Science, National Geographic, Discover, Time, and Newsweek,* John has taken astrophotography to the realm of an art form. While you will find his work in Fine Art exhibitions, you'll also find John busy as a serious astronomer—contributing to the Minor Planet Center and studying lunar transient phenomena. With Chumack Observatories located in both Dayton and Yellow Springs, John is an active member of the Miami Valley Astronomical Society and helps to fund his research through image sales. **www.galacticimages.com**

R. Jay GaBany

Born at the dawn of the space age, R. Jay GaBany has grown up and matured during a time when mankind's fascination with the great mysteries beyond our home planet has surged. His interest in astronomy started at an early age, sparked by the Apollo Moon Landing program. When Neil Armstrong and Buzz Aldrin were bouncing on the lunar surface, Jay was in his backyard observing the Moon through his first small refractor. But it was Carl Sagan's vision that ignited his adult enthusiasm for astronomy when Cosmos debuted and shortly thereafter he acquired his first 8" Schmidt-Cassegrain telescope.

Many other telescopes followed, as did two years learning how to image with a 35 millimeter camera in time for the passing of Halley's Comet in 1986. Family, kids, career, and expenses, however, turned him into a spectator as amateur astronomy converted from film to CCD imagery during the 1990s. After moving from Connecticut to San Jose, California, Jay began designing Web-based travel reservation systems during the day but at night began taking deep-space pictures, inspired by the work of Robert Gendler. Learning to produce images of the night sky with a CCD camera proved to be the most challenging, rewarding, and addictive activity he had ever undertaken. Today, images are taken both from his light-polluted backyard using a portable 12" telescope and remotely, using Internet control, with a 20" reflector from a dark location in the south-central mountains of New Mexico. **www.cosmotography.com**

Wes Higgins

Wes' interest in space and astronomy started when he was in the second grade and watching the first US manned space shot on television. He continued to follow with great interest the NASA space program all the way thorough the Apollo Moon landings. While growing up, he yearned for a telescope, but the thought lay dormant through college, marriage, and the start of his own business. Eleven years ago his dream came true and as he says, "I am sure that for the rest of my life I will be out observing and imaging every chance I get." **http://higginsandsons.com/astro**

International Occultation Timing Association (IOTA)

Many thanks go to the International Occultation Timing Association for their OCCULT Ephemeris 3.6.0 freeware which provided a look into the year's solar system events. **http://www.lunar-occultations.com/iota/**

Greg Konkel

Greg has many interests and two which occupy a great deal of his time are astronomy and photography. Having recently made the transition from film to digital cameras, he's enthused about the potential of this new technology and has focused his attention lately on integrating these two interests. The purpose of his website is twofold: to make a contribution regarding the technical issues surrounding digital astrophotography and to share some of the best images he's acquired—both astronomical and general photographic. **www.nwgis.com/greg**

Terry Mann

Terry Mann is President of the Astronomical League, a JPL Solar System Ambassador, and has served as President and Vice-President of the Miami Valley Astronomical Society. She has received the R. G. Wright Award and the Kepler Award, as well as an award from the Ohio House of Representatives for her dedicated research on the solar system. She has written articles for the Astronomical League's newsletter, the *Reflector*, and for local newspapers. Her astrophotography has appeared in local art galleries, newspapers, and TV newscasts. Her service and dedication to astronomy outreach is beyond comparison.

MMS/AATU

From the "land down under" comes the Melbourne Meade Scopes/Advanced Australian Telescope Users! My many thanks go to Stephen Boyd, Greg Bradley, Stuart Thompson, Mike Sidonio, and Peter 'Aussie Pete' Ward for sharing some of their incredible images of the Southern Hemisphere skies.

National Aeronautics and Space Administration (NASA)

NASA explores. NASA discovers. NASA seeks to understand. But most of all NASA shares. I would like to thank those good folks for providing all the wonderful resources available to amateur astronomers, and I express my personal thanks for the use of many archival photographs, illustrations, and other materials seen here. www.nasa.gov

Joe Orman

As a hiker, biker, climber, caver, rockhound, and ghost-towner, Joe has been exploring and photographing the special places of the southwest for more than 30 years. His sky shots are truly where photography and astronomy meet-in the middle ground between art and science. His list of credits is amazing, and the name Joe Orman has accompanied photographs in literally dozens of prestigious publications...and it is with great pride that his work is included here. http://joeorman.shutterace.com/Gallery.html

Palomar Observatory/Caltech

I cannot adequately express my gratitude toward Palomar Observatory and Caltech for the use of the POSS II Sky Survey images that you see in hundreds of places throughout this book. Although amateur photography has come a long way over the years, no one can surpass this huge database of images. I would also like to thank Linda Bustos for helping me obtain permission to use and present them to you. May their use as illustrations inspire you, as much as their research has inspired me. My many, many thanks... www.astro.caltech.edu/palomar

United States Naval Observatory (USNO)

Some of the solar system information found here is credited to the United States Naval Observatory's Multiyear Interactive Computer Almanac (MICA 3.0). Many thanks for their kind permission to use this software for this book. www.usno.navy.mil

Ken Vogt

There is no one to whom I owe a deeper debt of gratitude than to my consulting editor, Ken. By conventional standards, he swears his life has not been that successful—but he's far more talented than he will speak of. Living modestly in southern Indiana, he was able to retire from various menial employments in 1991 at age 45—but he's far from "retired."

Since that time, Ken pursued his love of music (playing keyboards) and fascination (and frustration) with computers. Although the bright sky in his home town has prevented any serious sky watching, he is a devotee of astronomy. Ken is also a passionate advocate of Distributed Computing, helping out with the creation of the BAUT BOINC team for projects like Einstein@Home. As he says, "At this stage of life, I'm very happy to help out around the Internet in any way I can."

Ken's "help" has been critical to the publication of this book. He has graciously spent uncounted hours reading and re-reading text and offering advice. His encouragement through some rough times has been instrumental in its completion, and his support has kept me on track. He is truly one of the brightest "stars" in my sky.

Roger Warner

Roger Warner lives in the UK in a town called Basildon, located in the county of Essex. Being both a father and a grandfather, his interest in astronomy began 7 years ago, but the last 2 years have been dedicated to learning the art of imaging. The Moon and the planets were his starting point—images taken with a low-cost camera, which is still used to this day.

The Moon became Roger's huge challenge—waiting for the moment of good seeing and grabbing those hidden secrets within. He began to get in close to capture those jaw-dropping pictures of the Moon's craters, valleys, and mountains. With the introduction of a modified webcam, he soon moved on to deep sky—learning the process all over again. His greatest wish is that the images he has produced will inspire others "to progress as well in this wonderful hobby." **www.lupas.pwp.blueyonder.co.uk/rwnewastro/lunar.htm**

My Special Thanks...

Go to the wonderful John Watson for having faith in me. To Harry Blom and Christopher Coughlin for having patience with me. And to Sharmila Krishnamurthy and the Integra India team for turning a vision into a reality.

My deepest appreciation goes once again to all of these fine folks and astronomers. May we all shine on...

About the Author

Tammy Plotner

An Ohio native, author Tammy Plotner has spent many years with various telescopes and binoculars, both learning and teaching the night sky. She is President of Warren Rupp Observatory, a founding member of Astronomy for Youth, a presenter and coordinator for the Night Sky Network, and a JPL/NASA Space Place editor and outreach speaker. Tammy is both a friend and a mentor to many astronomy groups, as well as an author of popular books and articles, and she is a contributing author for the Astronomical League's *Reflector* magazine.

While she loves viewing the distant Cosmos with the Observatory's 31" telescope, she's equally at home with a small refractor, and every size and type of telescope in-between. As an observer, she has won numerous awards, and was the first woman astronomer to achieve Comet Hunter's Gold status.

Although she finds writing about herself in the third person rather spooky, you'll find her a friendly face at star parties, at every public outreach program, speaking at colleges and libraries, traveling to distant astronomy clubs, and right here in her own backyard doing what she loves best...looking at the stars. http://theastronomer.tripod.com

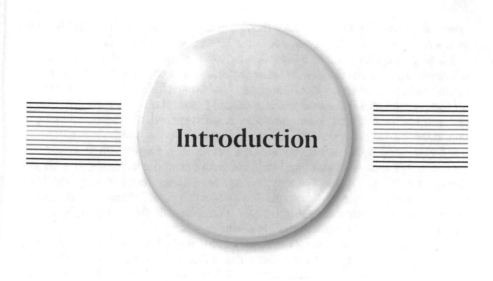

Introduction

Welcome, companion... For the next year we will take a journey together across the night sky. In these pages you will find lunar features, planets, meteor showers, single and multiple stars, open and globular clusters, as well as distant galaxies. There will be astronomy history to explore, famous astronomers to meet, and science to learn. You'll find things here for those who enjoy stargazing with just their eyes, binoculars, or even the largest of telescopes!

While these observing tips are designed with all readers in mind, not everyone lives in the same time zone, or the same hemisphere—and certainly no one has clear skies every night. But no matter where you live, or who you are, it is my hope that somewhere here you find something of interest to keep you looking up!

Learning the Night Sky

If you are new to astronomy, perhaps the most daunting part about beginning is learning all those stars. Relax! It's a lot easier than you think. Just like moving to a new city, everything will seem unfamiliar at first, but with a little help from some maps, you'll soon be finding your way around like a pro. Once you become familiar with the constellations and how they appear to move across the night sky, the rest is easy.

If you do not have maps of your own, try visiting your local library or online sites which can generate them. These show object positions in even greater detail, and most have a key of Greek letters to help you read them. This helps greatly to understand star hop instructions! Keep in mind that constellation charts are oriented just as if you held the map over your head, while star charts are printed in the order in which the sky moves—north is up and east is left.

These pages also contain a simple way of helping you understand relative positions and sizes in the sky. In some instances, you will find the directions to certain objects, use terms like "handspan" or "fingerwidth"—but what exactly do these terms mean? A *handspan* is measured from the tip of your little finger to the tip of your thumb when your arm is outstretched. For most, this simple measurement covers about 20' of sky. A *fistwidth* is the width of your closed hand at arm's length, and covers around 10', while a *fingerwidth* is around 2'. Although this type of instruction is not foolproof (and not everyone's hands are the same size), you'll find the simple directions given here will help you find the right area to begin. Just remember an object's relative direction will always remain the same. For example, southwest of Alpha will always be southwest of Alpha—no matter if a constellation is rising or setting.

While most astronomy terms are clear, others might seem to be contradictory. For example, why would a meteor shower or a Full Moon date appear differently from what is shown on a local calendar? The explanation is simple enough: Because astronomy is practiced worldwide and events often occur on either side of the International Date Line, a Universal standard is employed. Expressed as UT (Universal Time), specific dates are also "universal." A day behind for some—a day ahead for others! There are many online tools to help you calculate UT for your location, and with just a little practice you'll understand easily.

Another oddity comes in the form of what seem to be contradictory numbers. Why is a 9 millimeter eyepiece more "powerful" than a 32 millimeter? By dividing the focal length of the objective by the focal length of the eyepiece, you'll get magnifying power. In short, the higher the number (expressed in millimeters), printed on the eyepiece, the lower the magnification. Another seemingly contradictory number is the brightness—or magnitude—of an object. Why can you see a magnitude 1 star unaided, but not an 8? Again, the answer is that the scale is defined "in reverse"—the higher the magnitude number, the fainter something is. With each addition of one magnitude, the overall brightness drops by roughly two and a half times. Before your head starts to swim, remember it's impossible to know everything all at once. What might seem confusing at the start will become so routine as time passes that you won't even think of it. Just ask the experienced amateur astronomer who skips over the beginner's introduction!

Observing Equipment

While I would love to tell you which binoculars or telescope would be perfect for you, I can't. The choice in equipment differs as one's taste in automobiles. Every person uses equipment differently and under different circumstances.

Let's Start with Binoculars...

Anyone even vaguely interested in the night sky should own a pair of binoculars—even inexpensive ones. While they will never reveal the heavens in quite as much detail as even a small telescope, their availability, ease of use, and portability make them the perfect night sky companion. It takes no effort at all to aim them toward the stars and begin! But which ones? The very best size for astronomy are 7×50 and 10×50 models of the porro prism design, yet even modest 5×30 models will show a wealth of sky objects. While it might be tempting to get a "monster" model, remember the weight and how difficult it would be to steady them! Very large binoculars truly require a mount of some type. There is no recommended brand, but I suggest when purchasing to keep in mind they might get dropped or lost.

Don't be afraid to ask to try models out before purchasing. They should be comfortable to hold, and when you look through the eyepieces, the field should be evenly illuminated. When you look into the main lens, white reflections mean poorly coated optics, while deep purple and green reflections indicate higher quality. Better models will have a right diopter adjustment, and check to see that both sets of lenses are well collimated. No matter which you choose, I guarantee you'll enjoy these hand-held "twin" telescopes!

Now Let's Move on to Telescopes...

There are three basic designs. The refractor uses a lens to gather light, the reflector uses a mirror, and the catadioptric uses both. No matter which style you prefer, the goal is light gathering ability—not magnification. The larger the aperture (diameter), the more light the scope gathers and the more power it has to resolve objects. Steer clear of small telescopes you see in department stores and most camera shops: they are almost invariably of low quality, and for about the same amount of money you can end up with a fine telescope which will give years of use.

The *refracting* telescope is favored by those who enjoy high-power views of terrestrial subjects, as well as the Moon and planets—and provides suitable light gathering ability for plenty of deep sky study. Because the eyepiece is located at the end of the scope, it is often necessary to use a right-angle attachment to put the eyepiece in a comfortable location. While many claim a refractor is superior for seeing details, keep in mind that this style may put you in some very uncomfortable positions!

The *reflecting* telescope is the instrument of choice for deep sky observing. Large aperture is far more affordable and the performance on lunar and planetary objects is more dependent on the quality of the optics and seeing conditions, rather than the design. With the eyepiece located on the top side of the body of the telescope, this type of telescope is primarily used by observers who prefer to stand. Even the smallest (4.5") reflector will provide enough deep-sky studies to keep the average sky watcher entertained for a lifetime! Do they have a flaw? Yes. With large aperture also comes large size, and portability may become an issue. Do not let the word "collimation" frighten you. It is just the act of occasionally adjusting the primary mirror, and is no different from tuning a guitar.

The *catadioptric* design should thereby fulfill the best of both worlds—shouldn't it? The answer is yes—but it still doesn't come without drawbacks. This style telescope is very expensive and prone to dew up without preventative measures.

Now Let's Talk about mounts...

Here again we have three basic designs—the altazimuth, the equatorial, and the dobsonian. The *altazimuth* swings left/right and up/down, and requires manual adjustment to track the sky. These inexpensive and easy to use mounts are best suited to small refractors. The *equatorial* design moves in right ascension and declination—the proper movement and angle on the sky. When aligned to the pole, they only require a slight turn of a slow motion control to track and are capable of being fitted with automatic tracking devices. They come in a variety of weights and sizes suitable for telescopes of any type, and almost all come equipped with setting circles. The last type of mount is the *dobsonian*. Much like the altazimuth, it moves up, down, and side to side...but requires no tripod. It is nothing more than a simple, well-balanced rocker box. With small aperture dobs, this inexpensive design gives you total freedom to travel with your scope—but plan on having a backache while using it. Conversely, larger models have a more comfortable viewing position but lack portability.

So, How to Choose?...

More than anything, you must ask yourself what type of observing you enjoy the most and what scope will meet your needs—and your budget. There's no point in buying a large dobsonian if you need to travel to a dark-sky location to use it. You'll never be happy with a small refractor if you have dark skies outside your back door, an itch to galaxy hunt. I own models of every type and there's a reason for each one.

The small refractor (4") and its easy, lightweight mount is the perfect companion for travel. It's simple to carry, simple to set up, and provides great views—but not for everything I want to see. Its GoTo function is a wonderful aid when speed is of the essence, and its tracking feature makes it invaluable for doing outreach programs.

The small reflectors (4.5") are the workhorses of my fleet. They provide great lunar, solar, and double star views—along with the capability of capturing all the Messier objects and a goodly portion of the NGC targets. They are lightweight, portable enough, and I usually have at least two of the three I own always fully assembled and ready to be set outside at a moment's notice. They are all fine performers, but...I could still wish for just a little bit more.

The mid-size catadioptric (6") scope is a more tedious set up, but gives outstanding lunar and planetary views. It resolves tough double stars and provides crisp resolution on most star clusters. If astrophotography were a goal of mine, this would be the scope I would choose. But, given the telescope's expense, it's usually carefully packed away and seldom used on a whim.

The larger equatorially mounted reflector (8") is a superb tool for low surface brightness objects. With its big light grasp and resolution, this is a very fine scope to spend an evening (or many!) with. It's fairly easy to transport, a little difficult to set up, but the views are quite worth the time and effort. If I need portable aperture, this is the one I choose. But it still can't quite reach those faint galaxies...

The dobsonian model (12.5 inches) is my study scope. Far too large to be even remotely considered portable, it spends its life on a self-built mobile transporter nicknamed "The Grasshopper." While newer models are much lighter, this scope needed a way of moving it from storage to an outdoor observing area without killing its owner. Once I had designed a way of getting it to the backyard with ease, the whole vista of the heavens opened up. Now I was able to totally resolve clusters, study very faint galaxies, home in on lunar and planetary features, and split intense double stars. The dobsonian design left me with total freedom of movement, and after more than a decade of use, I have still not seen everything this scope is capable of. Of course, I have ridiculously large aperture at my disposal as well, but the 31 inch is housed in a professional observatory and what it "sees" is almost unfair compared to backyard equipment! And now for the next major issue.

To "GoTo" or not to "GoTo"...

Again, this is a matter of personal preference. It is my honest opinion that you do yourself a disservice by not learning to manually aim a telescope at an object. There is great joy to be had in studying the sky and finding a distant galaxy using nothing more than a map and your own two hands. So many folks have these wonderful systems gathering dust because they found out they require perfection in positioning as well as basic sky knowledge to use. Regardless of the claims of how many objects a database contains, only experience will tell you how many of these objects can be seen with your scope and under your sky conditions!

But do not be angry. There is also a beauty in these systems. For those with limited time, it only takes a little learning to use. Many such systems are also able to identify objects by their coordinates, so they do have their good points.

And so We Come to Eyepieces...

The bottom line is, you get what you pay for. A highly expensive eyepiece will not turn a bad telescope into a good one, but it will turn a good telescope into an awesome one. My best advice is to start with mid-priced optics and a limited variety of sizes. A 32 millimeter eyepiece is great for wide-field views, and a 25 or 17 millimeter is fantastic for most work. Eyepieces of 12, 10, and 5 millimeters are the powerhouses—but on a telescope without a drive unit, one of these often provides so much magnification as to be uncomfortable. Like fishing lures in a tackle box, you'll find yourself collecting a variety of eyepieces over the years—and each will favor certain uses. Only experimentation will provide the eyepieces that are right for you and for your telescope.

Now for Accessories...

These are the fun things to have, but none of them are necessary to practice astronomy: a sturdy case, a barlow lens, a set of basic color filters, a Moon filter, a polarizing filter, and a nebula filter. Optics cleaning kits are great—but a word to the wise—don't stress about sparkling clean optics. Unless it is more than 20% obstructed you will probably do more harm than good in cleaning. There are so many little things that make observing fun, and fun to have in your "kit." But what do you really need?

The very best accessories I recommend are a comprehensive Moon map, an easy to read sky chart, a watch, a pad of paper, a mechanical pencil, and a red flashlight. These are the most important things you will ever use. While basic photographic lunar maps are provided here, you may find charts of your own that you prefer. While note keeping is not a necessity, you'll find more often than not that you'll need it. Basic sketches are easy and you might find that you'll refer to your own notes often. Keep track of what you do, what you use, and when you see it. There are many fine observing programs offered through the Astronomical League and other observing clubs which provide awards for your study. And who knows? You might discover something new...

Ready to Observe

The Moon...

Let's start with one of the easiest to find yet most rewarding objects to study—the Moon. Its rugged craters, high mountains, and vast seas offer some of the finest details to be found in any astronomical target. It changes every night as the terminator—the line between sunset and shadow—progresses over the surface, revealing new details.

Unlike a star chart, Moon-feature instructions are based on lunar topography and not on our Earthly cardinal directions. While these pages outline what features should be visible on any given night, the position of the terminator may be slightly different for viewers in various time zones. Let's start by discussing how and when the Moon can be seen...

The Moon and the Earth both rotate at exactly the same speed, so we will always see the same "side"—yet its elliptical orbit causes a kind of "wobble" that we refer to as libration. This means there may be times when you can see just a bit more along the Moon's limb—the visible edge. When the Moon's orbit carries it between the Earth and the Sun, this is referred to as New Moon. It's still there...but we cannot see it.

As its orbit progresses, the Moon will slowly move to its first position, appearing in the night sky just after the Sun sets. The sunlight on the lunar surface will begin its march across the surface, progressing from lunar east to lunar west. At either pole of the Moon is the area called the cusp—the tip of the curve where the terminator ends. This progression of light is called the "waxing" phase.

When the Moon has reached its second position, it is directly opposite the Sun in our sky and the surface is totally illuminated—Full Moon. At this time, the Earth is between the Sun and the Moon. Most of the time, the Moon's orbit will carry it either north or south of the Earth's shadow, but about every six months it will slip inside that shadow and a lunar eclipse will occur. When it passes only partially into the cone of shadow it is known as a penumbral eclipse, and when it is directly aligned it is called a "full" or umbral eclipse.

Now the Moon is heading toward its third position and moving back toward the Sun. It will rise later each night and the terminator will now progress across the surface in the same direction—east to west—but this time the features will be seen at lunar sunset instead of sunrise! This is known as the "waning" phase. The Moon will become slimmer each night as it heads toward the rising Sun.

At first, the lunar landscape will look quite confusing—but keep in mind that lunar north has fewer craters than lunar south. As you study the Moon from month to month, craters will become more familiar to you and it won't be long until you know their names and can often tell what features will be visible—without even looking!

The Planets...

Just like the Sun and the Moon, the planets dance along an orderly path in the sky known as the ecliptic plane. Their progression against the background stars will seem slowest when they are the furthest away. During an observing season, it's possible to watch as the Earth overtakes a planet, much like running past an object on a race track. As we approach it, it will seem to slow down, stand still, and then move backwards as we go by. Once we have passed, it will then appear to resume forward motion.

Since Mercury and Venus are on the inside track of our race course, they move much more quickly around the Sun than the outer planets do. They will always appear just ahead of the rising Sun, or just after the setting Sun—and like our Moon will go through phases as they progress in their orbits.

Mars, Jupiter, and Saturn are three of the most highly observed planets and, at times, can offer up wonderful details to the telescope. But do not be disappointed if you do not see fantastic things on your first night out! There are many things to be considered when viewing these planets. Stop first to consider their distance at any given time, and the effect of our own atmosphere on observing conditions. Do not be discouraged! Large binoculars and the smallest of telescopes will reveal Jupiter's equatorial bands and clockwork movement of the four Galilean Moons...even the rings of Saturn! As aperture—the size of the optics—increases, so does the amount of detail that can be seen. But even the largest of telescopes cannot compensate for poor viewing conditions! The outer planets—Uranus and Neptune—also follow the ecliptic plane. Both Uranus's and Neptune's movements can be followed with binoculars, but even large telescopes offer little detail due to their great distance.

Night Lights...

Of course, there are other things within our own solar system that can also be easily studied—such as asteroids, comets, and satellites. Given the nature of this book—which was created with limited use of a planetarium program—these types of studies are best undertaken with the aid of either software or magazines such as *Sky and Telescope* or *Astronomy*. There are also online tools available to assist you and you'll find reference to many of these in the Resources section.

You'll also enjoy meteor showers throughout this observing year as well. While the dates that we pass through these cometary debris streams are predictable, the fall rate—the estimated amount that can be seen in a given time—is not. As a rule of thumb, you can see any given meteor shower from either hemisphere if you can see the constellation of the radiant—the general area from which they appear to originate. Keep in mind that ambient light plays a huge role in how many meteors can be seen—and the darker the skies, the better your chance for success.

Deep Sky...

This is the term given to objects which reside outside of our solar system. These include single stars, multiple star systems, open or "galactic" star clusters, globular clusters, nebulae, and distant galaxies. While many of these objects are within reach of small binoculars, just as many reside at the outer limits of the capabilities of a large telescope. It wouldn't be such a delightfully challenging hobby if everything were easy! Let's begin...

There are a few things to keep in mind as you begin exploring deep sky and the most important item is sky conditions. Even the largest of telescopes will have difficultly catching a faint galaxy through light-polluted skies or during poor atmospheric conditions (bad seeing). Nothing is a better teacher than experience, and it won't take long before you learn what your equipment is capable of.

For example, from a dark-sky site with favorable atmosphere, it is entirely possible to see the majority of the Messier catalog of objects with a pair of 5×30 binoculars...or even some objects with just the unaided eye. But binoculars are not the Hubble telescope and you need to understand what you might expect to see! A nebula will appear as a faint glowing cloud, a globular cluster as a round contrast change, and a star cluster as a "patch" of concentration. None of these will be particularly large given the fact that binoculars offer a wide field of view and little magnification. But how wonderful it is to use such simple equipment to view things so many light-years away!

A small-to-large telescope will increase the light grasp and allow you to see progressively fainter and fainter objects—and in greater detail. Instead of a round "smudgie" when looking at a globular cluster, individual stars will appear. A single bright star will reveal its tiny companion, nebulae will unfold, and the light of distant galaxies suddenly will become much closer... But all of this is dependent on one single thing—sky conditions. Experience is the greatest teacher. You cannot expect a small telescope to reveal an 11th magnitude galaxy, yet under the right conditions it is possible for it to reach beyond its theoretical limitations. Do not stop trying because you've had a few disappointments—learning comes with time. The eye must be trained to pick up on very faint things, and the best advice I can offer you is my three Ps for success: Practice, Patience, and Persistence.

You can learn to identify lunar features. You can learn how to read a star chart and find the positions in the sky. If you are having problems with something? Improvise. If the equatorial mount on your telescope puts you in an awkward position? Pick it up and turn it around. I assure you that the "polar alignment police" will not come to get you. If your small dobsonian telescope is hard on your back? Set it up on something! If the legs on the tripod are too high or you can't stand for a long period of time? Lower them and find a stool. If you live in a light-polluted area, it won't stop your love of the Moon and the planets and perhaps there will be a time that will allow you to get away to a dark-sky location.

The telescope that is loved the most is the one that gets used. If at all possible, consider keeping it fully assembled where it may be set outside at a moment's notice—such as in a garage, an outdoor shed, or near a door. You will be far more likely to spend 15 minutes with the Moon, or a half hour out of your busy schedule with deep sky if you do not have to go through complicated set up procedures. Forget the stress factor. Unless you are using a GoTo model, a drive system, setting circles, or planning on photography, it is not necessary to have everything perfectly aligned to enjoy the night sky or your telescope! If you reach the end of a slow motion cable's extent? Turn it back and reset the scope on the object. The only hard and fast rule in practicing amateur astronomy is to enjoy what you are doing.

If it is not practical in your circumstances to keep a telescope assembled then consider even an inexpensive pair of binoculars. If you think you can't afford them then think again. Many very suitable pairs of binoculars can be had for about the price of an extra large pizza! These small, handheld twin telescopes will increase your love of astronomy and whet your appetite for more.

Ready... Set... Go!

Now we have our equipment, our maps, and our notes... What's next? Start off with some common sense... Dress appropriately for the outdoors and allow your optics to reach ambient temperature. The next is probably the most important step for any observer: allow your eyes time to adapt dark. Picture yourself walking into a dark room. You can see nothing when you first enter, but after several minutes your eyes begin to adjust and you can "see" the sofa you are about to stumble over! This is even more true when viewing the night sky. When you go out from a brightly lit room to the outdoors, at first you will only see the brighter stars, but after several minutes you will see that many more begin to appear.

Everyone's eyes adjust at different rates. For some, it may only be a matter of minutes before you are able to spot faint objects, but for others it may take a lot longer. This is why you might find a faint galaxy, only to drag a friend or family member away from the television and discover they can't see it! Avoid bright lights, television, and computer screens as much as possible before observing. Darken the room or wear sunglasses. While this may sound silly, you'll find it will greatly increase your chances of finding that difficult galaxy or faint comet!

This is also why astronomers use red flashlights or red lighting... It helps to preserve night vision while reading a map, taking notes or preparing to observe. If you do not have one, take a look around you at what might suffice. Keychain lights often come in red and are sufficiently bright to read a map. Even a bit of red cellophane rubberbanded over a penlight will do!

There are also two techniques that will help increase your observing abilities. The first is called averted vision. While it might seem strange to ask you to view an object without looking directly at it, you'll find you're avoiding the low-sensitivity patch on your retina and faint objects will appear in greater contrast. The second is tapping the eyepiece gently. Movement, even slight, will make a low-contrast target easier to spot. Be sure to keep both eyes open to avoid eyestrain!

So, what else do we need before we begin to observe? While the art of note taking or sketching is not for everyone, it's something I highly recommend. Even if it's nothing more than a pocket notebook and a pencil stub, you are on your way. Imagine seeing Jupiter and its moons for the very first time! By having a handy way of taking notes, it takes very little effort to draw a circle and a few dots, and indicate on your notes which direction they move across the eyepiece field. Just this simple bit of information is enough to let you later identify which moons were in what position!

The same will hold true of everything you observe. By writing down dates and times—along with seeing conditions—it won't be very long until you are able to accurately assess what can be seen on any given night. You'll quickly discover that the first night you could see M44 unaided was also the first night you spotted M33 in binoculars! These notes are yours, and no one will come to "grade" them. They are all a part of learning... And you can learn!

Now let's be comfortable with ourselves, who we are, what we have, and what we know. There's a whole wonderful night sky filled with things to explore! Our equipment is ready and we are ready. So let's head out under the stars...

Together.

CHAPTER ONE

January, 2008

Tuesday, January 1

Greetings, fellow skywatchers! As we begin our observing time together, why not start before dawn this morning with a look at bright Spica and the Moon? You'll find the pair separated by around a fingerwidth as they rise together ahead of the first day of the New Year.

On this night in 1801, the skies were clear in Italy and astronomer Giuseppe Piazzi was busy at the eyepiece (Figure 1.1). He had made a discovery—the first known asteroid. Unlike modern communications, Piazzi's mailed observations failed to reach others for confirmation before his new discovery had moved too close to the Sun, and the object was lost until its return in September. What we now know as minor planet Ceres was relocated thanks to Gauss and his method of calculating orbits. Ceres was identified again on the last day of 1801 and reconfirmed on this date in 1802. Why not try your own hand at locating Ceres? All it takes is an accurate locator chart readily available online. Even smaller telescopes can easily spot this bright solar system member this evening as it cruises about a fistwidth northeast of Alpha Ceti (Menkar).

But Piazzi wasn't the only astronomer to start the New Year with a discovery: Sir William Herschel was also at the eyepiece on this night in 1789! Turn your eyes toward Orion and the easternmost star in the "belt"—Zeta Orionis. Named Alnitak, it resides at a distance of some 1600 light-years, but this 1.7 magnitude beauty contains many surprises for the telescope user. It is a triple star system. Fine optics, high power and steady skies will be needed to reveal its members.

Figure 1.1. Giuseppe Piazzi (widely used public image).

Figure 1.2. NGC 2024: The Flame Nebula (Credit—Palomar Observatory, courtesy of Caltech).

Look about 15' east and you will see that Alnitak also resides in a fantastic field of nebulosity which is illuminated by our tripartite star. NGC 2024 (RA 05 41 43 Dec –01 50 30) is an outstanding area of emission discovered by Herschel on this night and cataloged as H V.28 (Figure 1.2). At roughly magnitude 8, it is viewable by smaller scopes but will require a dark sky.

So what's so exciting about a fuzzy patch? Within it are hot, young stars producing the radiation to illuminate the gaseous nebula. It is being compressed and the atomic density is increasing. It has begun to fragment, forming clumps. Over thousands of years it will collapse under the weight of its own gravity, superheating, and forming protostars. As the "Flame" heats up, fusion occurs and new stars are born. Depending on the fragmentation process, they may be small, red dwarves—or giant blue stars exerting their influence on the gases around them. Because of the density of NGC 2024, studies must be performed in infrared to see what lies below the surface, but larger telescopes will deeply appreciate this nebula's many dark lanes, bright filaments, and unique shape.

Wednesday, January 2

Today in 1959, the USSR launched the very first Moon probe. Named Luna 1, it became the first extraterrestrial traveler. The spacecraft carried no propulsion systems of its own, but after having reached escape velocity, its third stage released a payload of sodium gas. This left a glowing, 6th magnitude "trail" allowing astronomers to trace it. Luna 1 made outstanding contributions to science—including the first confirmation that the Moon had no magnetic field. The probe was designed to impact the surface, and although it failed to do so, it did achieve another first with its flyby (Figure 1.3).

Tonight let's return to Orion's "belt" and starting with just our eyes look about a thumb's length south to discover an asterism of stars referred to as the "sword." On a clear, dark night away from city lights you can spot a glowing cloud of dust and gas surrounding Theta which has long held a place in astronomy history. It was first noted only one year after Galileo first used his telescope, and its discovery is credited to Nicholas Peiresc in 1611.

Now take out your binoculars and have a look at Messier 42—the "Great Orion Nebula" (Figure 1.4). Here, stars are still being born in a dense cloud and hundreds of them are less than a million years old. Compared to our own Sun's age of over 4 billion years, these would seem almost new! But think again at what you are looking at... The light you see tonight left this area around 1900 years ago.

So magnificent are the many details to be seen in the Orion Nebula, that chapter upon chapter could be devoted to its riches. For now, feast your eyes upon this 30 light-year expanse of dust, neutral and ionized hydrogen, and doubly ionized oxygen, all illuminated by the ultraviolet starlight of this stellar nursery. It is more than 20,000 times larger than our own solar system and its mass could form 10,000 stars like our own Sun!

Figure 1.3. Luna 1 (Credit—NASA).

Figure 1.4. M42: The Great Orion Nebula (Credit—R. Jay GaBany).

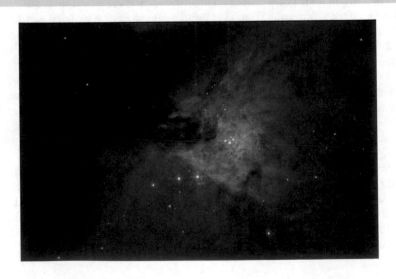

Thursday, January 3

Today is the birthdate of Russian astronomer Grigori Neujmin (1886). His important discovery was the rotating asteroid Gaspra. This is also the date on which Stephen Synnot discovered Juliet and Portia, two additional moons belonging to Uranus.

Tonight let's return to the Orion nebula for a much closer look. Only in 1656, when Christian Huygens sketched it, did it become well known for containing a "heart of stars"—an area now known to astronomers as the Trapezium. Considered the holy grail of multiple star systems, this is the fueling core of M42, Theta Orionis (Figure 1.5). Are you ready to walk into "the Trap?" Even the smallest of telescopes can reveal the four bright stars which comprise the quadrangle, and the seasoned veteran knows there are eight stars in this region. But what can you really see? The journey we are about to undertake requires both aperture and steady skies.

Figure 1.5. Theta Orionis: The Trapezium (Credit—Hubble Space Telescope, press release image).

All four primary stars are easily split, and appear to be in a dark notch. A mid-sized telescope will reveal two additional 11th magnitude stars, but excellent skies could mean that even smaller aperture could detect them as red companions to the blue–white primary stars. The remaining two components average about magnitude 16, putting them within reach of large amateur scopes. Power up!

While at first glance with a small telescope, the background region in this area might appear a black void, but it is not. The nebula continues here, but changes form. Instead of seeing "smoke-like" filaments, the region around the Trapezium is scalloped, like fish scales. A moment of steady seeing will easily reveal the prized G and H stars along with many more. Studies have revealed about 300 such stars within 5' of the Theta Orionis complex brighter than magnitude 17. According to K. A. Strand, their expansion rate puts them at an approximate age of 30,000 years, making the Trapezium the youngest star cluster known.

Regardless of the size of the telescope you use, you owe it to yourself to take the time to power up on the Trapezium. No matter how many stars you are able to resolve out of this region, you are looking into the very beginnings of starbirth!

Friday, January 4

On this universal date in 2004, history was made as the first Mars Rover—Spirit—landed safely on the surface of the Red Planet and opened its eyes on this vast, new world (Figure 1.6). Followed 21 days later by its twin geologist Opportunity, the two rovers have firmly established their place among the first

Figure 1.6. Mars Rover (Credit—JPL/NASA).

robotic missions to make a huge impact on the way we look at Mars. As of late 2006, the pair had been successfully operating for almost two years!

In the early morning hours, be sure to keep a watch for members of the Quadrantid meteor shower. Its radiant belongs to an extinct constellation once known as Quadran Muralis, but any meteors will seem to come from the general direction of bright Arcturus and Boötes. This is a very narrow stream, which may have once belonged to a portion of the Aquarids. As Jupiter's gravity continues to perturb the stream, another 400 years may mean this shower will become as extinct as the constellation for which it was once known. Today is also the birthdate of Wilhelm Beer (1797), an amateur astronomer who with Johann Mädler created an exhaustive map of the Moon—*Mappa Selenographia*. It was the first of its kind.

Tonight our study region is to the northeast of the Great Orion Nebula (M42) and has a designation of its own—M43 (Figure 1.7). Discovered by De Mairan in the latter half of the 18th century, this emission nebula appears to be separate from M42, but the division known as "the Fish Mouth" is actually caused by dark gas and dust within the nebula itself. At the heart of it is the 7th magnitude

Figure 1.7. M43 (Credit—Palomar Observatory, courtesy of Caltech).

"Bond's Star"—and wouldn't 007 be proud? This unusually bright OB star is creating a matter-bound Strömgren sphere!

Translated loosely, this star is actually ionizing the gas near it, making an orb-shaped area of glowing hydrogen gas. Its size is governed by the density of both the gas and the dust surrounding Bond's Star. This "exciting" star of our show is more properly known as Nu Orionis and near it is a dense concentration of neutral material known as the "Orion Ridge." It is this combination of dust—mixed with gases—that makes for a well-balanced area of star formation.

And besides... It's just cool!

Saturday, January 5

Woke up before dawn? Then you're in for a special treat as the thin crescent of the waning Moon pairs with Antares. Separated by less than half a degree, this could mean an occultation event for some observers. Be sure to refer to IOTA (International Occultation Timing Association) for accurate information.

Figure 1.8. A section of Barnard's Loop (Credit—-Palomar Observatory, courtesy of Caltech).

Let's return again to Orion tonight, but preferably with binoculars, since we will be studying a very large region known as "Barnard's Loop" (Figure 1.8). Extending in a massive area about the size of the "bow," you will find Barnard's namesake to the eastern edge of Orion, where it extends almost half the size of the constellation between Alpha and Kappa.

Because the Orion complex contains so many rapidly evolving stars, it stands to reason a supernova should have occurred at some time. Barnard's Loop is quite probably the shell leftover from such a cataclysmic event. If taken as a whole, it would encompass 10 degrees of the sky! For the most past, the nebula itself is very vague, but the eastern arc (where we are observing tonight) is relatively well defined against the starry field. Although it is similar to the Cygnus Loop (the Veil Nebula), our Barnard Loop is far more ancient. If you have transparent, dark skies? Enjoy! You can trace several degrees of this soft, wispy ancient remnant using just binoculars.

Sunday, January 6

Today in 1949, the first atomic clock built on theoretical work by Isidor Rabi and Norman Ramsey went into operation (Figure 1.9). This model used ammonia as its "pendulum," but only eight years later the first cesium beam device was built. Its primary standards are now keeping time to about one-millionth of a second

Figure 1.9. First atomic clock (widely used image).

per year. Like clockwork, objects we can see in the sky also keep incredibly accurate time. Tonight return again to Orion's belt as we have a closer look at its westernmost star—Mintaka (Figure 1.10).

Located around 1500 light-years away, Delta Orionis usually holds a magnitude of 2.20, but orbiting it at a clockwork rate of 5.7325 days is a nearly equal-magnitude star around 8 million kilometers away. Mintaka is a prime example of an eclipsing binary star, and although visually you won't really notice a 0.2 magnitude drop in light from its spectroscopic companion, let's take a closer look with binoculars. As one of the few easy double challenges, Mintaka will reveal its 6.7 magnitude visual companion star to its north. For over 100 years, this pair has been closely watched and no movement between the half light-year apart physical pair has been detected. For those with larger telescopes, power up and see if you can discover the 13th magnitude C star southwest of the primary. No matter how you look at Mintaka, this fascinating star has been a part of history. It was the very first to display stationary spectral lines which proved the existence of interstellar matter!

For those with larger scopes wishing a challenge, wait until the constellation of Sextans has well risen. Using Gamma as your guide, you will find NGC 2974 roughly two-thirds the distance north along an imaginary line between it and Iota Hydrae (RA 09 42 33 Dec –03 41 58) (Figure 1.11). Discovered on this night in 1785 by William Herschel and cataloged as H I.85, this near 11th magnitude elliptical galaxy has a bright core region...and with good reason. Studies have shown this E4 class galaxy may very well contain an embedded stellar disc!

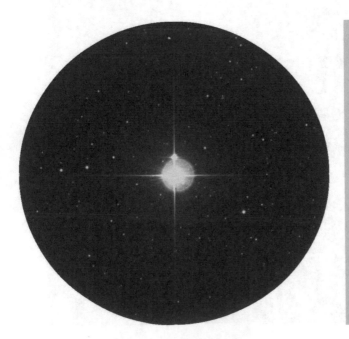

Figure 1.10. Delta Orionis: Mintaka (Credit—Palomar Observatory, courtesy of Caltech).

Figure 1.11. NGC 2974 (Credit—Palomar Observatory, courtesy of Caltech).

Monday, January 7

On this night in 1610, Galileo discovered three of Jupiter's four largest satellites using his simple telescope. This revelation changed the face of an Earth-centered Universe. This morning before dawn, why not take out your binoculars and see if you, too, can discover the "Galilean moons" with this simple equipment. You'll find the largest of all the planets in our solar system making a wonderful morning apparition along the very last remaining curve of the Moon. Because it is so near to the Sun, take precautionary measures while viewing it—but don't worry if you can't spot it now. It will race further and further away from our nearest star in the weeks ahead!

Tonight let's return to Orion's sword to have a look for something you might have missed. Starting with M42 and M43, be sure to log these two Messier catalog studies for your binocular or small telescope records, but have a much closer look about one degree north (RA 05 35 12 Dec –04 24 00).

NGC 1981 is a 4th magnitude open cluster which looks like a stellar member of the Orion group to the unaided eye (Figure 1.12). In small binoculars, it is easily resolved into a dozen or so members, with its brightest star weighing in at around magnitude 6. In the small telescope, as many as 20 individual members are resolved in chains and small groups. The region of NGC 1981 has been studied for rotational movement in the Orion arm of our galaxy and it was found that the stars in this cluster are actually rotating around our galactic center faster than the stars in the Perseus arm.

Figure 1.12. NGC 1981 (Credit—Palomar Observatory, courtesy of Caltech).

Well suited to even urban skies, NGC 1981 is also an Astronomical League Binocular Deep Sky Object (DSO) you will very much enjoy. For larger telescopes looking for a real challenge, double star Struve 750 is part of this entertaining and easy galactic cluster!

Tuesday, January 8

On this day in 1942—precisely 300 years after the death of Galileo—Stephen Hawking was born (Figure 1.13). Despite physical limitations, the British theoretical astrophysicist went on to become one of the world's foremost leaders in cosmological theory and his book *A Brief History of Time* remains one of the best written on the subject. Also born on this day in 1587 was Johannes Fabricius, son of the discoverer of variable star Mira, David Fabricius. Like many father and son teams, the pair went on to study astronomy together, and some of their most frightening work dealt with viewing sunspots through an unfiltered telescope—a practice which eventually blinded Galileo!

Figure 1.13. Stephen Hawking (widely used public image).

Figure 1.14. 42 Orionis Region (Credit—Palomar Observatory, courtesy of Caltech).

Tonight let's honor them by returning to the area just south of last night's study as we enjoy a New Moon challenge. Lace up your Nikes and let's head out to find "The Running Man"...

Located just a half a degree north of M43 and bordering on last night's study, this tripartite nebula consists of three separate areas of emission and reflection nebulae which seem to be visually connected. NGC 1977, NGC 1975 and NGC 1973 would probably be pretty spectacular if only they were a bit more distant from their grand neighbor! This whispery soft, conjoining nebula's fueling source is multiple star 42 Orionis (Figure 1.14). To the eye, a lovely "triangle" of bright nebulae with several enshrouded stars makes a wonderfully large region for exploration. Can you see the shape of the "Running Man" within?

Wednesday, January 9

Today in 1839, Scottish astronomer Thomas Henderson was the first to measure the distance to a star, while stationed at the Cape of Good Hope. Using geometrical parallax, Alpha Centauri became the first stellar standard other than our own Sun. Although Henderson began as a lawyer's clerk, his impressive list of 60,000 star positions led to his appointment as the first Astronomer Royal in Scotland.

Figure 1.15. NGC 1980 (Credit—Palomar Observatory, courtesy of Caltech).

With the Moon absent during the early evening, our goal for tonight is Iota Orionis. Known to the Arabs as "the Bright One of the Sword," we know it as the southernmost star in its asterism's namesake. Iota is estimated to be around 2000 light-years away and is about 20,000 times brighter than our own Sun. In the small telescope you will find Iota to be an easy and charming triple star. The bluish B star is relatively close at 11" in separation, but is a bright 6.9 in magnitude. Much more distant at 50" is the disparate, magnitude 11, reddish C star. Iota itself is a spectroscopic binary and you will note another "white" double (Struve 747) unrelated to Iota about 8' to the southwest.

If you look closely, you will see that Iota is involved in a great region of faint emission nebulosity, along with a small open cluster known as NGC 1980 (Figure 1.15). Congratulations! You've just spotted an object on the Herschel "400" list—H V.31. To be sure, the area is vague, as are all low-surface brightness nebulae, but do look closely around Iota where a much brighter, roundish area makes an unmistakable appearance!

Thursday, January 10

Robert W. Wilson was born this day in 1936. Wilson is the co-discoverer of the cosmic microwave background, and in 1978 along with Arno Penzias won the Physics Nobel Prize. While we're "listening in," on this day in 1946 the US Army's Signal Corps became the first group to successfully bounce radar waves

Figure 1.16. Project Diana (widely used public image).

off the Moon. Although this might sound like a minor achievement, let's look just a bit further into what it **really** meant.

At the time, scientists were hard at work to find a way to pierce the Earth's ionosphere with radio waves—a feat widely believed to be impossible. Headed by Lt. Col. John DeWitt and a handful of full-time researchers, Project Diana was a modified SCR-271 bedspring radar antenna set up in the northeast corner of Camp Evans and aimed at the rising Moon (Figure 1.16). A series of radar signals were broadcast and, in each case, the echo was picked up in exactly 2.5 seconds—the time it takes for light to travel to the Moon and back. The significance of Project Diana cannot be overestimated: its discovery that communication was possible through the ionosphere opened the way to space exploration. Although it would be another decade before the first satellites were launched into space, Project Diana paved the way for these achievements, so don't forget to salute the Moon as it makes a brief appearance on the western horizon tonight!

Tonight let's continue with our "radio study" of an object associated with a molecular cloud as we venture to a region known as Lynds' Dark Nebula (Lynds 1630), and have a look at NGC 2071 (RA 05 47 05 Dec +00 21 19) (Figure 1.17).

Figure 1.17. NGC 2071 (Credit—Palomar Observatory, courtesy of Caltech).

At its core is the smallest protoplanetary disc yet to be detected. Revolving around a young star, this "disc" could have the potential to form a solar system, and its size is very similar to that of Neptune's orbit. Located 1300 light-years away, it contains compact clusters of water molecules which allow researchers to study its motion through radio emissions. Known as masers, these regions amplify radio emissions, and the entire area has also been subject to jet activity. Although we cannot see the disc itself, in an average telescope you may be able to detect a faint nebula associated with a 9th magnitude star. As with many types of objects, sometimes high magnification is not the answer. Try staying with minimal power to spot it.

Friday, January 11

Tonight in 1787, Sir William Herschel discovered two of Uranus' many moons— Oberon and Titania. As luck would have it, a moon and an outer planet will figure high on our observing list as well, because the tender crescent of our own Moon is less than half a degree north of Neptune! An appearance this close could mean an occultation event, so be sure to visit IOTA for accurate information for your area.

Since we cannot deny the Moon, let's begin our series of journeys designed to acquaint you with specific craters. Around midway on the terminator tonight, you will spot a conspicuous old crater named for Belgian engineer and mathematician Michel Florent van Langren. Better known as Langrenus, this handsome old crater stretches out over 132 kilometers in diameter with walls rising up to 1981 meters high (Figure 1.18). The deepest part of the crater reaches down 4937 meters and

Figure 1.18. Langrenus and Vendelinus (Credit—Alan Chu).

Figure 1.19. NGC 2169 (Credit—Palomar Observatory, courtesy of Caltech).

could swallow Ecuador's Mt. Cotacachi. Is the Sun rising over its brilliant east wall? If so, look closely and see if you can spot Langrenus' central mountain peak. Rising up 1950 meters, it's as high as the base elevation in Jackson Hole, Wyoming!

Similar to Sir William's observation of this night, let's try one of his challenge 400 objects. Wait until Orion has well risen and our lunar companion has ducked west. Our mark will be the third vertex of a triangle with Xi and Nu and point back in the direction of Betelgeuse. Its name is Collinder 83.

It is believed it may have been observed by Hodierna before 1654, but its discovery is credited to William Herschel in 1784 and was cataloged by him as H VIII.24. Located 3600 light-years away and best known as NGC 2169 (Figure 1.19), this 6th magnitude open cluster is very well suited to even smaller binoculars. A diffuse nebulosity accompanies this 50-million year old cluster, but a small telescope should be able to resolve out its 30 or so stellar members. No matter what optics you choose to use, one bright asterism will stand out—the number '37.' Enjoy…and write down your observations!

Saturday, January 12

Today celebrates the founding in 1830 of what—in 1831—would become the Royal Astronomical Society. The RAS was conceived by John Herschel, Charles Babbage, James South, and several others. The RAS has published its Monthly Notices continuously since 1831. Believed to have been born today in 1907 was Sergei Pavlovich Korolev (Figure 1.20). While few people recognize Korolev's name, he was a Soviet rocket engineer whose contributions to the science made him as important to the Russian space program as Robert Goddard or Wernher von Braun were to the United States. His work led to Sputnik, Vostok, Voskhod, and eventually the Soyuz program.

Tonight let's return to our own lunar mission as we travel to the northeast quadrant of the Moon and identify the emerging Mare Crisium (Figure 1.21). The "Sea of Crises" stretches out about 400–500 kilometers—an area about the size of the state of Washington. Mare Crisium is not only unique for its lack of connection with any other maria, but it's home to a gravitational anomaly called a mascon. This "mass concentration" might possibly consist of fragments of the asteroid or comet whose impact with the lunar surface created the basin buried beneath the lava flow. The mascon creates an area of high gravity and causes changes in the orbits of lunar probes. This excess gravity has even been known to cause low-orbiting lunar satellites to either crash on land or be flung out into space!

Now let's turn our attention toward Orion and a binocular and small telescope cluster known as Collinder 69. While many of us have looked at Orion's triangular head before, what we may not realize is the area surrounding 3rd magnitude Lambda is an open cluster (Figure 1.22). Containing approximately 19 stars

Figure 1.20.
Sergei Korolev
(Credit—NASA).

Figure 1.21. Mare Crisium (Credit—Greg Konkel).

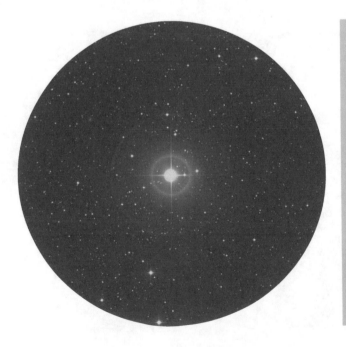

Figure 1.22. Lambda Orionis (Credit—Palomar Observatory, courtesy of Caltech).

ranging from 5th to 9th magnitude, look for a southward-extending chain which gives this collection its signature. The brightest of these stars is named as Meissa and the cluster itself is considered young. Probably formed no more than 10 million years ago, try again on a dark night and see if you can spot some nebulous filaments remaining from its birth!

Sunday, January 13

We begin our explorations tonight by looking near the center of the lunar terminator to identify Mare Fecunditatis (Figure 1.23). Stretching out across an area about equal in size to the state of California, the Sea of Fertility's western edge is lined with blocked areas of crust which have dropped below the thinly covered surface. Called "grabens," they are also bordered by parallel fault lines and are quite similar to such terrestrial features as Death Valley. Look along the eastern shore for a 125 kilometer long "bay" known as Sinus Successus. Congratulations on your own success at identifying these features!

Now turn your eyes toward the northeast corner of Orion and behold a variable star and a distant sun; astronomers have even observed large "hot spots" on the surface. It's designated Alpha Orionis—but is more commonly known as Betelgeuse (Figure 1.24). This star is so massive that if it were to replace our own Sun, it would fill our solar system out to the distance of the orbit of Jupiter, and is so distant that resolving it would be like aiming a telescope at a car headlight

Figure 1.23. Mare Fecunditatis (Credit—Greg Konkel).

Figure 1.24. Alpha Orionis: Betelgeuse (Credit—John Chumack).

from 9656 kilometers away. It is an irregularly pulsing, red supergiant which goes through its cycle about every 5.7 years and can drop in intensity by as much as a magnitude. Betelgeuse is also a well-known multiple star system, with four companions ranging from 11th to 14th magnitude, but it is believed its variability is caused by internal changes rather than by an eclipsing companion.

As you view this giant star tonight, keep in mind how much of its hydrogen has been expended and how many times it has expanded and contracted in the 425 years it took this light to reach your eyes. When it finally does go supernova, it will be almost half a century before we know it!

Monday, January 14

Let's begin our studies this evening on the lunar surface and have our first look of the year at crater Posidonius (Figure 1.25). Located on the northeastern shore of Mare Serenitatis, this huge, old, mountain-walled plain in considered a

Figure 1.25. Posidonius (Credit—Greg Konkel).

class V crater. Spanning 84 by 98 kilometers, you can plainly see Posidonius is shallow—dropping only 2590 meters below the surface. Tonight it will resemble a bright, elliptical pancake on the surface and we'll return to study it later in the year. For now, let's go warm up, wait for the Moon to set and Canes Venatici to rise...

Our destination is just slightly northwest of Beta CnV (RA 12 30 37 Dec +41 38 27). On this night in 1788, an astronomer was hard at work cataloging sky objects when he discovered the galactic pair of NGC 4490 and NGC 4485 to its north (Figure 1.26). At magnitude 10, the largest of this pair (4490) is an easy catch for the mid-sized scope, revealing an unusually shaped spiral sometimes called the "Cocoon Galaxy." Far more faint at magnitude 12, and considerably smaller, NGC 4485 is only 3' north, but its small size and low surface brightness make it more suited to larger aperture. Take a close look, however, for this diminutive study is also classed as an irregular elliptical.

Located around 25 million light-years from our own galaxy, this galactic pair has also been studied by the Chandra X-ray Observatory in the search for intermediate mass black holes. NGC 4490 has an active galactic nucleus, and both show evidence of ultra-luminous X-ray sources known as ULXes. Thanks to modern spectral photography, five sources of ULX have been identified to date.

If you think it's too cold to observe, then imagine Sir William Herschel braving the environment without modern clothing. Not only did he catalog NGC 4490

Figure 1.26. NGC 4490 and NGC 4485 (Credit—Palomar Observatory, courtesy of Caltech).

as H I.198 and NGC 4485 as H I.197, but he went on to discover an additional 27 objects on this same night! Be sure to make a note for your Herschel "400" object list... Sir William would be very proud.

Tuesday, January 15

On this date in 2006, the Stardust Mission safely returned its payload of cometary dust! The capsule of aerogel released over the Utah desert landed success-fully, allowing study of particles belonging to Comet Wild 2. Launched in 1999, the mission reached its target in 2004 and two years later provided the amateur community an opportunity to contribute to science firsthand through the "Stardust@Home" project (Figure 1.27).

Tonight on the lunar surface, all of Mare Serenitatis and Mare Tranquillitatis will be revealed, and so it is fitting we should take a look at both the "Serene" and the "Tranquil" seas (Figure 1.28).

Formed some 38 million years ago, these two areas of the Moon have been home to most of mankind's lunar exploration. Somewhere scattered on the basalt landscape on the western edge of Tranquillitatis, a few remains of the Ranger 6 mission lie tossed about, perhaps forming a small impact crater of their own. Its eyes were open, but blinded by a malfunction...forever seeing

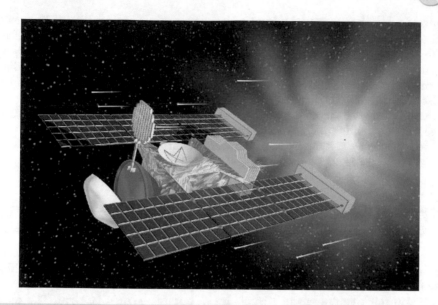

Figure 1.27. Artist's concept of Stardust (Credit—NASA).

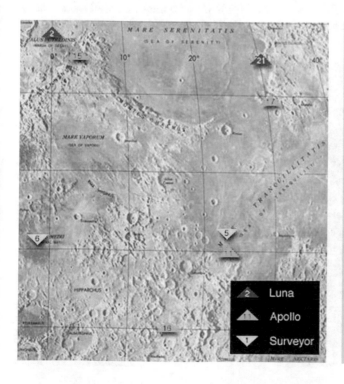

2	Luna
1	Apollo
1	Surveyor

Figure 1.28.
Serenity–Tranquility
region (Credit—NASA).

nothing. To the southwest edge lie the remnants of the successful Ranger 8 mission which sent back 7137 glorious images during the last 23 minutes of its life. Nearby, the intact Surveyor 5 withstood all odds and made space history by managing to perform an alpha particle spectrogram of the soil while withstanding temperatures considerably greater than the boiling point. Not only this, but it also took over 18,000 pictures!

Look closely at the map and you will find this area is also home to many of the Apollo landing sites, as well as the landing places of several unmanned craft. It is a place you can deeply appreciate for its historical significance...and is an Astronomical League Lunar Challenge!

Wednesday, January 16

Tonight if you'd like to try for the strange and unusual, your destination is about a thumb's width northwest of Betelgeuse. Open wide and say hello to Cederblad 59 (RA 05 46 00 Dec +09 00 00) (Figure 1.29)!

If you think you're seeing a whole lot of nothing—you'd be right. The main thing in this area is dark nebula Barnard 35, but our focus is on a peculiar little yellow star, FU Orionis. Is it a slow nova or is it a variable? One thing we do know is that we are looking at something through very heavy layers of obscuring dust which is at least several thousand light-years away. It has been proposed that FU may actually be a star just beginning to turn on!

As the prototype for this kind of study, FU Orionis may very well have a large accretion disc, and even an equally large planet—possibly the underlying cause

Figure 1.29. Cederblad 59, B35, and FU Orionis (Credit—Palomar Observatory, courtesy of Caltech).

of the huge mass shifts which trigger its activity. Although we aren't certain right now what causes the shifts in brightness, it's certainly an area of the sky worthy of your time and attention.

Thursday, January 17

Let's begin the evening with lunar study as we have a deeper look at the "Sea of Rains." Our mission is to explore the disclosure of Mare Imbrium (Figure 1.30), home to Apollo 15.

Stretching out 1123 kilometers over the Moon's northwest quadrant, Imbrium was formed around 38 million years ago when a huge object impacted the lunar surface creating a gigantic basin. The basin itself is surrounded by three concentric rings of mountains. The most distant ring reaches a diameter of 1300 kilometers and involves the Montes Carpatus to the south, the Montes Apenninus to the southwest, and the Caucasus to the east. The central ring is formed by the Montes Alpes, and the innermost has long been lost except for a few low

Figure 1.30. Mare Imbrium (Credit—NASA).

Figure 1.31. NGC 1907 (Credit—Palomar Observatory, courtesy of Caltech).

hills which still show their 600 kilometer diameter pattern through the eons of lava flow.

Originally, the impact basin was believed to be as much as 100 kilometers deep. So devastating was the event that a Moon-wide series of fault lines appeared as the massive strike shattered the lunar lithosphere. Imbrium is also home to a huge mascon, and images of the far side show areas opposite the basin where seismic waves traveled through the interior and shaped its landscape. The floor of the basin rebounded from the cataclysm and filled in to a depth of around 12 kilometers. Over time, lava flow and regolith added another five kilometers of material, yet evidence remains of the ejecta which was flung more than 800 kilometers away, carving long runnels through the landscape.

If you're feeling adventuresome, head off to Auriga for NGC 1907 (Figure 1.31). Situated just southwest of the grand M38 (RA 05 28 06 Dec +35 19 30), this pretty, magnitude 8 open cluster was discovered on this night by Herschel in 1787 and is on the "400" list as H VII.39!

Friday, January 18

Very early this morning the Moon will be very near the Pleiades. This will be the first of a series of possible occultations of the Seven Sisters this year. IOTA

Figure 1.32. Sinus Iridum (Credit—Greg Konkel).

has a page specifically devoted to occultations by the Pleiads (mostly for North America), so be sure to visit their site for details and maps of all of them.

Tonight we will return again to the lunar surface to have a look through binoculars or telescopes at another tremendous impact region—Sinus Iridum (Figure 1.32).

Sinus Iridum is one of the most fascinating and calming areas on the Moon. At around 241 kilometers in diameter and ringed by the Juras Mountains, it's known by the quiet name of the Bay of Rainbows, but was formed by a cataclysm. Astronomers speculate that a minor planet around 200 kilometers in diameter impacted our Moon at a glancing angle, and the result of the impact caused "waves" of material to wash up to a "shoreline," forming this delightful C-shaped lunar feature. The impression of looking at an earthly bay is stunning as the smooth inner sands show soft waves called "rilles," broken only by a few small impact craters. The picture is completed by Promontoriums Heraclides and LaPlace, which tower above the surface, at 1800 meters and 3000 meters respectively, and appear as distant "lighthouses" set on either tip of Sinus Iridum's opening. Enjoy this serene feature tonight... It's a Lunar Club challenge!

Now challenge yourself to a 6th magnitude open cluster just northwest of the top star in Orion's bow (RA 04 49 24 Dec +10 56 00) as we have a look at NGC 1662 (Figure 1.33). Discovered on this night in 1784 and cataloged as H VII.1 by Sir William Herschel, it won't make the popular lists because it's nothing more than a double handful of stars...or is it? Studied extensively for proper motion,

Figure 1.33. NGC 1662 (Credit—Palomar Observatory, courtesy of Caltech).

this galactic cluster may have once held more stars earlier in its lifetime. Enjoy its bright blue and gold members and mark your notes for locating a binocular deep-sky object!

Saturday, January 19

Somewhere out there before dawn in the year 1779, Charles Messier was still awake and at the eyepiece discovering M56! And Johann Bode (Figure 1.34) was born today in 1747. He was the publicizer of the Titus-Bode law, a nearly geometric progression of the distances of the planets from the Sun. Also born today, but in 1851, was Jacobus Kapteyn (Figure 1.35). Kapteyn studied the distribution and motion of half a million stars and created the first modern model of the size and structure of the Milky Way Galaxy.

Tonight I have a very special treat in store for you. For observers everywhere the Moon and Mars will make a wonderful close appearance. But, for a great many of you, this will be a chance to watch the Moon occult a bright planet! Be sure to check IOTA for accurate timing information in your area.

Now let's head to the lunar north as we explore another challenge region— Sinus Roris (Figure 1.36). "The Bay of Dew" is actually a northern extension of the vast region of the Oceanus Procellarum. Extending for about 202 kilometers,

Figure 1.34. Johann Bode (widely used public image).

many lunar maps aren't quite true to Sinus Roris' dimensions. Its borders are not exactly clear, given the curvature on which we see this feature, but we do know the eastern edges join Mare Frigoris. This area is much lighter than most features of this type. If you seek answers, then look further north as Roris' high albedo can be attributed to the ejecta from many nearby impacts.

It also holds a fanciful place in history, as seen in this excerpt from the science fiction story "Man on the Moon" by Wernher van Braun:

There's one section of the moon that meets all our requirements, and unless something better turns up on closer inspection, that's where we'll land. It's an area called Sinus Roris,

Figure 1.35. Jacobus Kapteyn (widely used public image).

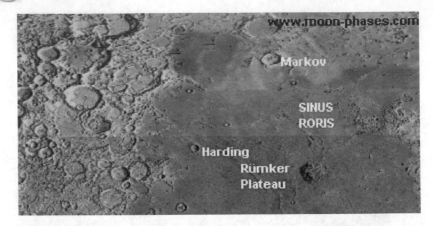

Figure 1.36. Sinus Roris (Credit—Alwyn Botha).

or "Dewy Bay," on the northern branch of a plain known as Oceanus Procellarum, or "Stormy Ocean" (so called by early astronomers who thought the moon's plains were great seas). Dr. Fred L. Whipple, chairman of Harvard University's astronomy department, says Sinus Roris is ideal for our purpose—about 1000 kilometers from the lunar north pole, where the daytime temperature averages a reasonably pleasant 40 degrees and the terrain is flat enough to land on, yet irregular enough to hide in.

Journey there tonight... And look for the Lunar Club Challenge, the "Man In The Moon!"

Sunday, January 20

Simon Mayr was born today in 1573 (Figure 1.37). Although Mayr's name is not widely recognized, we all recognize the names he's given to some satellites of Jupiter. Mayr observed the moons of Jupiter at nearly the same time as Galileo, and he was the one who assigned them the Greek names in use today. If you're up before dawn, look for Jupiter and see if you can spot Europa, Io, Ganymede, and Callisto and for yourself! They will all be to the west of Jupiter this morning, strung out in the order listed. Europa is closest, just 36" away, and Callisto is the furthest, 6' 42" west of the mighty planet.

While you're out, take a few minutes to watch the skies for the peak of the Coma Berenicid meteor shower. Although the activity for this one is fairly weak, with an average fall rate of about 7 per hour, it still warrants study.

So what makes this particular shower of interest? Noted first in 1959, the stream was eventually tied in 1973 to another minor shower in the same orbit, known as the December Leo Minorids. As we know, meteoroid streams are traditionally tied to the orbit of a comet, and in this case the comet was unconfirmed! Observed in 1912 by B. Lowe, an Australian amateur astronomer, the comet was officially designated as 1913 I and was only seen four times before losing it to sunrise.

Figure 1.37. Simon Mayr (widely used illustration).

Using Lowe's observations, independent researchers computed the comet's orbit, but it was basically forgotten until 1954. At that time, Fred Whipple was studying meteoroid orbits and made the association between his photographic studies and the enigmatic comet Lowe. By continuing to observe the annual shower, it was deduced that the orbital period of the comet was about 75 years, but the two major streams occurred about 27 and 157 years apart. Thanks to the uneven dispersion of material, it may be another decade before we see some real activity from this shower, but even one meteor can make your day!

Monday, January 21

John Couch Adams was born today in 1792 (Figure 1.38). Adams predicted the existence of Neptune. While Neptune is close to the western horizon at sunset, even unaided sky watchers might stand a chance at catching the innermost planet, Mercury, as it reaches its greatest eastern elongation. Look for it on the western skyline just as the Sun sets!

Also born today in 1908 was Bengt Strömgren (Figure 1.39)—the developer of the theory of ionization nebulae (H II regions). Tonight we'll defy the bright Moon and take a look at an ionization nebula as we return for an in-depth look at M42 (Figure 1.40).

Known as "The Great Orion Nebula," M42 is actually a great cloud of glowing gases whose size is beyond our comprehension. More than 20,000 times larger than our own solar system, its light is mainly fluorescent. For most people under dark skies, it will appear to have a slight greenish color—the result of doubly ionized oxygen. At the fueling heart of this immense region is the "Trapezium," its four easily seen stars are perhaps the most celebrated multiple system in the night sky. The Trapezium itself belongs to faint cluster of stars which are

Figure 1.38. John Couch Adams (public domain image).

now approaching the main sequence stage in an area known as "Huygenian Region."

Buried in this cloud of mainly hydrogen gas, there are many star-forming regions amidst the bright ribbands and curls. Appearing like "knots" in the structure, these are known as "Herbig-Haro objects" and are believed to be stars in their earliest states. There are also a great number of faint reddish stars and erratic variables—very young stars which may be of the accreting T Tauri

Figure 1.39. Bengt Strömgren (public domain image).

Figure 1.40. M42: The Great Orion Nebula (Credit—Palomar Observatory, courtesy of Caltech).

type. Along with these are "flare stars," whose rapid variations mean amateur astronomers have a good chance to witness new activity.

While you view M42, note that the region appears very turbulent. There is a very good reason. The Great Nebula's many different regions move at different speeds in both recession and approach. The expansion rate at the outer edges of the nebula point to radiation from the very youngest stars known. Although M42 may be as old as 23,000 years since the Trapezium brought it to "light," it is entirely possible new stars are still forming there.

Tuesday, January 22

Tonight is the Full Wolf Moon. Its name comes from the North American Indians who would hear the wolves howling in search of food in the cold, snow-covered, and barren landscape. In Europe it was referred to as the Moon after Yule, and 388 years and 18 days ago, Galileo Galilei changed the face of astronomy when he observed it. Pointing his newly developed telescope at our nearest celestial neighbor, his observation of mountains and craters on the surface opened the world's eyes to what lies just beyond the range of human sight. Said Galileo, "It is a beautiful and delightful sight to behold the body of the Moon." Tonight discover

Figure 1.41. Tycho's Ray System (Credit—Roger Warner).

for yourself what Galileo saw. Using any type of optical aid, trace the bright lunar rays extending from the brilliant Tycho or the deep impact of Copernicus (Figure 1.41). Let's return to Orion and have a much closer look at the blue–white giant Beta Orionis. The seventh brightest star in the sky is known by the name Rigel. Very little is known about its true distance from Earth, but the widely accepted value is 900 light-years. This white-hot star has a surface temperature of about 12,000 Kelvin and is thousands of times more powerful than our own Sun. If it were as close to us as Sirius, it would shine with a light as bright as 20% of the full Moon! But look closely at the brilliant star... Intermediate-sized telescopes under good conditions will find a 6.7 magnitude blue companion. Although it is not always an easy double star, you'll find it on the list for many challenges. But, chances are, you will never see the C star which accompanies the B.

Even if you just view Rigel with your eyes tonight, marvel at this young and powerful star. When the light you see left this star, the Crusades had begun...the Vikings were sailing to discover America...the Mayan Empire was beginning to crumble...paper was a new concept...and the very numbers we use today were just beginning to catch on!

But, if you hear a wolf howl...perhaps it might be the "Dog Star" on the rise. Alpha Canis Majoris, better known as Sirius, the fifth nearest star known, has played an important role throughout astronomy history (Figure 1.42). It was seen by Ptolemy, Homer, and Plutarch. The ancient Egyptians revered it and the

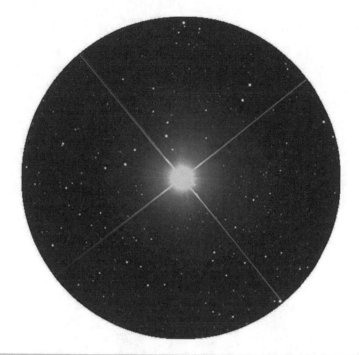

Figure 1.42. Sirius (Credit—John Chumack).

Greeks and Romans feared it. Watch its dazzling appearance while it is still fairly low and flashing all the colors of the rainbow. Enjoy it tonight and we'll be back to study...

Wednesday, January 23

With very little time before moonrise tonight, let's travel a fingerwidth northeast of Zeta Orionis and right on the celestial equator (RA 05 46 47 Dec +00 00 50) for a delightful area of bright nebulosity known as M78 (NGC 2068). Discovered by Méchain in 1789, M78 is part of the vast complex of nebulae and star birth comprising the Orion region. Fueled by twin magnitude 10 stars, the nebula almost appears to binoculars to resemble a "double comet." Upon close scrutiny with a telescope, observers will note two lobes separated by a dark band of dust. Each lobe bears its own designation: NGC 2067 to the north and NGC 2064 to the south. While studying, you will notice the entire area is surrounded by a region of absorption, making the borders appear almost starless. Filled with T Tauri-type stars and residing 1600 light-years away, this reflection nebula is a cloud of interstellar dust which reflects the light of these young stars, the brightest of which is HD 38563A. In 1919, Vesto Slipher was the first to discover its reflective nature. As of 1999, 17 Herbig-Haro objects have been associated with

Figure 1.43. M78 (Credit—Palomar Observatory, courtesy of Caltech).

M78, and are believed to be jets of matter being expelled from newly forming stars (Figure 1.43).

Tonight I ask you to journey to this vague region for the very sake of amateur astronomy...

On January 23, 2004 a young backyard astronomer named Jay McNeil was checking out his new 3" telescope by taking some long exposures of M78

Figure 1.44. Jay McNeil (press release photo).

Figure 1.45. McNeil's Nebula (Credit—Adam Block/NOAO/AURA/NSF).

(Figure 1.44). Little did Jay know at the time, but he was about to make a huge discovery! When he later developed his photographs, there was a nebulous patch there which had no designation. When he reported his findings to the professionals, they confirmed its novelty and realized that Jay had stumbled onto something quite unique! It is believed that Jay's discovery was a variable accretion disc around a newborn star—IRAS 05436-0007 (Figure 1.45). Little is known about the region, but it seems it had been caught in a photo once in the past but never studied. Even the Digital Sky Surveys had no record of it!

Although Jay's discovery might not be bright enough tonight to be seen just south of M78, it is a variable and circumstances play a big role in any observation. Before you assume being a backyard astronomer has no real importance to science, remember a teenager in a Kentucky backyard with a 3" telescope...catching what professionals had missed!

Thursday, January 24

Today is the birthday of American solar astronomer Harold Babcock (Figure 1.46). Born in 1882, Babcock proposed in 1961 that the sunspot cycle was a result of the Sun's differential rotation and magnetic field. Would you like to have a look at the Sun? Although solar observing is best done with a proper filter, it is perfectly safe to use the "projection method."

First off, *never* look at the Sun directly with the eye or with any unfiltered optical device, such as binoculars or a telescope! I'm not joking when I say this

Figure 1.46. Harold Babcock (widely used public image).

will **blind** you. Exposed film, mylar, and smoked glass are also *unsafe*. But don't be afraid, because I'm here to tell you how you, too, can enjoy the Sun. A *safe* way to observe sunspots is to "project" an image of the Sun through a telescope or binoculars onto a screen. This can be a simple as cardboard, a paper plate, a wall, or whatever you have handy. If you're using a telescope, be sure the finderscope is securely covered. If you'd like to try binoculars, just keep the cover on one of the two tubes. By using the shadow method, you will see a bright circle of light

Figure 1.47. "The Sounds of Earth" (Credit—NASA).

on your makeshift screen. This is the solar disc. Adjust the focus by moving the distance of the screen from your scope or binoculars until it is about the size of a small plate. If the image is blurry, use your manual focus until the edges of the disc become sharp. Even though it might take a little practice, you'll soon become proficient at this method and you'll be able to see a surprising amount of detail in and around sunspot areas. Happy and *safe* viewing to you all!

Today in 1986, the United States Voyager 2 was the first spacecraft to fly by Uranus, providing us on Earth the most outstanding photographs and information on the planet to date. After 11,114 days of successful operation, Voyager 2 still continues on toward the stars carrying "The Sounds of Earth" (Figure 1.47).

Later tonight, watch as the Moon and Regulus rise together. For most viewers, the pair will be around a half degree apart, but for some fortunate astronomers this could be an occultation event. Be sure to check IOTA for precise timing information in your area.

Friday, January 25

Today is the birthday of Joseph Louis Lagrange (Figure 1.48). Born in 1736, this French mathematician made important contributions to the field of celestial mechanics. No, we aren't talking about wrenches in space, but Lagrange points (Figure 1.49). What exactly are they? Let's have a look...

Lagrange envisioned a location in space where gravity and orbital motion would balance. As we know, an orbiting body is drawn to the center of the mass around which it revolves—be it the Earth and Moon, or the Earth and Sun. Five such places exist for each combination, but anything which orbits closer to the Sun revolves faster and would eventually pass the Earth. Thanks to Lagrange's calculations, a spacecraft placed about a hundredth of the distance between the Earth and Sun requires very little correction to maintain its orbit and keeps the same pace as the Earth. This position is known as Lagrange Point 1—a position enjoyed by the most prolific solar "observer" of all...the SOHO satellite!

Figure 1.48. Joseph Lagrange (widely used public image).

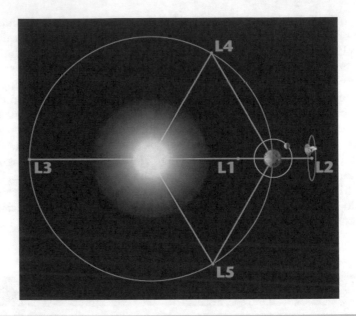

Figure 1.49. Lagrange points (Credit—NASA).

Tonight let's do some binocular studies of open clusters which belong to alternative catalogs. The first three are known as Dolidzes and your marker star is Gamma Orionis.

The first is an easy hop of about one degree northeast of Gamma—Dolidze 21. Here we have what is considered a "poor" open cluster. Not because it isn't nice—but because it isn't populous. It is home to around 20 or so low wattage stars of mixed magnitude with no real asterism to make it special. The second is about one degree northwest of Gamma—Dolidze 17. The primary members of this bright group could easily be snatched with even small binoculars, and would probably be prettier seen in this fashion. Five very prominent stars cluster together with some fainter members which are again, poorly constructed, but it includes a couple of nice visual pairs. Low power is a bonus on this one to make it recognizable.

The last of our Dolidzes is about two degrees north of Gamma—Dolidze 19. Two well-spaced, roughly 8th magnitude stars stand right out with a looping chain of far fainter stars between them and a couple of relatively bright members dotted around the edges. With the very faint stars added in, there are probably three dozen stars all told, and this one is by far the largest of this "Do" trio.

Now let's have a look at a deceptive open cluster located in Barnard's Loop around two degrees northeast of M78 (RA 05 53 45 Dec +00 24 36). Billed at a magnitude of roughly 8, NGC 2112 might be a binocular object, but it's a challenging one (Figure 1.50). This open cluster consists of around 50 or so stars of mixed magnitudes and only the brightest can be seen in small aperture. Add

Figure 1.50. NGC 2112 (Credit—Palomar Observatory, courtesy of Caltech).

a little more size in equipment and you'll find a moderately concentrated, small cloud of stars, fairly clearly distinguishable against the stellar background. Also known as Collinder 76, this unusual cluster resides in the galactic disc—an area of very old, metal poor stars. It is believed NGC 2112 is of an intermediate age, based on recent photometric and spectroscopic data.

Saturday, January 26

Today in 1962, the US space program launched a lunar probe named Ranger 3 (Figure 1.51). Its mission was to image the Moon right up until impact, land a seismometer, study gamma rays, and report on surface reflectivity of radar... But, it didn't happen. Two days after launch, the ill-fated Ranger 3 was on a runaway course toward the lunar surface when it received an erroneous command and lost contact with Earth. As a result, it overshot its mark by 36,800 kilometers and still remains in heliocentric orbit. If you're up before dawn this morning, have a look at the Moon. Not only is Ranger 3 still up there, but you'll find Saturn quite close to it as well!

Tonight let's jump to the border of Taurus and about a degree northeast of 15 Orionis as we have a look at two open clusters, NGC 1817 (RA 05 12 12 Dec +16 41 00) and NGC 1807 (RA 05 10 42 Dec +16 32 00) (Figures 1.52 and 1.53). The

Figure 1.51. Ranger 3 (Credit—NASA).

northernmost (1817) is a galactic cluster around the same age as the Hyades and, when studied for stellar motion, was found to contain several hundred members which truly belong to the cluster itself. Far less populous is the southern open (1807), which photometric measurements show may not truly be a cluster at all.

While both clusters are roughly the same magnitude, binoculars can spot NGC 1807 easily, but NGC 1817 remains a hazy patch. Even larger telescopes have difficulty resolving out this very rich cluster, which recent studies show contains as many as a dozen Delta Scuti–type stars. Be sure to mark your notes, for this pair is an urban observing challenge, a Herschel object, and a binocular deep sky target!

Sunday, January 27

On this day in 1967, tragedy struck at Pad 34. During a training exercise atop a Saturn IB rocket, astronauts Command Pilot Virgil I. Grissom, Senior Pilot Ed White, and Pilot Roger B. Chaffee gave their lives to further human exploration of space as fire swept through their module. Named "Apollo 1," stop for a moment tonight to remember these brave souls (Figure 1.54).

Figure 1.52. NGC 1817 (Credit—Palomar Observatory, courtesy of Caltech).

Figure 1.53. NGC 1807 (Credit—Palomar Observatory, courtesy of Caltech).

Figure 1.54. Apollo 1 crew (Credit—NASA).

They gave their lives in service to their country in the ongoing exploration of humankind's final frontier. Remember them not for how they died but for those ideals for which they lived.

From the launch pad commemorative plaque

And another man with ideals was Sir William Herschel, who was observing on this night in 1786. During this evening so long ago, he discovered seven new objects and tonight we'll take a look at just one—NGC 2186 (Figure 1.55).

Located just a little more than a degree east of 61 Orionis (RA 06 12 06 Dec +05 28 00), this magnitude 8 open cluster is within the realm of larger binoculars, but truly takes a little aperture to fully appreciate. Cataloged as H VII.25 this delightful small concentration of stars is bordered on either side by far brighter members. While it holds no special study appeal, it is on the well-known "400"

Figure 1.55. NGC 2186 (Credit—Palomar Observatory, courtesy of Caltech).

list and shows just the kind of dedication Sir William displayed in finding these small galactic jewels for future generations.

Monday, January 28

Today take the time to honor shuttle commander Dick Scobee, pilot Mike Smith, astronauts Ellison Onizuka, Judy Resnik, Ron McNair, and Greg Jarvis, and teacher Christa McAuliffe. They were the crew onboard the Challenger when it exploded on this day in 1986 (Figure 1.56). Godspeed...

We will never forget them, nor the last time we saw them, this morning, as they prepared for the journey and waved goodbye and slipped the surly bonds of Earth to touch the face of God.

President Ronald Reagan

Today also celebrates the birth of Johannes Hevelius in the year 1611. So what, you say? Then think on this... Hevelius was using a telescope to view the Moon's surface and produced the very first detailed maps which were published as *Selenographia* in 1647. That's 361 years ago! The Polish astronomer then went on to name a constellation which still remains in use today—Lynx. When asked to explain how he came up with the name, he said an observer needed to have eyes like a lynx just to see it!

Figure 1.56. Challenger crew (Credit—NASA).

Figure 1.57. Jones-Emberson 1 (Credit—Palomar Observatory, courtesy of Caltech).

If your Lynx-like eyes allow you to see the constellation, perhaps you'd like to try your hand at a very difficult planetary nebula? Perek/Kohoutek (PK) 164+31.1 (RA 07 57 51 Dec +53 25 17) is also known as Jones-Emberson 1 (Figure 1.57) after the discoverers of its central star... But don't think for one minute that you'll discover it with ease! Although it holds a respectable visual magnitude of 13, this ancient planetary is of such low surface brightness that it's often considered to be more like magnitude 17. Like all extreme objects, it's worth a try if you have large aperture and a nebula filter. The large ring structure can just be hinted at with averted vision on a good, dark night.

Tuesday, January 29

Over the next two nights let's embark on a very different type of study, as we have a look at 1 Geminorum and the region around it. If you have reasonably dark, clear skies then you can spot it unaided as the "toe" of the Gemini Twin Castor's "foot." While you might not think this 151 light-year distant, G spectral–type star is interesting, take a closer look with a telescope and you'll discover it's a very tight, similar-magnitude double star.

If I had to point you somewhere in the sky tonight with binoculars or a small telescope, I could not think of a finer place than M35 (Figure 1.58), located about a fingerwidth northeast of 1 Geminorum. Under very dark skies it will appear as a faint, fuzzy patch to the unaided eye and explode into a magnificent

Figure 1.58. M35 and NGC 2158 (Credit—John Chumack).

field of stars with optical aid. More than 200 individual stars grace this grand 2200 light-year distant galactic cluster, and sharp-eyed observers will note many different spectral types. While low power is essential to appreciate M35's starry span, larger telescopes will see another gift in the field as well—open cluster NGC 2158. Located just a breath away to the southwest of its stellar neighbor, this very rich, small cluster is quite near the outer edges of our galaxy, and was once thought to be a globular cluster candidate.

Wednesday, January 30

Tonight we'll continue our studies in Gemini with an open cluster which can be spotted in binoculars, but is more fully appreciated with the telescope. Your mark lies about a degree west of 1 Geminorum (RA 06 01 06 Dec +23 19 00) and its name is NGC 2129 (Figure 1.59). In smaller optics at low power, this tiny, sparse cluster might not appear to be too much when you first glimpse it, but power up. Located in our own local spiral arm, studies have shown it to be a very young cluster—not more than 10 million years old. Most optics can easily resolve the optical double in the center of this one! Be sure to mark your notes with H VIII.26.

For larger telescopes looking for a challenge, try IC 2157 around two degrees northeast of 1 Geminorum (RA 06 05 00 Dec +24 00 00) (Figure 1.60). Much like galactic cluster NGC 2158, this one is very distant and difficult to resolve with the average telescope. At a little brighter than magnitude 9, it's a bit sparser than its

Figure 1.59. NGC 2129 (Credit—Palomar Observatory, courtesy of Caltech).

Figure 1.60. IC 2157 (Credit—Palomar Observatory, courtesy of Caltech).

neighbor, but makes up for what it lacks with a concentration of hot, OB stars. Welcome to a cluster first studied by T. E. Espin!

Thursday, January 31

Today in 1961, Mercury-Redstone 2 was launched, carrying Ham the chimpanzee into a suborbital flight and to fame (Figure 1.61). In 1966, Luna 9 was launched. In 1958, the first US satellite—Explorer 1—was launched and met a milestone as it proved the Earth was surrounded by intense bands of radiation which we now refer to as the Van Allen Belts. In 1971, Apollo 14 was headed toward the Moon on this day...and we'll be back to study its landing area next month.

On this night in 1862, Alvan Graham Clark Jr. was at the eyepiece and made an unusual discovery (Figure 1.62). While watching Sirius, Clark uncovered the intense star's faint companion while testing an 18" refractor being built for Dearborn Observatory. The scope itself was built by Clark, his father, and his brother. Imagine his excitement when it turned up the white dwarf—Sirius B! Friedrich Bessel had proposed its existence back in 1844, but this was the first time it was confirmed visually.

Tonight, why not try your own hand at turning up this difficult double star? Also known as the "Scorching One," Alpha Canis Majoris is the brightest of

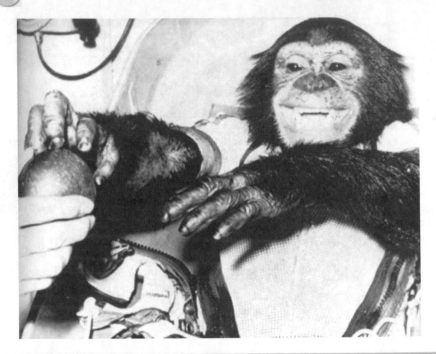

Figure 1.61. Ham (Credit—NASA).

Figure 1.62. Alvan Clark (widely used public image).

the fixed stars at an amazing magnitude of -1.42. With the exception of Alpha Centauri, Sirius is the closest of all the stars we can see unaided at only 8.7 light-years distant—but it's not standing still. As part of the Ursa Major Stream of moving stars, it has changed its position by 1.5 times the apparent width of the Moon in just 2000 years!

In the telescope, this main sequence star is a dazzling white tinged with blue. But thanks to our atmosphere, Sirius' light will produce all the colors of the rainbow as it sparkles in our eyes. For many of us this beautiful iridescence is all we will ever see of Sirius, but for those with 10" and larger telescopes, a perfectly steady sky will reveal Alpha Canis Majoris' secret—the white dwarf companion! Although this 8.5 magnitude star is well within the range of even small scopes, the blinding glare of the primary makes it a very elusive target. In another 20 years it will have reached its maximum separation of 11.5", but keep a watch to the southeast as you view the "Scorching One" tonight—perhaps you'll spot Sirius B!

If you had problems finding the companion, then don't worry. Back in 1948 on this same night, the first test photos using the Hale 5 meter (200 inches) telescope at Mt. Palomar were underway. Believe it or not, problems with the configuration and mounting of the mirror meant it was almost 2 years before the first official observing run was made by a scheduled astronomer!

CHAPTER TWO

February, 2008

Sky watcher alert! Be sure to set your alarm for before local dawn. This morning Venus and Jupiter will be separated by less than a degree and will make a stunning sight which requires no special equipment to enjoy. Not to be outdone, the Moon and Antares are also less than a degree apart, and this could be an occultation event. Be sure to check IOTA for accurate information in your area. Wishing you clear skies!

On this night in the year 1786, Sir William Herschel was quite busy at the eyepiece adding 18 new deep sky studies to his catalogs. One of these discoveries was IC 434, better known as "Barnard's Loop." If you missed your chance to view it last month, try again tonight. When you are finished, let's head toward Beta Eridani and take a two degree hop north (RA 05 06 54 Dec –03 20 30) to another object Herschel captured tonight—NGC 1788 (Figure 2.1).

Surrounding a handful of stars, this faint reflection nebula is easily captured in small- and mid-sized telescopes and will appear a bit more colorful to larger aperture since its gases and dust reflect mainly in the blue wavelengths. Much like the grand M42, this is a star-forming region—but it's an isolated molecular cloud—one containing low mass pre-main sequence members and brown dwarf candidates. But that's not all... Look closely at the southern region just outside its boundaries and tell me what you see? That's right: nothing. You are also looking at a dark dust obscuration region known as Lynds 1616. Although it is

Figure 2.1. NGC 1788 (Credit—Palomar Observatory, courtesy of Caltech).

only "somewhat" opaque as dark nebulae go, that's just another good reason to look this one up! Be sure to mark your Herschel "400" study list for H V.32.

Saturday, February 2

With a considerable amount of time before the Moon rises tonight, let's bundle up in parkas and snow boots as we head for a namesake planetary nebula so fitting for this time of year—the "Eskimo!" (Figure 2.2).

Fairly easily found by locating Delta Geminorum (Wasat) at the "waist" of Gemini, use the finderscope to locate the wide double 63 Geminorum to the east. You will find NGC 2392 is only two-thirds of a degree southeast (RA 07 29 11 Dec +20 54 42). Discovered by Sir William Herschel in 1787, the Hubble telescope revealed this planetary nebula in all its glory 213 years later. Containing a Sun-like central star, the outer gas shell is so complex that even today's science cannot fully understand it. Containing particle filaments in the inner disk which are being propelled outward by the wind of the central star, astronomers can only speculate what causes the outer disc to contain light-year long "knots" which may have originated at the same time as the parent star.

Easily distinguished in even small telescopes as a blue–green disc, this colorful planetary nebula is around 3000 light-years from us and requires large aperture to

Figure 2.2. NGC 2392: The Eskimo Nebula (Credit—Palomar Observatory, courtesy of Caltech).

truly appreciate. At high power the 10th magnitude central star is very apparent, and some of the features seen in photographs can be caught. If you have a nebula filter, use it: even more structure is revealed. While our eyes can never resolve the "Eskimo" as well as CCD imaging can, it is still possible to see the faint halo which surrounds the inner nebulosity, appearing like a "parka-style hood" around a human face. Who knows? If we stay out in the cold long enough, we might even see "polar" bears rising in the northeast! Welcome back, Ursa Major...

Sunday, February 3

Tonight we celebrate the success of Luna 9, also known as Lunik 9 (Figure 2.3). On this day in 1966, the unmanned Soviet lunar probe became the first to achieve a soft landing on the Moon's surface and successfully transmit photographs back to Earth. The lander weighed in at 99 kilograms, and the four petals, which formed the spacecraft, opened outward. Within five minutes of landing, antennae sprang to life and the television cameras began broadcasting back the first panoramic images of the surface of another world, proving that a lander would not simply sink into the lunar dust. Last contact with the spacecraft occurred just before midnight on February 6, 1966.

Figure 2.3. Luna 9 landing capsule (Credit—NASA).

If you are up before dawn this morning, get out your binoculars and have a look at the area of this first successful landing on the Moon—Oceanus Procellarum, the "Ocean of Storms." While it will be difficult to pick out small features, Procellarum is the long, dark expanse which runs from lunar north to south. On its western edge, you can easily identify the dark oval of Grimaldi. About one Grimaldi-length northward and on the western shore of Procellarum is where you would find the remains of Luna 9. While no Earth-bound optics could ever hope to achieve resolution of the mission's remains, it is still a wonderful way to improve your skills and enjoy a bit of history at the same time.

Since it's the weekend, why not hang around until midnight with William Herschel? He and Messier might be gone, but one of their discoveries is not. Tonight your goal is a little more than a fistwidth east of Sirius for a hazy patch of stars known as M47 (Figure 2.4).

M47 was known long before Messier's time because it approaches unaided eye visibility. When Charles discovered this 5th magnitude beauty in February 1771 he described it as a brighter neighbor of M46, but incorrectly logged its position! Thus was born the "missing Messier"—its mystery remained until 1934 when Oswald Thomas identified it. It's rather funny to note that because of this "messy mistake," William Herschel also saw it on this night in 1785, and cataloged it as H VIII.38. Both Herschel and Dreyer had problems with this one... But you won't have any problems as you view this bright cluster in either binoculars or telescope. It is a loose open cluster around 78 million years old which contains

Figure 2.4. M47 (Credit—John Chumack).

around 50 stars of various magnitudes in a region about the same size as the Full Moon. At roughly 1600 light-years away, you might even get a glimpse of an orange giant or two, along with beautiful double Sigma 1121 in its center!

Monday, February 4

This morning, discover some of the most beautiful celestial scenery you'll encounter all year as the crescent Moon, Jupiter, and Venus all dance together in the predawn skies. This one will be worth getting up early for!

Today is the birthday of Clyde Tombaugh (Figure 2.5). Born in 1906, Tombaugh discovered Pluto, just 24 years and 2 weeks after his birth. Have you ever wondered what part of the sky Tombaugh was looking at when last of our "classical" planets were identified? Then tonight turn your eyes to the skies and look about a fistwidth southwest of Castor and Pollux for fainter Delta Geminorum (Figure 2.6).

Known as Wasat, this 3.5 magnitude K-type star is around 53 light-years distant and has a visual companion easily seen with binoculars. Turn a telescope its way and you'll soon find that Wasat is a very beautiful binary with a disparate 8.2 magnitude red companion! The pair orbit each other extremely slowly—on the order of about 1200 years. It is suspected the primary could also be a very close

Figure 2.5. Clyde Tombaugh (widely used public image).

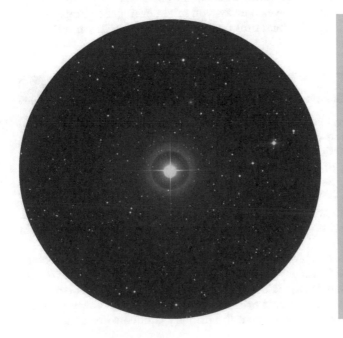

Figure 2.6. Delta Geminorum (Credit—Palomar Observatory, courtesy of Caltech).

Figure 2.7. NGC 2420 (Credit—Palomar Observatory, courtesy of Caltech).

binary. Oddly enough, we often associate Pluto with Ophiuchus, just as we do Uranus with Capricorn... But both objects were in the constellation of Gemini during their discovery!

For an 8th magnitude telescopic challenge, head about four degrees east of Wasat (RA 07 38 23 Dec +21 34 24) for the rich galactic cluster NGC 2420 (Figure 2.7). Residing about 3000 light-years above the galactic disk, this deceptive looking cluster's members have an average chemical composition much like our own Sun, yet it is only about 1.7 billion years old. Containing about 1000 stars which range from members of the main sequence to red giants and blue stragglers, NGC 2420 is truly a cosmological playground! Be sure to mark your notes for H VI.1...

Tuesday, February 5

On this day in 1963, Maarten Schmidt (Figure 2.8) measured the first quasar redshift, and in 1974 the first close-up photography of Venus was made by Mariner 10. If you're up before dawn, see if you can spot the brilliant return of Venus.

Tonight let's journey further south into Lepus as we take a look at Alpha. Its name is Arneb and it is a quality double star which resides around 900 light-years distant. Arneb's 11th magnitude disparate companion will take a larger scope

Figure 2.8. Maarten Schmidt (widely used public image).

to resolve. Its wide separation of 35.5" means it is probably not a true physical companion, but it is a challenge worthy of your time.

For binoculars and small scopes, hop due east of Alpha about a fingerwidth for a brilliant multiple star system which is also designated as an open cluster—NGC 2017 (Figure 2.9). First cataloged by Sir William Herschel at the Cape of Good

Figure 2.9. NGC 2017 (Credit—Palomar Observatory, courtesy of Caltech).

Hope, this interesting group of stars will show in the same field as Alpha Leporis in binoculars, but come to colorful life in the telescope. The stars in this small open cluster are all gravitationally bound to each other and are a well-studied source of both radio and infrared emissions. NGC 2017 produces a dense wind from a thin H II region hidden within it, which may be from a loose distribution of gas and dust. Power up. As aperture increases, so does resolution. Watch as the primary colorful members begin to split into disparate pairs as the combination of aperture and magnification increases. It's a much underrated jewel box!

Wednesday, February 6

On this day in 1971, astronaut Alan Shepard became the first "lunar golfer" as he teed off on the Moon's surface (Figure 2.10). Think the ball is still in orbit? Then think again as his shot made a successful "hole in one" in a crater just tens of meters away! Today we also celebrate the fiery return of the Soviet Space Station Salyut 7 (Figure 2.11). Launched into orbit in 1982, the station was plagued by electrical and maneuvering problems. Despite this, cosmonauts would remain onboard for as long as eight months before returning to Earth. The project was abandoned in 1986, but some of the equipment and supplies were transferred to the orbiting Mir. On this day in 1991, the Soviet space station Salyut re-entered our atmosphere and was lost.

Have you ever wondered if you can spot orbiting spacecraft? Yes, you can. Many objects are visible to the unaided eye if you know where and when to look. Try checking with www.heavens-above.com for highly accurate information

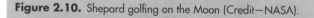

Figure 2.10. Shepard golfing on the Moon (Credit—NASA).

Figure 2.11. Salyut 7 (widely used public image).

Figure 2.12. NGC 2537 (Credit—Palomar Observatory, courtesy of Caltech).

for your specific area. Many events are wonderful to witness. Among the most spectacular sights is an Iridium flare—the Sun reflecting off the highly polished sides of a communications satellite. And an International Space Station fly-over is also a wonder to behold! Try it tonight...

Of course, it was a dark night for Sir William as well and although he scored eight objects (NGC 2204 made the "400" list), we're going to look for one more fitting for our dark night and large scopes—the "Bear Paw." Located about three degrees north-northwest of 31 Lynx (RA 08 13 15 Dec +45 59 30), NGC 2537 (Figure 2.12) is a part of a trio of three galaxies including the A component and IC 2233. At slightly fainter than magnitude 11, it's a superior catch for mid-sized scopes and begins to show signs of structure for larger aperture—such as bright star-forming regions. Because of its unusual characteristics, H IV.56 is also known as peculiar galaxy Arp 6!

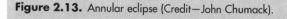

Thursday, February 7

On this day in 1889, the first American national astronomy organization was born—the Astronomical Society of the Pacific.

Today the place to be is in the Southern Hemisphere—especially Antarctica! An annular solar eclipse will be visible over part of Antarctica, while a partial eclipse will be viewable for all of New Zealand and a portion of eastern Australia (Figure 2.13). For these lucky observers, the next two nights will be the peak of the Centaurid meteor shower. Discovered by Michael Buhagiar of Australia, this stream has two radiants—Alpha and Beta. While both occur at roughly the same time and roughly from the same place, tonight's Alpha peak has a regular fall rate of around three per hour and an average magnitude of 2.4; while tomorrow's Beta stream produces up to 14 per hour, and they are far brighter at magnitude 1.6. Wishing our mates clear skies for these events!

For those of you a bit more to the north with small telescopes or binoculars—don't despair. Instead wait for Hydra to rise and try capturing what Herschel

Figure 2.13. Annular eclipse (Credit—John Chumack).

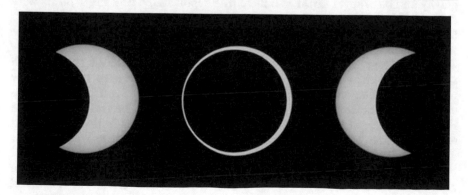

Figure 2.14. NGC 3242: Ghost of Jupiter (Credit—John Chumack).

did on this night in 1785! The "Ghost of Jupiter" planetary nebula is roughly magnitude 8 and can be found about a fingerwidth south of Mu Hydrae (Figure 2.14). Although it takes a larger telescope to resolve the eye-like features in this bright nebula, you can still expect to see the blue disc of H IV.27!

Friday, February 8

Today celebrates the discovery of the Sayh al Uhaymir 094 "Mars Meteorite" (Figure 2.15). Found this day in 2001, scientists had long known Mars' surface was home to many impact craters which may have caused space-born debris. It was only a matter of time before a bit of this debris would be captured by Earth's gravity and be brought down as a meteorite. Upon study, tiny gas deposits were discovered in it which nearly matched the atmosphere of Mars as measured by the Viking Landers, and its mineral composition also leads scientists to believe the meteor originated from Mars.

In mythology, Lepus the Hare is hiding in the grass at Orion's feet. As we will see, there are many objects of beauty hidden within what seems to be a very ordinary constellation. Before the Moon hides the "Rabbit" for this month, there are some objects worthy of our attention. If you look to the feet of Orion and the brightest star of Lepus, you will see they make a triangle in the sky. Tonight we are headed toward the center of that triangle (RA 05 19 45 Dec –25 03 50) for a singular deep sky object—the Spirograph Nebula (Figure 2.16).

Shown in all its glory through the eye of the Hubble Telescope, the light you see tonight from the IC 408 planetary nebula left in the year 7 AD. Its central

Figure 2.15. Mars Meteorite SaU 094 (Credit—NASA).

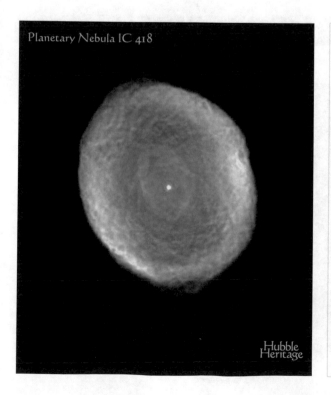

Planetary Nebula IC 418

Hubble
Heritage

Figure 2.16. The Spirograph Nebula (Credit—Hubble Heritage Team (NASA).

star, much like our own Sol, was in the final stages of its life at that time, and just a few thousand years earlier was a red giant. As it shed its layers into about a tenth of a light-year of space, only its superheated core remained—its ultraviolet radiation lighting up the expelled gas. Perhaps in several thousand years the nebula will have faded away, and in several billion years more the central star will have become a white dwarf—a fate which awaits our own Sun.

At magnitude 11, it is well within reach of a small- to medium-sized telescope. Like all planetary nebulae, the more magnification the better the view. The central star is easily seen against a slightly elongated shell, and larger telescopes bring an "edge" to this nebula which makes it very worthwhile studying. Spend some quality time with this object. With larger scopes, there is no doubt a texture to this planetary will delight the eye...and touch the heart!

Saturday, February 9

Did you spot the very young crescent of the Moon tonight along the western horizon? Be sure to mark your lunar observing notes for "the New Moon in the Old Moon's Arms." With Selene soon absent, it's time to get more serious about Lepus and do some galaxy hunting. Our first mark will be Mu Leporis, with NGC 1832 (Figure 2.17) in the same field to its north (RA 05 12 03 Dec –15 41 16).

Figure 2.17. NGC 1832 (Credit—Palomar Observatory, courtesy of Caltech).

Figure 2.18. NGC 1964 (Credit—Palomar Observatory, courtesy of Caltech).

At a rough magnitude of 12, this small galaxy isn't for the small scope, but is reasonably bright and easy to study with aperture. It is an ongoing object of study because of its spiral arm pattern, rotation rate, and star formation rate. And a supernova was discovered here in 2004 by the Lick Observatory Supernova Search and Federico Manzini. Look for a slightly oval shape oriented from north to south, which brightens toward the core. A faint star can be seen at the edge of the arm structure to the northeast. It is best at mid-magnifications.

Our second hop takes us about one degree southeast of Beta (RA 05 33 22 Dec –21 56 41) and into a stellar field for NGC 1964 (Figure 2.18). At a visual magnitude of 10.8, this Herschel "400" galaxy shows an oval disc elongated from the northeast to southwest with a bright core area. Several faint stars overlay the galaxy but are not involved with it. It can be spotted with scopes as small as 4.5", but truly requires larger aperture to appreciate.

Sunday, February 10

Tonight the slender first crescent of the Moon makes its presence known on the western horizon. Before it sets, take a moment to look at it with binoculars. The beginnings of Mare Crisium will show in the northeast quadrant, but look just a bit further south for the dark, irregular blotch of Mare Undarum—the

Sea of Waves. On its southern edge, and to lunar east, look for the small Mare Smythii—the "Sea of Sir William Henry Smyth." Further south of this pair and at the northern edge of Fecunditatis is Mare Spumans—the "Foaming Sea." All three of these are elevated lakes of aluminous basalt belonging to the Crisium basin.

When the Moon has set, let's continue on our tour of Lepus and the galaxy hunt. Tonight we'll go from one corner to the other as we begin with Iota and hop two degrees west for NGC 1784 (RA 05 27 Dec −11 52 21) (Figure 2.19).

At magnitude 11.8, this barred spiral can be spotted in mid-aperture scopes as a misty oval with a slightly brighter center. With larger telescopes and optimal conditions, the central bar structure can be revealed as an elongated brightness toward the core region, with some brighter knots noted in the arms. In studies done by Doug Ratay at radio wavelengths, NGC 1784 was mapped for its distribution of hydrogen gas both inside and outside the galaxy structure. His incredible findings showed an orbiting area of gas which could be part of a small galaxy located about 100 million light-years away.

Our second mark is slightly more than three degrees south-southwest of Epsilon—NGC 1744 (RA 04 59 58 Dec −26 01 36) (Figure 2.20). Despite seeming to be possible in a large scope—magnitude 12.3—this north–south-inclined barred spiral is anything but easy from the Northern Hemisphere—or the South! Very low surface brightness means this particular galaxy is a tough customer even for veterans, and at best will show as a thin, nebulous area with no definition.

Figure 2.19. NGC 1784 (Credit—Palomar Observatory, courtesy of Caltech).

Figure 2.20. NGC 1744 (Credit—Palomar Observatory, courtesy of Caltech).

Monday, February 11

Today in 2003, the world was stunned as it took its first look at the oldest light in the Universe. Launched on June 30, 2001, NASA's Wilkinson Microwave Anisotropy Probe (WMAP, Figure 2.21) hangs out in space about 1.6 million kilometers from Earth at Lagrange Point 2. The map shown here is one of the most important results of its study of the comic microwave background, and is strong evidence in favor of the "Big Bang Theory;" placing the age of our Universe at about 13.7 billion years.

On this day in 1970 Lambda 4S-5, the first Japanese satellite, was launched. Tonight our own satellite cannot be denied as the Moon lights up the early evening skies. Let's have a look, shall we? As it begins its westward journey after sunset in a position much easier to observe, the lunar feature we are looking for is at the north-northeast of the lunar limb and its view is often dependent on libration. What are we seeking? "The Sea of Alexander von Humboldt"...

Mare Humboldtianum is seen in this picture as fully revealed, yet sometimes it can be hidden from view because it is an extreme feature (Figure 2.22). Spanning 273 kilometers, the basin in which it is contained extends for an additional 600 kilometers and continues around to the far side of the Moon. The mountain ranges which accompany this basin can sometimes be glimpsed under perfect lighting conditions, but ordinarily are just seen as a lighter area. The mare was

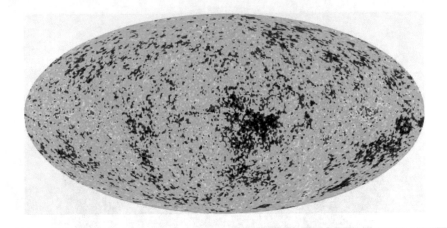

Figure 2.21. The cosmic microwave background from WMAP (Credit—NASA/WMAP).

Figure 2.22. Mare Humboldtianum (Credit—NASA).

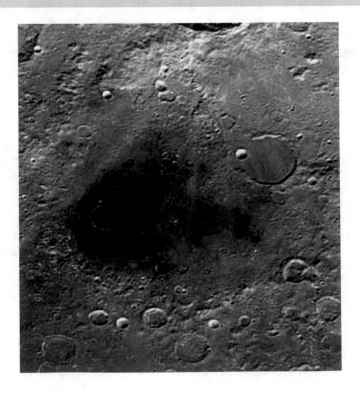

formed by lava flow into the impact basin, yet more recent strikes have scarred Humboldtianum. Look for a splash of ejecta from crater Hayn further north and the huge 200 kilometer strike of crater Bel'kovich on Humboldtianum's northeast shore.

Tuesday, February 12

Today is also the anniversary (2001) of NEAR landing on asteroid Eros (Figure 2.23). The Near Earth Asteroid Rendezvous mission was the first to ever orbit an asteroid, successfully sending back thousands of images. Although it was not designed to land on Eros, it survived the low speed impact and continued to send back data. And where is asteroid Eros? You'll find our 11.3 magnitude friend scooting along through the southern reaches of Pisces about 10 degrees north-northwest of Uranus. It's close to the western horizon at skydark, so don't wait too late!

With tonight's Moon in a much higher position to observe, let's begin with an investigation of Mare Fecunditatis—the Sea of Fertility (Figure 2.24).

Stretching 1463 kilometers in diameter, the combined area of this mare is equal in size to the Great Sandy Desert in Australia—and almost as vacant in interior features. It is home to glasses, pyroxenes, feldspars, oxides, olivines, troilite, and metals in its lunar soil, which is called regolith. Studies show the basaltic flow inside of the Fecunditatis basin perhaps occurred all at once, making its chemical composition different from other maria. The low titanium content means it is

Figure 2.23. NEAR image of Eros (Credit—NASA).

Figure 2.24. Fecunditatis region (Credit—Greg Konkel, annotation by Tammy Plotner). (1) Taruntius, (2) Secchi, (3) Messier and Messier A, (4) Lubbock, (5) Guttenberg, (6) Montes Pyrenees, (7) Goclenius, (8) Magelhaens, (9) Columbo, (10) Webb, (11) Langrenus, (12) Lohse, (13) Lame, (14) Vendelinus, and (15) the Luna 16 landing site.

between 3.1 and 3.6 billion years old! The western edge of Fecunditatis is home to features we share terrestrially—grabens. These dropped-down areas of landscape between parallel fault lines occur where the crust is stretched to the breaking point. On Earth, these happen between tectonic plates, but on the Moon they are found around basins. The forces created by lava flow increase the weight inside the basin, causing a tension along the border which eventually faults and creates these areas. Look closely along the western shore of Fecunditatis where you will see many such features.

Now let's take a walk across the Sea of Fertility and see how many lunar challenge features you can identify!

Wednesday, February 13

Today is the birthday of J.L.E. Dreyer (Figure 2.25). Born in 1852, the Danish-Irish Dreyer came to fame as the astronomer who compiled the New General Catalogue (NGC) published in 1878. Even with a wealth of astronomical catalogs to choose from, the NGC objects and Dreyer's abbreviated list of descriptions still remain the most widely used today.

Figure 2.25.
J.L.E. Dreyer (widely
used public image).

Figure 2.26. Nectaris region (Credit—Greg Konkel, annotation by Tammy Plotner).
(1) Isidorus, (2) Madler, (3) Theophilus, (4) Cyrillus, (5) Catharina, (6) Dorsum Beaumont,
(7) Beaumont, (8) Fracastorius, (9) Rupes Altai, (10) Piccolomini, (11) Rosse, (12) Santbech,
(13) Pyrenees Mountains, (14) Guttenberg, (15) Capella.

Since the Moon is now beginning to interfere, why not spend a few days really taking a look at the lunar surface and familiarizing yourself with its many features? Tonight would be a great time for us to explore "The Sea of Nectar."

At around 1000 meters deep, Mare Nectaris covers an area of the Moon equal to that of the Great Sandhills in Saskatchewan, Canada (Figure 2.26). Like all maria, it is part of a gigantic basin which is filled with lava, and there is evidence of grabens along the western edge of the basin. While Nectaris' basaltic flows appear darker than those in most maria, it is one of the older formations on the Moon, and as the terminator progresses you'll be able to see where ejecta belonging to Tycho crosses its surface. For now? Let's have a closer look at the mare itself and its surrounding craters... Enjoy these many features which are also lunar challenges—and we'll be back to study each later in the year!

Thursday, February 14

Happy Valentine's Day! Today is the birthday of Fritz Zwicky (Figure 2.27). Born in 1898, Zwicky was the first astronomer to identify supernovae as a separate class of objects. He also conceived of the possibility of neutron stars. Among his wide range of achievements, Zwicky catalogued galaxy clusters and even designed jet engines!

Tonight let your imagination sweep you away as we go mountain climbing—on the Moon! On the lunar surface all of Mare Serenitatis will be revealed, and along its northwestern shore lie some of the most beautiful mountain ranges you'll ever view—the Caucasus to the north and the Apennines to the south (Figure 2.28). Like its earthly counterpart, the Caucasus Mountain range stretches almost 550

Figure 2.27. Fritz Zwicky (widely used public image).

Figure 2.28. The Caucasus and Apennine Mountains (Credit—Greg Konkel).

kilometers and some of its peaks reach upwards of six kilometers—summits as high as Mt. Elbrus! Slightly smaller than its terrestrial namesake, the lunar Apennine Mountain range extends some 600 kilometers with peaks rising as high as 5 kilometers. Be sure to look for Mons Hadley, one of the tallest peaks you will see at the northern end of this chain. It rises above the surface to a height of 4.6 kilometers, making that single mountain about the size of asteroid Toutatis.

But this holiday wouldn't be complete unless I presented you with a Valentine—a star of singular beauty. Located about three fingerwidths southwest of Rigel, or a little more than a fingerwidth northwest of Mu, in the constellation of Lepus (RA 04 59 36 Dec −14 48 22), is R Leporis—better known as "Hind's Crimson Star" (Figure 2.29).

Discovered in October of 1845 by J. R. Hind, R Leporis will require optical aid to view since it is a Mira-type variable which moves from approximately magnitude 6 to as low as magnitude 11 in about 432 days. As a carbon star, this particular example is well worth viewing for its intense ruby color when near minimum. As R Leporis undergoes its changes, it produces amazing amounts of carbon. To understand what makes it dim, think of an oil lamp. As the carbon "soot" collects on the glass, like the star's outer atmosphere, the light decreases until the soot is sloughed off and the process is repeated. At a rough distance of approximately 1500 light-years, Hind's Crimson Star will become an observing favorite and is also a challenge on many lists. Enjoy!

Figure 2.29. R Leporis: Hind's Crimson Star (Credit—Palomar Observatory, courtesy of Caltech).

Friday, February 15

Born on this day in 1564 was the man who fathered modern astronomy—Galileo Galilei. Two and a half centuries ago, he became first scientist to use a telescope for astronomical observation and his first target was the Moon. Tonight your lunar assignments are relatively easy. We will begin by identifying "The Sea of Vapors" (Figure 2.30).

Look for Mare Vaporum on the southwest shore of Mare Serenitatis. Formed from newer lava flow inside an old crater, this lunar sea is edged to its north by the mighty Apennine Mountains. On its northeastern edge, look for the now washed-out Haemus Mountains. Can you see where lava flow has reached them? This lava has come from different time periods and the slightly different colorations are easy to spot even with binoculars.

Further south and edged by the terminator is Sinus Medii—the "Bay in the Middle." With an area about the size of Massachusetts and Connecticut combined, this lunar feature is the midpoint of the visible lunar surface. In 1930, experiments were being conducted to determine this region's surface temperature—a project begun by Lord Rosse in 1868. Surprisingly enough, results of the two studies were very close, and during full daylight temperatures in Sinus Medii can reach the boiling point, as shown by Surveyors 4 and 6—which landed near its center (Figure 2.31).

Figure 2.30. Mare Vaporum as seen from Apollo (Credit—NASA).

Figure 2.31. Surveyor 6 sampling Sinus Medii (Credit—NASA).

Figure 2.32. Palus Putredinus as seen from Apollo (Credit—NASA).

Now take a hop north of Mare Vaporum for a look at "The Rotten Swamp"—Palus Putredinus (Figure 2.32). A little more pleasingly known as the "Marsh of Decay," this nearly level surface of lava flow is also home to a mission—the hard landing of Lunik 2. On September 13, 1959 astronomers in Europe reported seeing the black dot of the crashing probe. The event lasted for nearly 300 seconds and spread over an area of 40 kilometers.

Saturday, February 16

On this day in 1948, Gerard Kuiper was celebrating his discovery of Miranda—one of Uranus' moons. Just 42 years earlier on this day, both Kopff and Metcalf were also busy: discovering asteroids! And today is the birthday of François Arago (Figure 2.33). Born in 1786, Arago was a pioneer in the illumination of the wave nature of light. He is also credited with the invention of the polarimeter and other optical devices.

Tonight let's celebrate Arago's achievements by having a look at a lunar feature named for him (Figure 2.34). Are you becoming more confident in your abilities? Then let's start by looking for the western shore of Mare Tranquillitatis. Arago crater is quite small, only 28 kilometers in diameter, but it will show as a small, bright ring almost centrally along the western edge. If you can't pick it up, don't

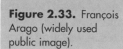

Figure 2.33. François Arago (widely used public image).

Figure 2.34. Albategnius region (Credit—Roger Warner, annotation by Tammy Plotner). (1) Flammarion, (2) Herschel, (3) Ptolemaeus, (4) Alphonsus, (5) Davy, (6) Alpetragius, (7) Arzachel, (8) Thebit, (9) Purbach, (10) Lacaille, (11) Blanchinus, (12) Delaunay, (13) Faye, (14) Donati, (15) Airy, (16) Argelander, (17) Vogel, (18) Parrot, (19) Klein, (20) Albategnius, (21) Muller, (22) Halley, (23) Horrocks, (24) Hipparchus, (25) Sinus Medii.

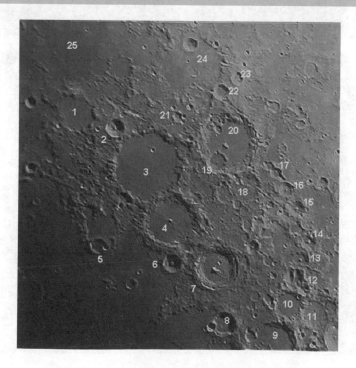

worry! There's plenty more we can have a look at tonight. Let's try looking just south of Sinus Medii and identify the features mentioned in Figure 2.34!

Even if you don't investigate with a telescope or binoculars, be sure to look at the Moon tonight. The "Red Planet"—Mars—will only be about a fingerwidth away to the south. Once you've identified Mars, it's easy to watch its motions across the ecliptic plane!

Sunday, February 17

For now, let's continue onwards with our lunar studies as we locate the emerging "Sea Of Islands." Mare Insularum will be partially revealed tonight as one of the most prominent of lunar craters—Copernicus—comes into view (Figure 2.35). While only a small section of this reasonably young mare is now visible southeast of Copernicus, the lighting will be just right to spot its many different-colored lava flows. To the northeast is a Lunar Club challenge: Sinus Aestuum. Latin for the Bay of Billows, this mare-like region has an approximate diameter of 290 kilometers, and its total area is about the size of the state of New Hampshire.

Figure 2.35. Copernicus region (Credit—Greg Konkel, annotation by Tammy Plotner). (1) Mons Wolf, (2) Eratosthenes, (3) Gay-Lussac, (4) Montes Carpatus, (5) Copernicus, (6) Reinhold, (7) Mare Insularum, (8) Gambart, (9) Apollo 14 landing site, (10) Frau Mauro, (11) Bonpland, (12) Parry, (13) Lalande, (14) Ptolemaeus, (15) Herschel, (16) Flammarion, (17) Mosting, (18) Sinus Medii, (19) Triesnecker, (20) Murchison, (21) Pallas, (22) Bode, (23) Ukert, (24) Sinus Aestuum, (25) Stadius.

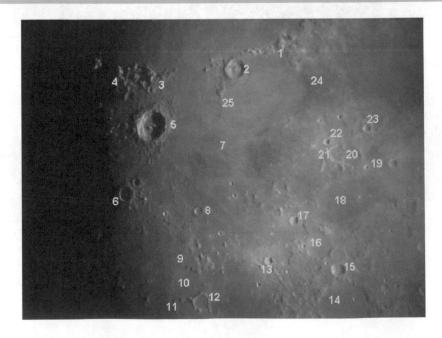

Containing almost no features, this area has a low albedo—providing very little surface reflectivity.

Now let's take a look and see what we can identify…and happy hunting!

Monday, February 18

On this day in 1930, a young man named Clyde Tombaugh was very busy checking out some photographic search plates taken with the Lowell Observatory's 13" telescope. His reward? The discovery of Pluto!

Tonight your reward will be taking a close look at the lunar surface as we investigate Mare Cognitum, "The Sea That Has Become Known." Also formed by an impact, the remains of the basin ring still exist as the bright semi-circle of the Montes Riphaeus which borders it to the northwest. Look for the very bright point of the crater Euclid to guide you. Just to its north is the Fra Mauro formation (Figure 2.36), the landing area for Apollo 14. Now let's talk about why exploration in this area was so important!

Named for the 80 kilometer diameter Fra Mauro crater, these highlands are hills believed to be the ejecta from the impact which formed Mare Imbrium. This debris may have come from as deep as 161 kilometers below the surface, and so studying this area would help us understand the physical and chemical nature of the area below the lunar crust.

Figure 2.36. Fra Mauro (Credit—Wes Higgins).

The Fra Mauro formation became more interesting to scientists when the Apollo 12 seismometer at Surveyor Crater (177 kilometers to the west) relayed to Earth signals of monthly moonquakes. These phenomena were believed to have originated in the Fra Mauro crater as the Moon passed through its perigee. Apollo 14 landed in the hills at the edge of the crater Fra Mauro near a newer impact region called Cone Crater—around 305 meters across and 76 meters deep. Astronauts Shepard and Mitchell took samples from the crater's outer walls and photographed the interior.

Tuesday, February 19

Today is the birthday of Nicolas Copernicus (Figure 2.37). Born in 1473, he was the creator of the modern solar system model which explained the retrograde motion of the outer planets. Considering this was well over 530 years ago, and in a rather "unenlightened" time, his revolutionary thinking about what we now consider natural is astounding.

With the Moon moving further east each night it has now passed Pollux and is headed toward Saturn. Even though it's not quite full yet, can you see the effect it has on nearby stars? Now that it is further from Orion and Taurus, those primary stars are beginning to appear again—yet there are still none visible to the unaided eye in Monoceros. Even 4.6 magnitude Beta doesn't show!

Tonight let's return again to the lunar surface to study how the terminator has moved and take a close look at the way features change as the Sun brightens the moonscape. Can you still see Langrenus? How about Theophilus, Cyrillus, and Catharina? Does Posidonius still look the same? Each night features further east become brighter and harder to distinguish—yet they also change in subtle and unexpected ways. We'll look at this in the days ahead, but tonight let's walk the terminator as one of the most beautiful features has now come into view—the "Bay of Rainbows."

Sinus Iridum's C-shape is easily recognizable in even small binoculars (Figure 2.38)—yet there are a wonderland of small details in and around the area

Figure 2.37.
Nicholas Copernicus (widely used public image).

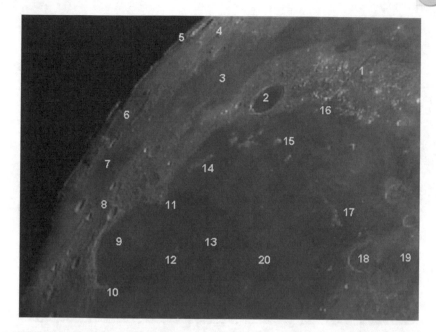

Figure 2.38. Iridum region (Credit—Roger Warner, annotation by Tammy Plotner). (1) Alpine Valley, (2) Plato, (3) Mare Frigoris, (4) Philolaus, (5) Anaximenes, (6) J. Herschel, (7) Sinus Roris, (8) Bianchini, (9) Sinus Iridum, (10) Promontorium Heraclides, (11) Promontorium LaPlace, (12) Helicon, (13) Leverrier, (14) Straight Range, (15) Mons Pico, (16) Mons Piton, (17) Montes Spitzbergen, (18) Archimedes, (19) Apollo 15 landing area, (20) Mare Imbrium.

for the small telescope which we'll study as the year goes by. Take the chart with you tonight and see how many of the features mentioned in Figure 2.38 you can identify and add to your lunar challenges!

Wednesday, February 20

On this day in 1962, John Glenn became the first American to orbit the Earth while aboard the Mercury capsule Friendship 7 (Figure 2.39). Today in history also celebrates the Mir space station's launch in 1986 (Figure 2.40). Mir (Russian for "peace") was home to both cosmonauts and astronauts as it housed 28 long-duration crews during its 15 years of service. To date it is one of the longest running space stations and a triumph for mankind. *Spaseba*!

Tonight turn your eyes to the Moon as we perform a "launch" of our own and identify a featured known as the "Cow Jumping over the Moon." It is strictly a visual phenomenon—a combination of dark maria which looks like the back, forelegs, and hindlegs of the shadow of that mythical animal (Figure 2.41). It's not only fun, but an Astronomical League Lunar Challenge as well!

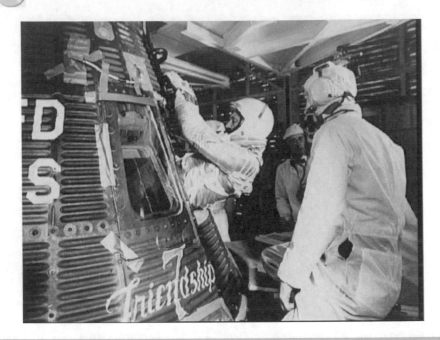

Figure 2.39. John Glenn and Friendship 7 (Credit—NASA).

Figure 2.40. Mir (Credit—NASA).

Figure 2.41. Half Moon (Credit—Roger Warner).

If you're up for another challenge, then it's time to identify the Oceanus Procellarum—The Ocean of Storms. Visible to the unaided eye as the vast gray sea which stretches along the terminator of the Moon's entire mid to northwest quadrant. Like many mythically based astronomy names, Oceanus Procellarum came from superstition associated with its appearance. Once upon a time, it was believed sighting this feature presaged bad weather, but backyard astronomers know its reappearance only forecasts the coming Full Moon! Formed by volcanic eruptions, this largest of all maria is little more than an expanse of long-hardened magma stretching from north to south a distance of around 2500 kilometers. Although to the eye it may seem featureless, it is home to some very unusual features and some historic missions which we will study in the future.

Thursday, February 21

Are you ready for tonight? Beginning at 00:34:59 UT, the Moon will begin to slide inside the Earth's penumbral shadow. You got it... It's time for a lunar eclipse! Although the beginning passage is the most subtle part of the event, it will be visible in western Asia, Europe, and Africa about the time the Moon sets. For observers in the Americas and the Pacific region, this gentle shading will end around the time the Moon rises. At 01:42:59 UT, the Moon will pass into the deep umbral shadow and not leave until 05:09:07—this is the time frame of partial

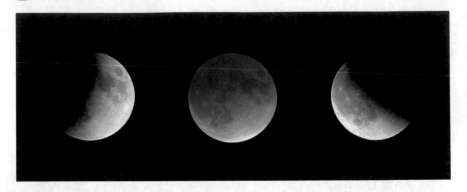

Figure 2.42. Lunar eclipse (Credit—John Chumack).

eclipse. Totality will occur for the Americas, Europe, Africa, the Middle East, and Western Asia, beginning at 03:00.34 UT with maximum at 03:26:05 and end at 03:51:32. Considering the umbral magnitude is estimated at 1.1, this should make for a deep orange color and will be quite worth taking the evening to observe (Figure 2.42).

Take the time to try your hand at photography and make notes. Estimate the eclipse by the Danjon scale: L=4 is an orange or coppery color with a blue tint where the umbral and penumbral shadow meets; L=3 is brick colored, with a gray or yellow rim; L=2 is a deep red and the Moon will be very dark in the center; L=1 is a dark eclipse. The Moon will appear brown or dark gray, while the surface features will be hard to see; L=0 is so dark that the Moon is nearly invisible. Power up with a telescope to watch a rare treat as the shadow advances over the craters, mountains and seas... And then the Sun rises over them again!

But don't get so wrapped up in the eclipse that you forget to take a look around the Moon. For many observers, bright Regulus will be less than a degree away from the lunar limb and a lucky few will also get to witness either a grazing event or an occultation on this same night! Be sure to scan around with binoculars to have a look at other objects as well. How often can you see a DSO on a night lit by the Full Moon?

Wishing you clear skies...

Friday, February 22

Today in 1966, Soviet space mission Kosmos 110 was launched. Its crew was canine, Veterok (Little Wind) and Ugolyok (Little Piece of Coal), both history-making dogs (Figure 2.43). The flight lasted 22 days and held the record for living creatures in orbit until 1974—when Skylab 2 carried its three-man crew for 28 days.

Figure 2.43. Veterok and Ugolyok (Courtesy of Alexander Chernov).

Figure 2.44. NGC 2186 (Credit—Palomar Observatory, courtesy of Caltech).

With the Moon absent for a short while tonight, let's take a look at two dogs as we identify the constellations of Canis Major and Canis Minor. It's time to study Monoceros! By using the red giant Betelgeuse, diamond-bright Sirius and the beacon of Procyon, we can see these three stars form a triangle in the sky with Sirius pointing toward the south.

The "Unicorn" is not a bright constellation, and most of its stars fall inside this area—with its Alpha star almost a handspan south of Procyon. Using the belt of Orion as a guide, look a handspan east, this is Delta. A fistwidth away to the southeast is Gamma; with Beta about two fingerwidths further along. About a palmwidth southeast of Betelguese is Epsilon. Although this might seem simple, knowing these stars will help you find many wonderful objects. Let's start our journey tonight two fingerwidths northwest of Epsilon…

NGC 2186 is a triangular open cluster of stars set in a rich field which can be spotted with binoculars, and it reveals 30 or more stars to even a small telescope (Figure 2.44). Not only is this a Herschel "400" object which can be spotted with simple equipment, but a highly studied galactic cluster containing circumstellar discs!

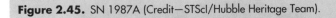

Saturday, February 23

In 1987, Ian Shelton made an astonishing visual discovery—SN 1987a (Figure 2.45). This was the brightest supernova in 383 years. Located near the Tarantula Nebula in the Large Magellanic Cloud, this beauty came in at close

Figure 2.45. SN 1987A (Credit—STScI/Hubble Heritage Team).

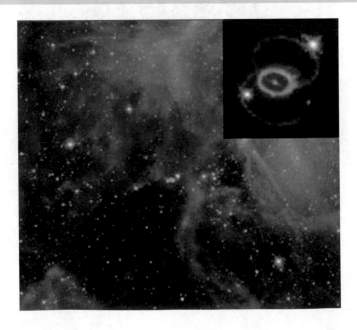

to magnitude 3 and was visible for some time to the unaided eye. Seven years later, the Hubble Space Telescope (HST) observed a few strange rings in the area which were believed to have been created before the event itself—this still remain unexplained.

Sir William Herschel was also on a voyage of discovery tonight as well... And his choice was in the constellation of Monoceros. Get your binoculars or telescopes out and let's head a little less than a fistwidth east-southeast of Alpha (RA 08 00 01 Dec –10 46 12) to check out NGC 2506 (Figure 2.46). On a dark night, this near 7th magnitude object is perhaps one of the most impressive of the Monoceros open clusters. Caught in a chain of stars, it displays a rich concentration—almost appearing like a globular cluster. Because of this, NGC 2506 has been used in the study of old, metal-poor galactic clusters. Its evolution has enriched its iron content, and despite its extreme age it is still a beauty! Be sure to mark your "400" notes with H VI.37.

Would you like a challenge? Try NGC 2236 (Figure 2.47), about two degrees southwest of 13 Monoceros (RA 06 29 42 Dec +06 50 00). Although it was discovered the same night as NGC 2506, Herschel had picked it out of the starry sky seven years earlier. At roughly magnitude 8, this little triangle of stellar points is far more difficult to recognize as a cluster, yet Sir William was a very sharp observer. It truly is a physical association of stars located on the inner edge of the outer Perseus spiral arm, and was logged on this night 224 years ago as H VII.5.

Figure 2.46. NGC 2506 (Credit—Palomar Observatory, courtesy of Caltech).

Figure 2.47. NGC 2236 (Credit—Palomar Observatory, courtesy of Caltech).

Sunday, February 24

Today in 1968, during a radar search survey, the first pulsar was discovered by Jocelyn Bell (Figure 2.48). The co-directors of the project, Antony Hewish and Martin Ryle, matched these observations to a model of a rotating neutron star, winning them the 1974 Physics Nobel Prize and proving a theory which J. Robert Oppenheimer had proposed 30 years earlier.

Would you like to get a look at a region of the sky containing a pulsar? Then look for guidestar Alpha Monocerotis to the south and bright Procyon to its north. By using the distance between these two stars as the base of an imaginary triangle, you'll find pulsar PSR 0820+02 at the apex of your triangle pointed east (Figure 2.49). In Figure 2.49, I wonder which "star" it is?

You won't have to wonder if you aim your binoculars or telescopes toward our next study. Tonight we'll use both Sirius and Beta Monocerotis as our guides to have a look at one fantastic galactic cluster for any optical aid—M50 (Figure 2.50). Hop about a fistwidth east-southeast of Beta, or northeast of Sirius...and be prepared!

Perhaps discovered as early as 1711 by G. D. Cassini, it was relocated by Messier in 1772 and confirmed by J. E. Bode in 1774. Containing perhaps as

Figure 2.48. Jocelyn Bell (widely used public image).

Figure 2.49. PSR 0820+02 (Credit—Palomar Observatory, courtesy of Caltech).

Figure 2.50. M50 (Credit—Palomar Observatory, courtesy of Caltech).

many as 200 members, this colorful old cluster resides almost 3000 light-years away.

The light of the stars you are looking at tonight left this cluster at a time when iron was first being smelted and used in tools. The Mayan culture was just beginning to develop, while the Hebrews and Phoenicians were creating an alphabet. Do you wonder if it looked the same then as it does now? In binoculars you will see an almost heart-shaped collection of stars, while telescopes will begin to resolve out color and many fainter members—with a very notable red one in its midst. Enjoy this worthy cluster and make a note... You've captured another Messier object!

Monday, February 25

Tonight it's time for us to go hunting some obscure objects which require dark skies. Our guidestar will be Epsilon Monocerotis and we'll head about three fingerwidths northeast for a vast complex of nebulae and star clusters.

To the unaided eye, 4th magnitude S Monocerotis is easily visible, and to small binoculars so are the beginnings of a rich cluster surrounding it. This is NGC 2264 (Figure 2.51). Larger binoculars and small telescopes will easily pick

Figure 2.51. NGC 2264—(Credit): Palomar Observatory, courtesy of Caltech.

out a distinct wedge of stars. This is most commonly known as the "Christmas Tree Cluster," whose name was given by Lowell Observatory astronomer Carl Lampland. With its peak pointing due south, this triangular group is believed to be around 2600 light-years away and spans about 20 light-years. Look closely at its brightest star—S Monocerotis is not only a variable, but also has an 8th magnitude companion. The group itself is believed to be almost 2 million years old. The nebulosity is beyond the reach of a small telescope, but the brightest portion illuminated by one of its stars is the home of the Cone Nebula. Larger telescopes can see a visible V-like thread of nebulosity in this area which completes the outer edge of the dark cone.

To the north is a photographic-only region known as the Foxfur Nebula, part of a vast complex of nebulae which extends from Gemini to Orion. Northwest of the complex are several regions of bright nebulae, such as NGC 2247, NGC 2245, IC 446, and IC 2169. Of these regions, the one most suited to the average scope is NGC 2245, which is fairly large, but faint, and accompanies an 11th magnitude star. NGC 2247 is a circular patch of nebulosity around an 8th magnitude star, and it will appear much like a slight fog. IC 446 is indeed a smile to larger aperture, for it will appear much like a small comet with the nebulosity fanning away to the southwest. IC 2169 is the most difficult of all. Even with a large scope a "hint" is all!

Enjoy your nebula quest...

Tuesday, February 26

Celestial Scenery Alert! Be sure to set your alarm for just before local dawn. This morning the two innermost planets—Venus and Mercury—will grace the eastern sky just about one degree apart.

Today is the birthdate of Camille Flammarion (Figure 2.52). Born in 1842, he became a widely read author in astronomy and conceived the idea that we were not alone—the concept of extraterrestrial life. Yet, Flammarion was just a little bit more than the great-grandfather of SETI. In 1877, he had an unusual chance most of us can only dream of. He had his hands on a personal copy and notes of the Messier Catalog. Using it as a reference he later revised the catalog, but before 1917 his studies led him to identify M102 with NGC 5866. By 1921, Flammarion had added M104—now known as NGC 4594—to the catalog as well, and it became the first of many additions.

Now return again to Epsilon Monocerotis. Our destination is about a finger-width east as we seek out another star cluster which has an interesting companion—a nebula!

NGC 2244 is a star cluster embroiled in a reflection nebula spanning 55 light-years and most commonly called "The Rosette" (Figure 2.53). Located about 2500 light-years away, the cluster heats the gas within the nebula to nearly 10,000°C, causing it to emit light in a process similar to that of a fluorescent tube. A huge percentage of this light is hydrogen-alpha, which is scattered back from its dusty shell and becomes polarized.

While you won't see any red hues in visible light, a large pair of binoculars from a dark-sky site can make out a vague nebulosity associated with this open cluster. Even if you can't see this, it is still a wonderful cluster of stars crowned by the yellow jewel of 12 Monocerotis. With good seeing, small telescopes can easily spot the broken, patchy wreath of nebulosity around a well-resolved symmetrical concentration of stars. Larger scopes, and those with filters, will make out separate

Figure 2.52. Camille Flammarion (widely used public image).

Figure 2.53. NGC 2244: The Rosette Nebula (Credit—John Chumack).

areas of the nebula which also bear their own distinctive NGC labels. No matter how you view it, the entire region is one of the best for winter skies.

After that, it's time to relax and enjoy the Delta Leonid meteor shower. Burning through our atmosphere at speeds of up to 24 kilometers per second, these slow travelers will seem to radiate from a point around the middle of Leo's "back." The fall rate is rather slow at around 5 per hour, but they are still worth keeping a watch for!

Wednesday, February 27

Today is the birthday of Bernard Lyot (Figure 2.54). Born in 1897, Lyot went on to become the inventor of the coronagraph in 1930. By all accounts, Lyot was a wonderful and generous man who, sadly, died of a heart attack when returning from a trip to view a total eclipse.

While we have early dark skies on our side, let's head for a handful of difficult nebulae in a region just west of Gamma Monocerotis. For binoculars, check out the region around Gamma, it is rich in stars and very colorful! You are looking at the very outer edge of the Orion spiral arm of our galaxy. For small scopes, have a look at Gamma itself—it's a triple system on many challenge lists. For larger scopes? It's Herschel hunting time...

NGC 2183 and NGC 2185 will be the first you encounter as you move west of Gamma (RA 05 10 50 Dec –06 12 01) (Figure 2.55). Although they are faint, just remember they are nothing more than clouds of dust illuminated by faint

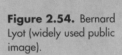

Figure 2.54. Bernard Lyot (widely used public image).

Figure 2.55. NGC 2183 and NGC 2185 (Credit—Palomar Observatory, courtesy of Caltech).

Figure 2.56. NGC 2182 and NGC 2170 (Credit—Palomar Observatory, courtesy of Caltech).

stars, and are on the edge of our galactic realm. The stars which formed inside them provided the light source for these wispy objects, and at their edges is intergalactic space.

To the southwest (RA 06 09 30 Dec –06 19 40) is the weaker NGC 2182, which will appear as nothing more than a faint star with an even fainter halo about it, with NGC 2170 more strongly represented in an otherwise difficult field (Figure 2.56). While the views of these objects might seem vaguely disappointing, you must remember not everything is as bright and colorful as seen in a photograph. Just knowing that you are looking at the collapse of a giant molecular cloud 2400 light-years away is pretty impressive!

Thursday, February 28

Even though Monoceros holds many more interesting objects, tonight let's head toward the upside-down Y of the constellation of Canis Major and pick up some studies while dark skies are in our favor. Our first destination lies about three fingerwidths south of brilliant Sirius and is viewable with little optical aid—and even without dark skies!

Messier object 41 was recorded as far back as 325 BC—in Aristotle's time (Figure 2.57). Since it resides at a distance of around 2350 light-years, the light you

Figure 2.57. M41 (Credit—Palomar Observatory, courtesy of Caltech).

Figure 2.58. NGC 2283 (Credit—Palomar Observatory, courtesy of Caltech).

see tonight in fact came from the time of Aristotle! First set to note by Flamsteed in February of 1702, and hosting around 100 true members of various magnitudes, this open cluster is very bright and totally resolved to larger telescopes. Its central star is a K-type red giant and many blue giants can also be seen. Not only is M41 a wonderland of resolution for almost every optical aid, but it's an Urban Observing Club challenge, a Binocular Messier, and just plain fun!

For the large telescope, head north about another three degrees to spot NGC 2283 (RA 06 45 53 Dec −18 12 38) (Figure 2.58) This small, faint, 13th magnitude spiral galaxy has a bright nucleus and is very difficult to spot because it's involved with a small field of stars. Because Sirius is slightly more than a degree north, very good sky conditions are needed to spot this tough Herschel (H III.271) object!

Friday, February 29

It's leap year day! Celebrate today by beginning with observations of the Moon and Antares. For most viewers, the pair will be around a half a degree apart, but for a lucky few this could mean an occultation or grazing event. Be sure to check with a reputable source, such as IOTA, for accurate times and locations.

It's not often we have an extra night to study, and we'll start by honoring Southern Hemisphere observers by exploring the fantastic NGC 3372—the Eta Carinae Nebula (Figure 2.59). As a giant, diffuse nebula with a visual brightness

Figure 2.59. Eta Carinae (Credit—Greg Bradley).

Figure 2.60. NGC 2359: Thor's Helmut (Credit—Palomar Observatory, courtesy of Caltech).

of magnitude one, it contains the most massive and luminous star in our Milky Way galaxy, Eta Carinae. It's also home to a small cluster, Collinder 228, which is one of the eight cataloged open clusters within the area of this huge star-forming region; the others are Bochum (Bo) 10, Trumpler (Tr) 14 (also cataloged as Cr 230), Tr 15 (= Cr 231), Cr 232, Tr 16 (= Cr 233), Cr 234, and Bo 11. The star Eta Carinae is involved in open cluster Trumpler 16. This fantastic nebula contains details which northerners can only dream about, such as the dark "Keyhole" and the "Homunculus" around the giant star itself. A fantastic region for exploration with both telescopes and binoculars!

Although it's far more humble, for northern observers let's take a look about a fistwidth east-northeast of Sirius for NGC 2359 (RA 07 18 30 Dec −13 13 48) (Figure 2.60). Sometimes known as "Thor's Helmut" or more commonly the "Duck Nebula" (H V.21), this unusual character is the result of central Wolf-Rayet star, its stellar winds, and the surrounding interstellar matter. As the high-velocity wind leaves this powerful star, it pushes the matter ahead of it, both compressing and expanding this ring-like shell. As it grows, it even collects more gas and dust from the interstellar medium, bringing its current mass to about 27 times that of our own Sun. While most planetary nebulae contain old stars nearing the end of their lives, the central Wolf-Rayet star in NGC 2359 is very young. Its ultraviolet photons are the fueling source of this delightful 8th magnitude emission nebula!

CHAPTER THREE

March, 2008

Saturday, March 1

In 1966 Venera 3 became the first craft to touch another planet as it impacted Venus (Figure 3.1). Although its communications failed before it could transmit data, it was a milestone achievement. If you're up before dawn, be sure to have a look at Venus and say *Spaseba!*

On this date in 2003, yet another in the *Deep-Sky Companion* series of books was released by the notable observer and author Stephen O'Meara (Figure 3.2). Known to all for his high-quality sketches of solar system objects and uncanny observing skills, O'Meara was the first to sight Comet Halley on its return in 1985, and the first to sketch the dark spokes in Saturn's rings (even before Voyager had imaged them). Still part of the editorial staff at *Sky and Telescope*, and a treasured lecturer, let's take just a moment to congratulate Steve-O on a lifetime achievement. *Aloha!*

George Abell was born on this day in 1927 (Figure 3.3). Abell was the man responsible for cataloging 2712 clusters of galaxies from the Palomar sky survey, a task he completed in 1958. Using these plates, Abell proposed that the grouping of such clusters delineated the arrangement of matter in the universe. He developed the "luminosity function," which shows the relationship between brightness and the number of members in each cluster, allowing you to infer their distances.

Figure 3.1. Venera 3 (Credit—NASA).

Figure 3.2. Stephen O'Meara (widely used public image).

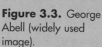

Figure 3.3. George Abell (widely used image).

Abell also discovered a number of planetary nebulae and developed the theory (along with Peter Goldreich) of their evolution from red giants. Abell was a fascinating lecturer and a developer of several television series dedicated to explaining science and astronomy in a fun and easy to understand way. He was also president and a member of the Board of Directors of the Astronomical Society of the Pacific, and he served in the American Astronomical Society, and was a member

Figure 3.4. A small portion of Abell 1367 (Credit—Palomar Observatory, courtesy of Caltech).

of the Cosmology Commission of the International Astronomical Union. And he accepted editorship of the *Astronomical Journal* just before he died.

To celebrate Abell's achievements, I can think of no finer area of the sky than one of his own galaxy clusters—A 1367—located about a degree southwest of 93 Leonis (RA 11 44 44 Dec +19 41 59) (Figure 3.4). Centered on 13th magnitude NGC 3842, this is an area of challenging intrigue spanning a degree of sky. It contains as many as two dozen small galaxies which can be seen with a large telescope, but will require very fine skies and patience to identify its many members. A 1367 is in an area of sky which many astronomers return to year after year, and it is a gemstone for galaxy collectors.

Sunday, March 2

On this date in 2004, the first deep space mission of the European Space Agency (ESA) was launched. Named Rosetta (Figure 3.5), its initial launch toward Comet 46/P Wirtanen was postponed, so the goal became Comet 67/P Churyumov-Gerasimenko. After traveling through space for a period of about 10 years, the mission should reach the comet sometime in the year 2014, when it will release

Figure 3.5. Concept of Rosetta and Philae (Credit—NASA).

a lander named Philae to explore the surface of the comet—giving us much more information about the composition and other properties of these distant travelers!

Since this is a weekend night, why not keep the big telescope handy and do a little galaxy hopping in the region south of Beta Canis Majoris?

Our first mark will be NGC 2207 (RA 06 16 22 Dec –21 22 20)—a 12.3 magnitude pair of interacting galaxies (Figure 3.6). Located some 114 million light-years away, this pair is locked in a gravitational tug of war. The larger of the pair is NGC 2207, and it is estimated an encounter began with the Milky Way–sized IC 2163 about 40 million years ago. Like the M81 and M82 pair, NGC 2207 will cannibalize the smaller galaxy—yet the true space between their stars is so large that actual collisions may never occur. While our eyes may never see as grandly as a photograph, a mid-sized telescope will make out the signature of two galactic cores with intertwining material. Enjoy this great pair!

Now shift further southeast for NGC 2223 (RA 06 24 36 Dec –22 50 17) (Figure 3.7). Slightly fainter and smaller than the previous pair, this round, low surface brightness galaxy shows a slightly brighter nuclear area and a small star caught on its southern edge. While it seems a bit more boring, it did have a supernova event as recently as 1993!

Figure 3.6. NGC 2207 (Credit—Palomar Observatory, courtesy of Caltech).

Figure 3.7. NGC 2223 (Credit—Palomar Observatory, courtesy of Caltech).

Monday, March 3

If you're up before dawn this morning, be sure to take a look at the close appearance of Jupiter and the Moon. While the pair is still well separated, they make a pleasing sight! Also look at the eastern skyline for Mercury. It's now reached its greatest elongation. Has its relative position near Venus changed over the last few days?

Tonight let's return to Canis Major with binoculars and have a look at Omicron 1, the westernmost star in the central Omicron pair (Figure 3.8). While this bright, colorful gathering of stars is not a true cluster, it is certainly an interesting group.

For larger binoculars and telescopes, hop to Delta and then northeast for Tau and the open cluster NGC 2362 (RA 07 18 36 Dec –24 59 00) (Figure 3.9). At a distance of about 4600 light-years, this rich little cluster contains about 40 members and is one of the youngest of all known star clusters. Many of the stars you can resolve have not even reached main sequence yet! Still gathering itself together, it is estimated this stellar collection is less than a million years old. Its central star, Tau, is believed to be a true cluster member and one of the most intrinsically luminous stars known. Put as much magnification on this one as skies will allow—it's a beauty!

Figure 3.8.
Omicron 1
Canis Majoris
(Credit—Palomar
Observatory, courtesy
of Caltech).

Figure 3.9. NGC
2362 (Credit—Palomar
Observatory, courtesy
of Caltech).

Tuesday, March 4

Today we celebrate the birth of Sir Patrick Moore (Figure 3.10). Author of 60 books, and since 1957 the host of "The Sky at Night," this gentleman-scholar has sparked the interest of several generations of skywatchers. So excellent were his observations that the Soviets used his lunar charts to correlate the first Lunik 3 pictures of the far side of the Moon—he was also involved with NASA's lunar mapping prior to the Apollo missions. After four decades of tireless service, we salute you!

In 1835, Giovanni Schiaparelli opened his eyes for the very first time and later opened ours with his accomplishments (Figure 3.11)! As the director of the Milan Observatory, Schiaparelli (and not Percival Lowell) was the fellow who popularized the term "Martian canals" somewhere around the year 1877. Far more importantly, Schiaparelli was the man who made the connection between the orbits of meteoroid streams and the orbits of comets almost 11 years earlier than his planetary observations!

Tonight, why not have a look at Mars? Even if you don't use a telescope, try comparing the Red Planet's color to both Betelgeuse and Aldebaran. Both binoculars and small telescopes will reveal the orb of our small neighbor, while steady skies, larger optics, and higher magnification allow a certain amount of detail to be seen.

Now let's have a look at HR 2447 Canis Majoris (Figure 3.12). While you won't find it labeled on most maps, you can easily identify it by hopping about two fingerwidths southwest of Epsilon to Kappa. About the same distance west, you will see roughly equal magnitude Lambda. Almost equidistant between these two stars (RA 06 37 47 Dec –32 20 23), look for a very faint 5th magnitude star...

There's a very good reason to take a look at HR 2447—for it has a planet orbiting it at about the same distance as Mars orbits our Sun. At a rough distance of 395 light-years from Earth, this class K star is a giant, nearing the end of its evolution. Unlike the situation with most planets discovered around dwarf stars, it's very difficult to ascertain the mass of such a dying star—and to accurately assess planetary size. While the planet could be anywhere from 5 to 10 times the

Figure 3.10. Sir Patrick Moore (widely used public image).

Figure 3.11.
Giovanni Schiaparelli
(widely used public
image).

Figure 3.12. HR 2447 Canis Majoris (Credit—Palomar Observatory, courtesy of Caltech).

size of Jupiter, we do know it orbits its parent star in slightly less than 2 years. From the surface of this planet, you would certainly enjoy a rather strange view of its sun: given the star's diameter and distance, in daylight you would see an orb which would be a least 10 times larger than Sol!

Wednesday, March 5

Celestial Scenery Alert! The place to be is outdoors just before local dawn to watching the beautiful apparition of the rapidly waning Moon, Venus, and Mercury. The inner planets are not only dancing very close together but are less than half a degree from the Moon's limb for almost all observers. For some, this is bound to be an occultation event, so be sure to check IOTA for accurate times and locations. Although daylight will rob the view, only hours later the Moon will occult Neptune as well!

Today is the birthday of Gerardus Mercator, famed mapmaker, who started his life in 1512 (Figure 3.13). Mercator's time was a rough one for astronomy, but despite a prison sentence and the threat of torture and death for his "beliefs," he went on to design a celestial globe in the year 1551.

Tonight let's hop about four fingerwidths east-northeast of Sirius (RA 07 17 42 Dec –15 38 00). Look for 5th magnitude SAO 152641 to guide you to a faint patch of stars in binoculars and a superb cluster in a telescope—NGC 2360 (Figure 3.14). Comprised of around eighty 10th magnitude and fainter stars, this particular cluster will look like a handful of diamond dust scattered on the sky. Discovered by Caroline Herschel in 1783, this intermediate-aged galactic cluster is home to red giants and is heavy in metal abundance. Mark your notes, because not only is this a Herschel object, but is known as Caldwell 58 as well!

Figure 3.13.
Gerardus Mercator (widely used public image).

Figure 3.14. NGC 2360 (Credit—Palomar Observatory, courtesy of Caltech).

Thursday, March 6

If you get a chance to see sunshine today, then celebrate the birthday of Joseph Fraunhofer, who was born in 1787 (Figure 3.15). As a German scientist, Fraunhofer was truly a trailblazer in modern astronomy. His field? Spectroscopy!

After having served his apprenticeship as a lens and mirror maker, Fraunhofer went on to develop scientific instruments, specializing in applied optics. While designing the achromatic objective lens for a telescope, he was watching the spectrum of sunlight passing through a thin slit and saw the dark lines which make up the "rainbow bar code." Fraunhofer knew some of these lines could be used as wavelength standards, so he began measuring. He labeled the most prominent of the lines he saw with the letters still used today!

His skill in optics, mathematics, and physics led Fraunhofer to design and build the very first diffraction grating which was capable of measuring the wavelengths of specific dark lines in the solar spectrum. Did his telescope designs succeed? Of course! His work with the achromatic objective lens is the design still used in modern telescopes!

In 1986, the first of eight consecutive days of flybys began as VEGA 1 and Giotto became the very first spacecraft to reach Halley's Comet (Figure 3.16). Although there may not be a bright comet for us to observe tonight, we can

Figure 3.15. Joseph Fraunhofer (widely used public image).

Figure 3.16. Giotto image of Comet Halley's nucleus (Credit—NASA).

Figure 3.17. Hubble's Variable Nebula (Credit—Palomar Observatory, courtesy of Caltech).

have a look at a wonderful comet-shaped object which displays all the "blues" of Fraunhofer's work. You'll find it about three fingerwidths northeast of Epsilon, about two degrees northeast of star 13 in Monoceros (RA 06 39 10 Dec +08 45 00).

This is NGC 2261, more commonly known as "Hubble's Variable Nebula" (Figure 3.17). Named for Edwin Hubble, this 10th magnitude object can be seen in smaller telescopes and is very blue in appearance to larger apertures. Its cometary shape isn't what's so unusual, but the variability of the nebula itself. The illuminating star, R Monocerotis, does not display a normal stellar spectrum and may host a protoplanetary system. R is usually lost in the high surface brightness of the structure of the nebula, yet the whole thing varies with no predictable timetable—perhaps due to dark masses shadowing the star.

Friday, March 7

Today John Herschel, the only child of William Herschel (the discoverer of Uranus), was born in 1792 (Figure 3.18). He became the first astronomer to thoroughly survey the Southern Hemisphere's sky, and he was discoverer of photographic fixer. Also born on this day, but in 1837, was Henry Draper (Figure 3.19)—the man who made the first photograph of a stellar spectrum.

Figure 3.18. John Herschel (widely used public image).

Figure 3.19. Henry Draper (widely used public image).

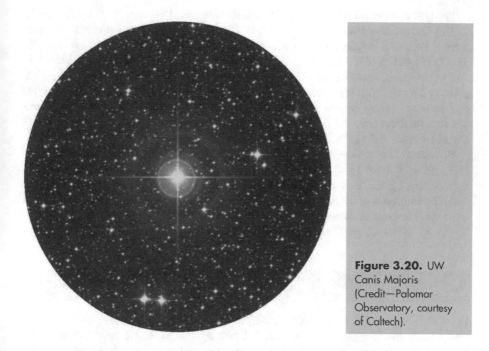

Figure 3.20. UW Canis Majoris (Credit—Palomar Observatory, courtesy of Caltech).

Figure 3.21. NGC 2354 (Credit—Palomar Observatory, courtesy of Caltech).

Tonight is New Moon and time to lather, rinse, and repeat. Start by returning to Tau Canis Majoris and NGC 2362. Just off the cluster's north-northeast corner, have a look at a single, unusual star—UW Canis Majoris (RA 07 18 40 Dec –24 33 31) (Figure 3.20). At magnitude 4.9, this super-giant spectroscopic binary is one of the most massive and luminous in our galaxy. Its two stars are separated by only 27 million kilometers (17 million miles), and revolve around each other at a frenzied pace—in less than 4.5 days. This speed means the stars themselves are flattened and would appear to be almost egg shaped. The primary itself is shedding material that's being collected by the secondary star.

Now drop about two degrees southwest of NGC 2362 for another open cluster— NGC 2354 (RA 07 14 07 Dec –25 44 24) (Figure 3.21). While at best this will appear as a small, hazy patch to binoculars, NGC 2354 is actually a rich galactic cluster containing around 60 metal-poor members. As aperture and magnification increase, the cluster shows two delightful circle-like structures of stars, similar to Figure 3.8. Be sure to make a note... You've captured another Herschel "400" object!

Saturday, March 8

On this day in 1977, the NASA airborne occultation observatory made a unique discovery—Uranus had rings! Uranus is near conjunction right now, but if you'd like to have a look at a ringed planet, why not check out the

Figure 3.22. M47 (Credit—Palomar Observatory, courtesy of Caltech).

grandest of all? Saturn is happily lighting up the night very near the backwards question mark of Leo.

Before the Moon begins to interfere with deep-sky studies, let's have a look at a small area of sky which contains not only three Herschel objects but two Messiers as well—M46 and M47. You'll find them less than a handspan east of Sirius and about a fistwidth north of Xi Puppis.

The brightest of the two clusters is M47 (RA 07 36 36 Dec –14 29 00) (Figure 3.22) and at 1600 light-years away it's a glorious object for binoculars. It is filled with mixed magnitude stars which resolve fully to aperture, and has the double star Struve 1211 near its center. While M47 is in itself a Herschel object, look just slightly north (about a field of view) to pick up another cluster which borders it. At magnitude 6.7, NGC 2423 isn't as grand, but it contains more than two dozen fairly compressed faint stars with a lovely golden binary at its center.

Now return to M47 and hop east to locate M46 (RA 07 41 42 Dec –14 49 00) (Figure 3.23). While this star cluster will appear to be fainter and more compressed in binoculars, you'll notice one star seems brighter than the rest. Using a telescope, you'll soon discover the reason. In its northern portion, the 300 million year old M46 contains a Herschel planetary nebula known as NGC 2438. The cluster contains around 150 resolvable stars and may involve as many as 500. The bright planetary nebula was first noted by Sir William Herschel and

Figure 3.23. M46 (Credit—Palomar Observatory, courtesy of Caltech).

then again by John. While it would appear to be a member of the cluster, the planetary nebula is just a little closer to us than the cluster. Be sure to mark your notes... There's a lot there in just a little area!

Sunday, March 9

Today is the anniversary of the Sputnik 9 launch in 1966 which carried a dog named Chernushka (Blackie). Also today we recognize the birth of David Fabricius (Figure 3.24). Born in 1564, Fabricus was the discoverer of the first variable star—Mira. Tonight the Moon will be no more than a slim crescent just after sunset, so let's start with an unusual variable star as we look at Beta Canis Majoris—better known as Murzim (Figure 3.25).

Located about three fingerwidths west-southwest of Sirius, Beta is a member of a group of stars known as quasi-Cepheids—stars which have very short-term, and also very small, brightness changes. First noted in 1928, Beta changes no more than 0.03 in magnitude, and its spectral lines will widen in cycles longer than those of its pulsations.

When you've had a look at Beta, hop another fingerwidth west-southwest for open cluster NGC 2204 (RA 06 14 09 Dec –18 38 36) (Figure 3.26). Chances are that this small collection of stars was discovered by Caroline Herschel in 1783, but it was added to William's list. This challenging object is a tough call for even large binoculars and small telescopes, since only around a handful of its dim members can be resolved. To the larger scope, a small round concentration can be seen, making this Herschel study one of the more challenging. While it might not seem like it's worth the trouble, this is one of the oldest of galactic clusters residing in the halo and has been used in the study of "blue straggler" stars.

Figure 3.24. David Fabricius (widely used public image).

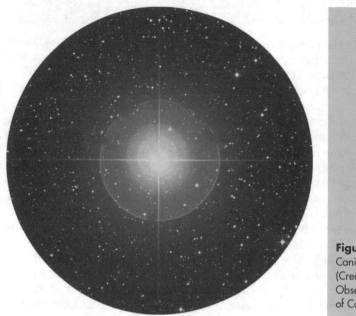

Figure 3.25. Beta Canis Majoris (Credit—Palomar Observatory, courtesy of Caltech).

Figure 3.26. NGC 2204 (Credit—Palomar Observatory, courtesy of Caltech).

Monday, March 10

Before we chase the stars tonight, take the time to do a little lunar observing. Don't forget to look for crater Langrenus about midway along the Moon's crescent, and shallower and featureless Vendelinus south of it. Return to the Mare Crisium area and look for conspicuous crater Cleomides to its north. All three are Lunar Club challenges!

For stargazers, take a look at Eta Canis Majoris while you're out this evening (Figure 3.27). You'll find it on the southeast corner of the upside down Y of the constellation. Its name is Aludra, and despite being some 3000 light-years from Earth, it outshines many other stars in the night. Far younger than our own Sol, Aludra is a blue supergiant and already in the final stages of its life. Continuously expanding, it may already be becoming a red supergiant—awaiting that final supernova event within a few million years.

Now take out binoculars and let's hop about a fistwidth north of bright Eta Canis Majoris and have a look at a "double cluster"—NGC 2383 (RA 07 24 42 Dec –20 56 42) and NGC 2384 (RA 07 25 12 Dec –21 01 24) (Figure 3.28). Just showing in binoculars as a faint patch, this pair will begin resolution with larger scopes. Studied photometrically, it would appear these fairly young clusters have contaminated each other by sharing stars: this has also occurred in some clusters located in the Magellanic Clouds. Enjoy this unusual collection of stars...

Figure 3.27. Eta Canis Majoris: Aludra (Credit—Palomar Observatory, courtesy of Caltech).

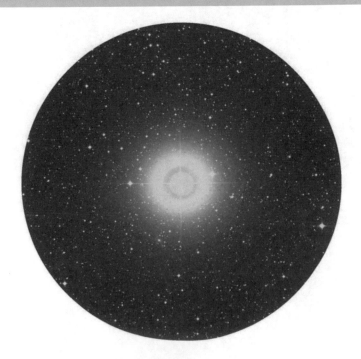

Figure 3.28. NGC 2383 and 2384 (Credit—Palomar Observatory, courtesy of Caltech).

Tuesday, March 11

Today is the birthday of Urbain Leverrier (Figure 3.29). Born in 1811, Leverrier predicted the existence of Neptune, leading to its discovery.

Tonight let's go to the Moon to discover an ancient and ruined crater which lies on the southern shore of Mare Nectaris. To binoculars, Fracastorius will look like a shallow, light colored ring, but a telescope will reveal its northern wall is missing—perhaps melted away by the lava flow which formed the mare (Figure 3.30). This is all that remains of a once grand crater which was more than 117 kilometers in diameter. The tallest of its eroded walls still stand at an impressive 1758 meters, placing them as high as the base elevation of Mt. Hood, yet in places nothing more than a few ridges and low hills still stand to mark the crater's remains. Power up and look for interior craterlets. Be sure to mark your lunar observing challenge notes with your observations!

Even with the Moon's interference, we can still have a look at a nice open cluster from an alternative catalog compiled by astronomers Harlow Shapley and P. J. Melotte. You'll find it less than a fingerwidth southwest of Alpha Monocerotis (RA 07 38 24 Dec –10 41 00).

Known as Mel 72 (Figure 3.31), this young open cluster can just barely be picked up with binoculars as a slight compression in the starfield, while larger

Figure 3.29. Urbain Leverrier (widely used public image).

telescopes will completely resolve it. Rich and condensed, older star catalogs may sometimes claim this region is called Collinder 467, but according to the work of Archinal and Hines this was a case of mistaken identity. Look for a beautiful stream of stars heading northward from this "off the beaten path" open cluster!

Figure 3.30. Fracastorius (Credit—Wes Higgins).

Figure 3.31. Melotte 72 (Credit—Palomar Observatory, courtesy of Caltech).

Wednesday, March 12

Tonight we take on another Lunar Club challenge as we look mid-way along the terminator at the west shore of Mare Tranquillitatis for crater Julius Caesar (Figure 3.32). This is also a ruined crater, but it met its demise not through lava flow but from a cataclysmic event. The crater is 88 kilometers long and 73 kilometers wide. Although its west wall still stands over 1200 meters high, look carefully at the east and south walls. At one time, something plowed its way across the lunar surface, breaking down Julius Caesar's walls and leaving them to stand no higher than 600 meters at the tallest.

While you're out, this would also be a good time to have a look at Epsilon Canis Majoris (Figure 3.33). Located at the southwestern corner of the upside down Y, Epsilon is actually the second brightest star in the constellation. Its name is Adhara and it's one of the hottest of the known bright stars. If it were as close as Sirius, it would outshine Venus by over 15 times! Adhara also has one of the most extreme ultraviolet spectra known, and it has been used to study the interstellar medium, and also as a model for the atmospheres of nearby stars. But there's more... Epsilon is also a great double star. While its companion is quite disparate at roughly magnitude 8, the pair can be easily separated with a small telescope and the visual companions with just binoculars!

Figure 3.32. Julius Caesar (Credit—Wes Higgins).

Figure 3.33. Epsilon Canis Majoris (Credit—Palomar Observatory, courtesy of Caltech).

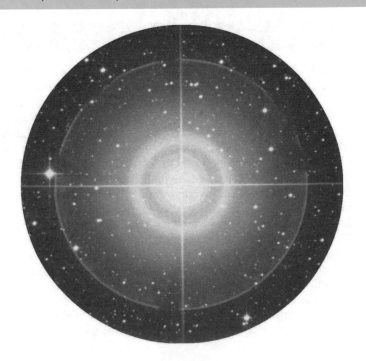

Thursday, March 13

On this day in 1781, Uranus was discovered by William Herschel. Also on this day, in 1855, Percival Lowell was born in Boston (Figure 3.34). Educated at Harvard, Lowell went on to found the observatory which bears his name in Flagstaff, Arizona, and spent a lifetime studying Mars. Tonight you can honor Lowell by looking at Mars yourself. While you might not see a great many details, think of how many strides have been made since Lowell's time and how advanced our knowledge of Mars has become!

Much like our crater from last night, tonight we'll explore the lunar surface for another surface scar caused by a glancing blow. Head to the north near the terminator as the incredible Alpine Valley now comes into view (Figure 3.35).

Figure 3.34. Percival Lowell (widely used public image).

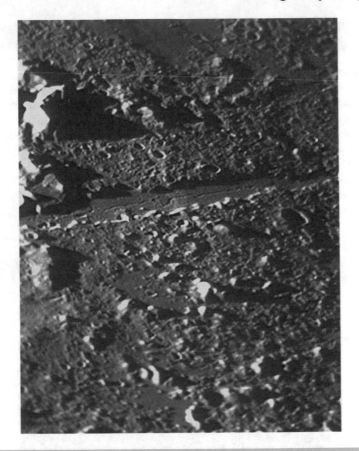

Figure 3.35. Valles Alpes (Credit—Wes Higgins).

Cutting its way through the lunar Alps with a width of 1.5 to 27 kilometers and stretching 177 kilometers long, it is possible this unusual feature was formed by a tectonic shift, but this is unlikely. Viewable through binoculars as a thin, dark line, telescopic observers at highest powers will enjoy a wealth of details in this area, such as a crack running inside its boundaries. No matter how it came to be, it's a very unusual feature, and a Lunar Club challenge. Catch it tonight!

Friday, March 14

Today is the birthday of Albert Einstein (Figure 3.36). Born in 1879, Einstein was one of the finest minds of our time. He conceived of gravity in terms of spacetime curvature—dependent on the energy density. Winner of the 1921 Physics Nobel prize, Einstein's work on the photoelectric effect is the basis of modern light detectors.

Tonight the northern area of the Moon will offer up wonderful details to help you on your way to lunar studies. Last month we reviewed the south at this phase,

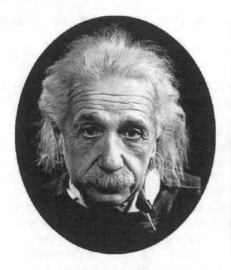

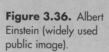

Figure 3.36. Albert Einstein (widely used public image).

so why don't we do the same for the north? Many of these features (marked in Figure 3.37) can be spotted in binoculars and are very easy with a telescope at mid-range magnifications. Let's have a look at them.

Figure 3.37. Plato region (Credit—Greg Konkel, annotation by Tammy Plotner). (1) Eudoxus, (2) Aristotle, (3) Caucasus Mountains, (4) Lunar Alps, (5) Valles Alpes, (6) Aristillus, (7) Autolycus, (8) Archimedes, (9) Mons Piton, (10) Mons Pico, (11) Straight Range, (12) Plato, (13) Mare Frigoris, (14) W. Bond, (15) Barrow, (16) Meton, (17) Cassini, (18) Alexander, (19) Montes Spitzbergen, (20) Mons Blanc.

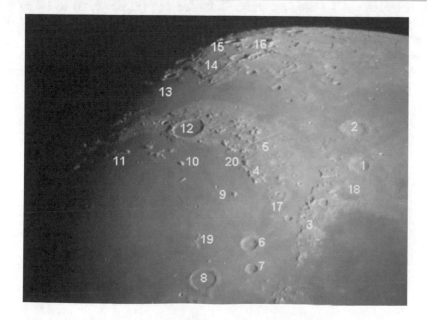

How many of these craters are Lunar Club challenges? Be sure to note your observations!

Saturday, March 15

Tonight Mars will be hauntingly close to the Moon, but let's celebrate the birth of Nicolas Lacaille (Figure 3.38). Born in on this day in 1713, Lacaille's measurements confirmed the Earth's equatorial bulge. Let's begin our look at Lacaille tonight on the lunar surface. Your first assignment will be to locate the crater named for him (Figure 3.39)!

Located northeast of Tycho, this 68 kilometer wide crater isn't the most impressive in the area, but it does drop down about 2775 meters below the Moon's surface. This depth is very comparable to that of Hawaii's Haleakala Crater, which reaches down almost an identical distance to the ocean's floor. For the most part, Lacaille's interior is very smooth, with only a few minor craterlets along the northern edge marring the smooth lava. Look carefully at its eastern wall, where a much newer impact has occurred, and to the south where the crater walls have eroded.

Now let's talk about the man. From 1750 to 1754, Abbé Nicolas du La Caille had the good fortune to study the Southern Hemisphere's stars from the Cape of Good Hope. He named 14 of the 88 constellations and studied 10,000 stars. He cataloged several nebulous star clusters and even discovered a galaxy! While this might seem pretty simple, think on this... Lacaille did it with a 0.5 inch refractor. Most of our modern finderscopes are bigger!

To further honor Lacaille tonight, let's continue by picking up an open cluster in a constellation he named. While his original catalog number II.2 put this object in Puppis, we now recognize Collinder 140 (Figure 3.40) to be part of Canis Major—just a few degrees south of Eta (RA 07 23 18 Dec –32 04 00). At around 3.5 magnitude, this rich open cluster is sufficiently large and bright to be enjoyed even through the moonlight! Around 1000 light-years away and estimated to be 22 million years old, even the smallest aperture will enjoy this

Figure 3.38. Nicolas Lacaille (widely used public image).

Figure 3.39. Crater Lacaille (Credit—Greg Konkel, annotation by Tammy Plotner).

Figure 3.40. Collinder 140 (Credit—Palomar Observatory, courtesy of Caltech).

large, jewel-like collection of stars which includes the double Dunlop 47. This one is best in binoculars!

Sunday, March 16

Today we celebrate the birthday (1850) of Caroline Herschel (Figure 3.41), William's sister. Devoted to her brother, she was much more than just his assistant—she was his friend. At least 14 of the objects in the Herschel catalog bear her signature and William's own note: "Lina found it." To preserve her brother's night vision, Carolyn took on the arduous task of setting down on paper William's directions and descriptions, yet she was just as avid at the eyepiece. From her memoirs she states:

> I knew too little of the real heavens to be able to point out every object so as to find it again without losing too much time by consulting the Atlas. But all these troubles were removed when I knew my brother to be at no great distance making observations with his various instruments on double stars, planets, etc., and I could have his assistance immediately when I found a nebula, or cluster of stars, of which I intended to give a catalogue...

By the end of her long life, Caroline Herschel had discovered six comets, received numerous honors, and is often referred to as "The First Woman of Astronomy."

Also on this day in 1926, Robert Goddard launched the first liquid-fuel rocket. But he was first noticed in 1907 when a cloud of smoke issued from a powder rocket fired in the basement of the physics building in Worcester Polytechnic Institute. Needless to say, the school took an interest in the work of this shy student. Thankfully they did not expel him, and thus began his lifetime of work

Figure 3.41. Carolyn Herschel (widely used public image).

Figure 3.42. Robert Goddard (Credit—NASA).

in rocket science. Goddard was also the first to realize the full implications of rocketry for missiles and space flight, and his lifetime of work was dedicated to bringing this vision to realization. While most of what he did went unrecognized for many years, tonight we celebrate the name of Robert H. Goddard (Figure 3.42). This first flight may have gone only 12 meters, but 40 years later on the date of his birth, Gemini 8 was launched, carrying Neil Armstrong and David Scott into orbit!

Tonight have a look at the Moon with either telescope or binoculars to reveal two excellent, and easily identifiable, Lunar Club challenges—Eratosthenes and Copernicus (Figure 3.43). Be sure to note both features and we'll return in time to have a closer look at both of these incredible craters.

When you are finished with your lunar observations, turn your eyes or scopes toward Rho Puppis (Figure 3.44), just slightly more than a fingerwidth northeast

Figure 3.43. Eratosthenes and Copernicus (Credit—Greg Konkel).

Figure 3.44. Rho Puppis: Tureis (Credit—Palomar Observatory, courtesy of Caltech).

of Eta Canis Majoris. Its name is Tureis, and it's a very special star. Located only 63 light-years away, 3rd magnitude Tureis is about twice the size of our own Sun, but produces six times more energy. It is one of the night sky's brightest Delta Scuti type variables, changing in luminosity by 10% in a bit more than three hours. Along with this rapid variation in brightness is an almost equally rapid pulsation which causes changes in Tureis' spectrum—a variation once believed caused by a spectroscopic companion. While many star charts still recognize Rho as a double star, the only evidence you will find with a telescope is a far-orbiting, 13th magnitude, M class red dwarf. With an estimated orbital period of 10,000 years, it is not yet established if this star is a true companion or not.

While outside, be on watch for the Corona-Australid meteor shower. While the fall rate is low at 5–7 per hour, our friends in the Southern Hemisphere might stand a chance with this one!

Monday, March 17

On this day in 1958, the first solar-powered spacecraft was launched. Named Vanguard 1, it was an engineering test satellite (Figure 3.45). From its orbital position, the data taken from its transmissions helped to refine the true shape of the Earth.

On the Moon, the terminator has now revealed the placid Sinus Iridum (Figure 3.46). The two features we will look closely at tonight are the Promonto-riums which guard the opening of Iridum like two lighthouses. The easternmost is LaPlace, named for Pierre LaPlace. Little more than 56 kilometers in diameter, it rises above the "Bay of Rainbows" some 3019 meters; almost identical in height

Figure 3.45. Vanguard 1 (Credit—NASA).

Figure 3.46. Sinus Iridum (Credit Roger Warner).

Figure 3.47. K Puppis (Credit—Palomar Observatory, courtesy of Caltech).

to Buttermilk Mountain near Aspen. Promontorium Heraclides covers roughly the same area, yet rises to little more than half of LaPlace's height. Both are telescopic Lunar Club challenges so be sure to mark your notes!

Tonight we'll have a look at another binary system about another fingerwidth north-northeast of Rho. Its name is K Puppis, but it is more commonly referred to on double star lists as Kappa (Figure 3.47). Even small telescopes will enjoy this widely spaced, close magnitude pair of white stars in a great field!

Tuesday, March 18

Today in 1965, the first ever spacewalk was performed by Alexei Leonov (Figure 3.48) outside the Soviet Voskhod spacecraft. The "walk" only lasted around 20 minutes and Alexei had problems in re-entering the spacecraft because his space suit had enlarged slightly. Imagine his fear as he had to let air leak out of his space suit in order to squeeze back inside. When they landed off-target in the heavily forested Ural Mountains, the crew of two had to spend the night in the woods surrounded by wolves. It took over 24 hours before they were located, and rescuers had to chop their way through the forest on skis to recover them. Brave men!

Tonight it's time to walk the Southern Highlands again as crater Schiller comes into view (Figure 3.49). While Schiller itself is rather recognizable, as the Sun lights up the moonscape many craters change appearance as new ones become

Figure 3.48. Alexei Leonov (Credit—NASA).

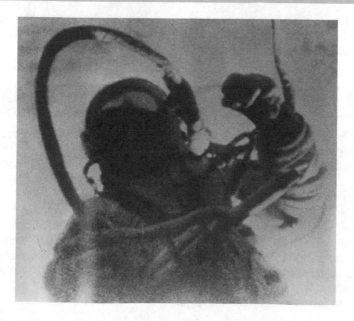

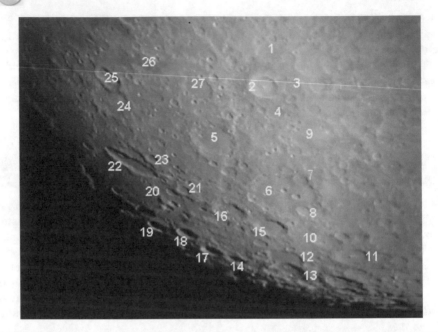

Figure 3.49. Region around Schiller (Credit—Greg Konkel, annotation by Tammy Plotner). (1) Sasserides, (2) Tycho, (3) Pictet, (4) Street, (5) Longomontanus, (6) Clavius, (7) Porter, (8) Rutherford, (9) Maginus, (10) Gruemberger, (11) Moretus, (12) Klaproth, (13) Casatus, (14) Wilson, (15) Blancanus, (16) Scheiner, (17) Kircher, (18) Bettinus, (19) Zucchius, (20) Segner, (21) Rost, (22) Schiller, (23) Bayer, (24) Mee, (25) Hainzel, (26) Lacus Timoris, (27) Wilson.

more dramatic. Let's take a look at what can be seen during this phase and how many you can identify. Best of luck and be sure to use this map whenever this area comes into view!

Wednesday, March 19

Tonight the gibbous Moon will be almost overpoweringly bright (Figure 3.50), but that doesn't mean there isn't something fun to do! For most observers, Regulus will be less than a degree away from the lunar limb and Saturn about twice as far. For a lucky few, Regulus could be either an occultation or grazing event, so be sure to check sources like IOTA for precise times and locations!

There will be no well-known objects visible tonight, so why not take a look at Xi Puppis? All you need to do is just hop again northeast from our last two studies! Located about 1300 light-years away, Xi is a true binary star with the name of Asmidiske (Figure 3.51). Its 13th magnitude companion may also be a binary—but a spectroscopic one. Also seen in the field are many other easily resolved stars. While they aren't quite as distant as Xi itself, they still make for a wonderful journey on a moonlit night!

Figure 3.50. Gibbous Moon (Credit—Greg Konkel).

Figure 3.51. Xi Puppis: Asmidiske (Credit—Palomar Observatory, courtesy of Caltech).

Thursday, March 20

Today, spare a moment for Robert Burnham Jr., who died on this day in 1993. While it would seem more fitting to honor Burnham on the date of his birth, the author wants to remind the reader of all Burnham was. Born on June 16, 1931, Robert Jr. was a reclusive man who had few friends, never married, and spent most of his time exploring the heavens with his self-made telescope. After discovering a comet, he became part of the Lowell Observatory staff, but one year after publishing the revered *Celestial Handbook* the funding ran out for his position. Rather than become an observatory janitor after such an illustrious career, Burnham spent the rest of his life alone and poor, selling paintings of his cats in the park. Despite discovering numerous comets and asteroids, Burnham died unnoticed and unsung at the early age of 61.

Mr. Burnham? We remember...

If you're up before dawn this morning, be sure to check out the eastern horizon for the return of the swift inner planet Mercury. Watch in the days ahead as it rises higher and higher each morning on its climb toward Venus!

Today is Vernal Equinox, one of the two times of the year when day and night become equal in length. From this point forward, the days will become longer—and our astronomy nights shorter! To the ancients, this was a time of renewal and planting—led by the goddess Eostre. As legend has it, she saved a bird whose wings were frozen from the winter's cold, turning it into a hare which

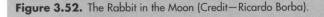

Figure 3.52. The Rabbit in the Moon (Credit—Ricardo Borba).

could also lay eggs. Do you have children or grandchildren? Tonight would be a great night to show them the "Rabbit in the Moon" (Figure 3.52)!

Since the dawn of mankind, we have been gazing at the Moon and seeing fanciful shapes in the lunar features. Tonight as the Moon rises is your chance to catch the "Rabbit," which is a compilation of all the dark maria. The Oceanus Procellarum forms the ear while the Mare Humorum makes the nose. The Rabbit's body is Mare Imbrium and the front legs appear to be Mare Nubium. Mare Serenitatis is the backside and the picture is complete where Mare Tranquillitatis and Mare Fecunditatis shape the hind legs—with Crisium as the tail. See the Moon with an open mind and open eyes…and find the Rabbit!

Friday, March 21

Tonight is the Full Moon. In many cultures, it is known as the "Worm Moon." As ground temperatures begin to warm and produce a thaw for the Northern Hemisphere, earthworms return and signify the return of robins. For Native Americans in the far north, this was also considered to be the "Crow Moon." The return of this black bird signaled the end of winter. Sometimes it has been called the "Crust Moon" because warmer temperatures melt existing snow fall during the day, leaving it to freeze to a crust during the night. Perhaps you may have also heard it referred to as the "Sap Moon." This marks the time of tapping maple trees to make syrup. To early American settlers, it was called the "Lenten Moon" and was considered to be the last Full Moon of winter. Let's hope!

Figure 3.53. Saturn (Credit—Wes Higgins).

No matter how clear the skies are tonight, we're not going to be able to escape the Moon, however you call it! So let's turn an eye toward an ever-changing planet—Saturn (Figure 3.53). Even a small telescope can resolve Saturn's rings, and at high magnification see significant details. Look for things like the broad Cassini division in the ring plane, as well as the shadow of the planet on the rings. Be sure to take note of Saturn's many moons as well! While Titan orbits well outside the rings and is bright enough to be noticed even in a small telescope, even apertures of 4" can pick up the smaller moons which orbit close to the ring system. No matter how you choose to look at it, Saturn is one of the most mysterious and fascinating of our solar system's members.

Saturday, March 22

Today in 1799 Friedrich Argelander was born (Figure 3.54). He was a compiler of star catalogues, studied variable stars and created the first international astronomical organization. To honor Argelander and his achievements tonight, let's have a look at an easy variable star which can be followed even without special equipment. Its name is Zeta Geminorum (Figure 3.55) and you'll find it on the southern side of Gemini, and it's the southernmost of the two stars at the "waist."Given the proper name of Mekbuda ("the lion's folded paw"), 1200 light-year distant Zeta is one of the brightest of Cepheid variables with a temperature similar to our own Sun... But it's there where any similarities end. Zeta is a supergiant which shines 5700 times brighter than our own star. Its short period makes it very easy to observe—it changes by about a half a magnitude every 10.2 days. It is a bit unusual, because its rise to maximum and fall to minimum occur in an almost equal amount of time. You can observe that it will gradually brighten to about the same magnitude as Kappa (south of Pollux) and then slowly fade to about the same brightness as Upsilon (southwest of Pollux). For the telescope, this charming star also has two companions (8th and 11th magnitude) which aren't physically related, but make a nice appearance in the eyepiece.

Figure 3.54.
Friedrich Argelander
(widely used public
image).

Figure 3.55. Zeta Geminorum (Credit—Palomar Observatory, courtesy of Caltech).

Although the Moon will greatly interfere, enjoy a spring evening with two meteor showers. In the Northern Hemisphere, look for the Camelopardalids. They have no definite peak, and a "screaming" fall rate of only 1 per hour. While that's not much, at least they are the slowest meteors—entering our atmosphere at speeds of only seven kilometers per second!

Far more interesting to both hemispheres will be the March Geminids which peak tonight. They were discovered and recorded in 1973 and then confirmed in 1975. With a much faster fall rate of about 40 per hour, these slower than normal meteors will be fun to watch! When you see a bright streak, trace it back to its point of origin: which did you see, a Camelopardalid or a March Geminid?

Sunday, March 23

Today in 1840, the first photograph of the Moon was taken (Figure 3.56). The daguerreotype was exposed by American astronomer and medical doctor J. W. Draper. Draper's fascination with chemical responses to light also led him to another first photograph of the Orion Nebula.

Before the Moon commands the sky, let's have a look at an object that's better suited for southern declinations—NGC 2451 (Figure 3.57). As both a Caldwell object (Collinder 161) and a southern skies binocular challenge, this 2.8

Figure 3.56. J. W. Draper's daguerreotype of the Moon (Credit—Courtesy of New York University).

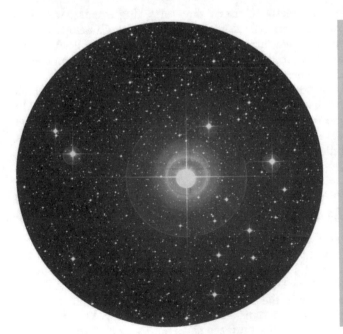

Figure 3.57. NGC 2451 (Credit—Palomar Observatory, courtesy of Caltech).

Figure 3.58. Raw image using described method (Credit—Tammy Plotner).

magnitude cluster was probably discovered by Hodierna. Consisting of about 40 stars, its age is believed to be around 36 million years. It is very close to us at a distance of only 850 light-years. Take the time to study this object closely—due to the thinness of the galactic disk in this region, it is believed we are seeing two clusters superimposed on one another.

And here comes the Moon... Why not try walking in Draper's footsteps tonight and photographing it? You will be very surprised at how the simplest of equipment can achieve some astonishing results! For this experiment, all you need is a telescope with a 1.25 diameter eyepiece and a disposable camera. Always begin by snapping a picture of a fully lighted area so the developers will know where to cut the film. Set the scope on the Moon and adjust the focus for 20/20 vision. Carefully hold the camera directly to the eyepiece (almost all cameras' lenses will mate perfectly) and wrap your fingers around it to hold it steady. Now shoot... While this method won't win any awards, the results will far exceed Draper's experiments (Figure 3.58)!

Monday, March 24

If you're up before dawn this morning, grab your binoculars and have a look at the intricate dance of orbits as Mercury passes Venus in the brightening skies!

Today is the birthday of Walter Baade (Figure 3.59). Born in 1893, Baade was the first to resolve the Andromeda galaxy's individual stars using the Hooker

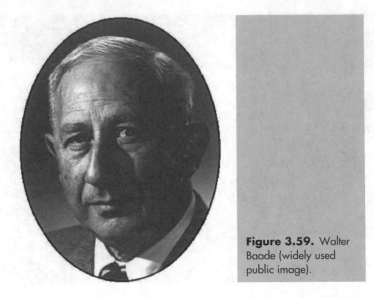

Figure 3.59. Walter Baade (widely used public image).

telescope during World War II blackout times, and he also developed the concept of stellar populations. He was the first to realize there were two types of Cepheid variables, thereby refining the cosmic distance scale. He is also well known for

Figure 3.60. M105 (Credit—Palomar Observatory, courtesy of Caltech).

discovering an area toward our galactic center which is relatively free of dust, now known as "Baade's Window."

On this night, 212 years before Baade was born, another great astronomer was at the eyepiece of his telescope making a new discovery. That astronomer was none other than Messier's assistant—Pierre Méchain—and the distant galaxy he turned up that night was eventually cataloged as M105 (Figure 3.60).

Located almost midway between 52 and 53 Leonis, and about a palm's width east of Regulus (RA 10 47 49 Dec +12 34 54), Méchain's discovery lies in what most astronomers refer to as the Leo I galaxy group. It is the brightest of the ellipticals and, thanks to the work of Baade, we know it to be around 38 million light-years distant. As one of the most evenly distributed of elliptical galaxies in terms of surface brightness, at it core beats a heart which exceeds our own Sun's mass by over 50 million times. With a rough magnitude of 9, this giant beauty is within range of large binoculars and small telescopes, but aperture is required to spot its companions—NGC 3384 and NGC 3389.

Tuesday, March 25

Today in 1655, Titan—Saturn's largest satellite—was discovered by Christian Huygens (Figure 3.61). He also discovered Saturn's ring system during this same year. After 350 years, the probe named for Huygens stunned the world as it reached the surface of Titan and sent back information on this distant world.

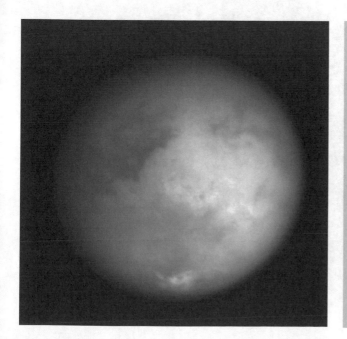

Figure 3.61. Titan (Credit—NASA).

On this day in 1951, 21 centimeter wavelength radiation from atomic hydrogen in the Milky Way was first detected. These 1420 megahertz H I studies continue to form the basis of a major part of modern radio astronomy. If you would like to have a look at a source of radio waves known as a pulsar, then aim your binoculars slightly more than a fistwidth east of bright Procyon. The first two bright stars you encounter will belong to the constellation of Hydrus and you will find pulsar CP0 834 just above the northernmost—Delta.

Hang on to those binoculars and hop southwest of Regulus to Alpha Hydrae (Figure 3.62). About two degrees southwest (RA 09 20 29 Dec –09 33 20) in the same field you will find an interesting star—27. This is a great double star for small optics, easily separated and easily found. The 5th magnitude primary appears to be slightly yellow in tint compared to the pure white of the 7th magnitude companion. Not only do they share proper motion, but a place on many doubles observing lists!

For an outright big telescope challenge, hop another degree southeast for NGC 2863 (RA 09 23 36 Dec –10 26 01) (Figure 3.63). At near magnitude 13, it's not the brightness of this galaxy which makes it difficult—but the size. Nestled between the stars, this tiny galaxy was discovered on this date in 1786 by Sir William Herschel and cataloged as H III.520. Good luck!

Figure 3.62. 27-P Hydrae (Credit—Palomar Observatory, courtesy of Caltech).

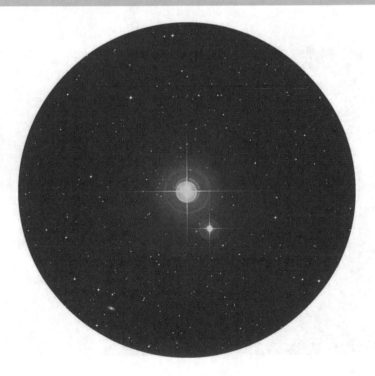

Figure 3.63. NGC 2863 (Credit—Palomar Observatory, courtesy of Caltech).

Wednesday, March 26

Tonight with the Moon absent for awhile, we'll start in northern Puppis and collect three more Herschel studies as we begin at Alpha Monocerotis and drop about four fingerwidths southeast to 19 Puppis (RA 08 10 42 Dec –12 50 00). NGC 2539 averages around 6th magnitude and is a great catch for binoculars as an elongated hazy patch with 19 Puppis on the south side (Figure 3.64). Telescopes will begin resolution on its 65 compressed members, as well as split 19 Puppis— a wide triple. This open cluster has been studied as a role model for lithium depletion and is roughly the same age as the Hyades.

Shift about five degrees southwest (RA 07 55 06 Dec –17 43 00) and you'll find NGC 2479 directly between two finderscope stars (Figure 3.65). At magnitude 9.6, it is for telescopes only and will show as a smallish area of faint stars at low power. Head another degree or so southeast (RA 08 00 43 Dec –19 04 19) and you'll encounter NGC 2509 (Figure 3.66)—a fairly large collection of around 40 stars which can be spotted in binoculars and small telescopes. This pretty open cluster has recently been the object of 2MASS studies for age, distance, and metallicity.

Be sure to mark your observing notes because all three of these open clusters are on the Herschel "400" list!

Figure 3.64. NGC 2539 (Credit—Palomar Observatory, courtesy of Caltech).

Figure 3.65. NGC 2479 (Credit—Palomar Observatory, courtesy of Caltech).

Figure 3.66. NGC 2509 (Credit—Palomar Observatory, courtesy of Caltech).

Thursday, March 27

Set your alarm clock for the early morning hours as red Antares and the waning Moon slow dance together on the ecliptic plane less than a degree apart! A union this close usually means an occultation for some part of the world, so be sure to check sources such as IOTA for accurate information in your area.

So what's special on the agenda tonight? Just a discovery—an extraordinarily beautiful one at that. On this night in 1781, the unsung astronomy hero Pierre Méchain happened on an incredible galaxy in Ursa Major. Located about three fingerwidths northeast of Mizar and Alcor (RA 14 03 13 Dec +54 20 53), this near 8th magnitude galaxy was added as one of the last on the Messier list, but it ranks as one of the first to be identified as a spiral.

While M101 is huge and bright (Figure 3.67), binoculars will spot only the bright central region—yet the average beginner's scope (4.5") will begin to reveal arm structure with aversion. As aperture increases, so does detail and some areas are so bright that Herschel assigned them their own catalog numbers. Even Halton Arp noted this one's lopsided core as number 26 ("Spiral with One Heavy Arm") on his peculiar galaxies list!

At a distance of 27 million light-years, M101 might be somewhat disappointing to smaller scopes, but photographs show it as one of the most fantastic spirals in the Cosmos. Dubbed the "Pinwheel," it heads up its own galactic group consisting

Figure 3.67. M101: The Pinwheel Galaxy (Credit—R. Jay GaBany).

of NGC 5474 to the south-southeast and NGC 5585 to the northeast, which are visible to larger scopes. It is estimated there may be as many as six more members as well! Be sure to take the time to really study this galaxy. The act of sketching often brings out hidden details and will enrich your observing experience.

Friday, March 28

Born today in 1749, Pierre LaPlace was the mathematician who conceived both the metric system and the nebular hypothesis for the origin of the solar system (Figure 3.68). Also born on this day 1693 was James Bradley, an excellent astrometrist who discovered the aberration of starlight (1729) and the nutation of the Earth. And, in 1802, Heinrich W. Olbers discovered the second asteroid, Pallas, in the constellation Virgo while making observations of the position of Ceres, which had only been discovered 15 months earlier. Five years later on this same date in 1807, Vesta—the brightest asteroid—was discovered by Olbers in Virgo, making it the fourth such object found.

Your assignment, should you choose to accept it, is to locate Vesta. You'll find it just a bit south of the union of Uranus, Venus, and Mercury about 30 minutes before local dawn. Pallas is too close to the Sun right now for safe viewing. While asteroid chasing is not for everyone, both Vesta and Pallas are often bright enough to be identified with just binoculars. In the coming months, each will rise higher each morning in the predawn sky. Use an online resource to get accurate locator charts and keep a record of spotting these solar system planetoids!

Indeed, this was a date of discovery as the prolific Sir William found yet another object for future generations to marvel at. Your destination is around a degree east of Alpha Lyncis, and is in the field with a 7th magnitude star in Leo Minor (RA 09 24 18 Dec +34 30 48). It's name? NGC 2859 (Figure 3.69).

Figure 3.68. Pierre LaPlace (widely used image).

Located about 23 million light-years away, this handsome barred spiral was cataloged on this night in 1786 as H I.137. At around magnitude 11, it's within the reach of average telescopes and the observer will first note its bright core region. But don't stop there: while there's nothing unusual about barred structure, this galaxy appears to have a detached halo around it. Often known as the "Ring Galaxy," this structure could perhaps be caused by gravitational forces reacting with gases along certain points in the bar structure, and so creating a resonance. Oddly enough, each of the four companion galaxies of NGC 2859 contains a

Figure 3.69. NGC 2859 (Credit—Palomar Observatory, courtesy of Caltech).

compact object or quasar-like phenomenon, and they all have similar redshifts. Be sure to add this "space oddity" to your observing notes!

Saturday, March 29

Today celebrates the first flyby of Mercury, made by Mariner 10 in 1974. Mariner 10 was unique (Figure 3.70). It was the first spacecraft to use a gravity assist from the planet Venus to help it travel on to Mercury. Due to its trajectory, Mariner 10 was only able to study half the surface, but its 2800 photographs showed that Mercury looks very similar to our Moon, has an iron-rich core, a magnetic field, and a very thin atmosphere. For skywatchers, the swift inner planet is now getting too close to the Sun for comfortable observation. It will be about another month before Mercury is visible in the evening skies, so begin watching the western skyline at the beginning of May.

Before this month ends, let's finish up our tour of Puppis. For observers living in high northern latitudes, you'll never see all of this constellation, but there will be some things for you to explore, as well as a great deal for our friends in the Southern Hemisphere. On our list tonight is a Herschel object which lies directly on the galactic equator around five degrees north-northwest of Xi (RA 07 36 12 Dec –20 37 00).

NGC 2421 is a magnitude 8.3 open cluster which will look like an exquisitely tiny "Brocchi's Cluster" in binoculars (Figure 3.71), and it will begin to show

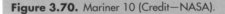

Figure 3.70. Mariner 10 (Credit—NASA).

Figure 3.71. NGC 2421 (Credit—Palomar Observatory, courtesy of Caltech).

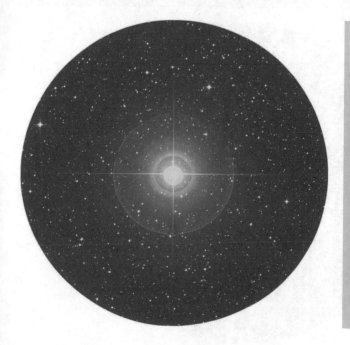

Figure 3.72.
Sigma Puppis
(Credit—Palomar
Observatory, courtesy
of Caltech).

good resolution of its 50 or so members to an intermediate telescope, in an arrowhead-shaped pattern. It's bright, it's fairly easy to find, and it's a great open cluster to add to your challenge study lists!

For the southern observer, try your hand at Sigma Puppis (Figure 3.72). At magnitude 3, this bright orange star holds a wide separation from its white 8.5 magnitude companion. Sigma's B star is a curiosity... While it resides at a distance of 180 light-years from our solar system, it would be about the same brightness as our own Sun if placed one Astronomical Unit from Earth!

Sunday, March 30

If you're out late or up before sunrise, be sure to take a look at the Moon and Jupiter making a pleasing pairing along the ecliptic. No special equipment is needed!

With early evening dark skies, let's finish our study of the Herschel objects in Puppis. Only three remain, and we'll begin by dropping south-southeast of Rho and center the finder on a small collection of stars to locate NGC 2489 (RA 07 56 18 Dec −30 04 00) (Figure 3.73). At magnitude 7, this bright collection is worthy of binoculars, but only the small patch of stars in the center is the cluster. Under aperture and magnification you'll find it to be a loose collection of around two dozen stars formed in interesting chains.

Figure 3.73. NGC 2489 (Credit—Palomar Observatory, courtesy of Caltech).

Figure 3.74. NGC 2571 (Credit—Palomar Observatory, courtesy of Caltech).

Figure 3.75. NGC 2567 (Credit—Palomar Observatory, courtesy of Caltech).

The next are a north–south-oriented pair around five degrees due east of NGC 2489. You'll find the northernmost—NGC 2571—at the northeast corner of a small finderscope or binocular triangle of faint stars (RA 08 19 00 Dec –29 45 00) (Figure 3.74). At magnitude 7, it will show as a fairly bright hazy spot with a few stars beginning to resolve. Around 30 mixed magnitude members will be revealed to aperture.

Less than a degree south is NGC 2567 (RA 08 18 32 Dec –30 38 00)(Figure 3.75). At around a half magnitude less in brightness, this rich open cluster has about 50 members to offer the larger telescope, which are arranged in loops and chains. Congratulations on completing these challenging objects!

Monday, March 31

Today in 1966, Luna 10 was on its way to the Moon. The unmanned, battery-powered Luna 10 was a USSR triumph (Figure 3.76). Launched from an Earth orbiting platform, the probe became the first to successfully orbit another solar

Figure 3.76. Luna 10 (Credit—NASA).

Figure 3.77. M93 (Credit—Palomar Observatory, courtesy of Caltech).

system body. During its 460 orbits, it recorded infrared emissions, gamma rays, and analyzed lunar composition. It monitored the Moon's radiation environment, and also discovered what eventually would be referred to as "mascons"—mass concentrations—below maria surfaces which gravitationally influence orbiting bodies.

Take your telescopes or binoculars out tonight and look just north of Xi Puppis (RA 07 44 36 Dec –23 52 00) for a "mass concentration" of starlight known as M93 (Figure 3.77). Discovered in March of 1781 by Charles Messier, this bright open cluster is a rich concentration of various magnitudes which will simply explode in sprays of stellar fireworks in the eyepiece of a large telescope. Spanning 18 to 22 light-years of space and residing more than 3400 light-years away, it contains not only blue giants, but lovely golds as well. Jewels in the dark sky...

As you view this cluster tonight, seize the moment to remember Messier, because this is one of the last objects he discovered personally. He described it as "A cluster of small stars without nebulosity"—but did he realize the light he was viewing at the time left the cluster during the reign of Ramses III? Ah, yes...sweet time. Did Charles have a clue this cluster of stars was 100 million years old? Or realize it was forming about the time Earth's land masses were breaking up, dinosaurs ruled, and the first mammals and birds were evolving? Although H. G. Wells "Time Machine" is a work of fiction, each time we view through a telescope we take a journey back across time itself. Enjoy the mystery!

CHAPTER FOUR

April, 2008

Tuesday, April 1

Today in 1960, the first weather satellite—Tiros 1—was launched. While today we think of these types of satellites as commonplace, the Television InfraRed Observation Satellite was quite an achievement. Weighing in at 120 kilograms, it contained two cameras and two magnetic tape recorders—along with onboard batteries and 9200 solar cells to keep them charged. While it only operated successfully for 78 days, for the first time ever we were able to see the face of the Earth's changing weather (Figure 4.1).

Tonight let's begin a new adventure as we move into the constellation of Cancer and view an ancient "predictor" of the weather. This will be an ideal time to familiarize yourself with its dim stars and one very bright open cluster. Try using both Pollux and Procyon to form the base of an imaginary triangle. Now aim your binoculars or finderscope north near the apex to discover M44—the Beehive (RA 08 40 24 Dec +19 41 00) (Figure 4.2).

According to ancient lore, this group of stars (often called the Praesepe) foretold a coming storm if it was not visible in otherwise clear skies. Of course, this came from a time when combating light pollution meant asking your neighbors to dim their candles. But, once you learn where it's at, it can be spotted unaided even from suburban settings. Hipparchus called it the "Little Cloud," but not until the early 1600s was its stellar nature revealed.

Believed to be about 550 light-years away, this awesome cluster consists of hundreds of members—with at least four orange giants and five white dwarfs.

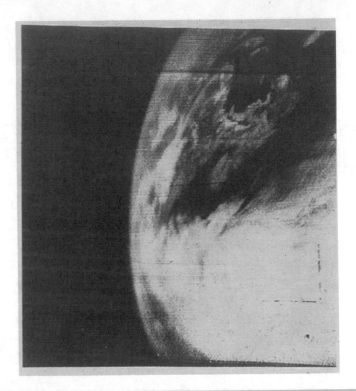

Figure 4.1. First image of Earth from Space: TIROS 1 (Credit—NASA).

Figure 4.2. M44: The Beehive Cluster (Credit—John Chumack).

M44's age is similar to that of the Pleiades, and it is believed both clusters have a common origin. Although you won't see any nebulosity in the Beehive, even the very smallest of binoculars will reveal a swarm of bright stars, and large telescopes can resolve as many as to 350 faint stars. Capture it tonight!

Wednesday, April 2

Today in 1889, the Harvard Observatory's 13" refractor arrived at Mt. Wilson. Just one month later, it went into astronomical service at Lick Observatory, located at Mt. Hamilton (Figure 4.3). It was at Mt. Wilson where the largest telescopes in the world resided from 1908 to 1948: the 60" for the first decade, followed by the 100" . This latter mirror is still the largest solid piece ever cast in plate glass and weighs 4.5 tons. Would you believe it's just 13 inches thick?

Today in 1845, the first photograph of the Sun was taken (Figure 4.4). While solar photography and observing is the domain of properly filtered telescopes, no special equipment is necessary to see some effects of the Sun—only the correct conditions. Right now Earth's magnetosphere and magnetopause (the point of contact) are positioned correctly to interact with the Sun's influencing interplanetary magnetic field (IMF)—the plasma stream which flows past us as solar winds. During the time around equinox, this leaves the door wide open for one of the most awesome signs of spring—aurora (Figure 4.5)! Visit the Geophysical Institute to sign up for aurora alerts and use their tools to help locate the position of the Earth's auroral oval.

Figure 4.3. The Hooker Telescope (Credit—NASA).

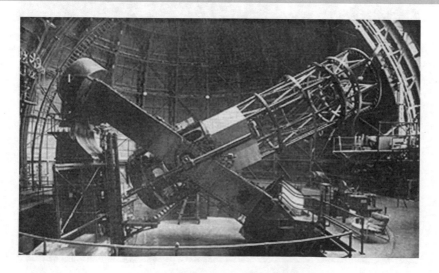

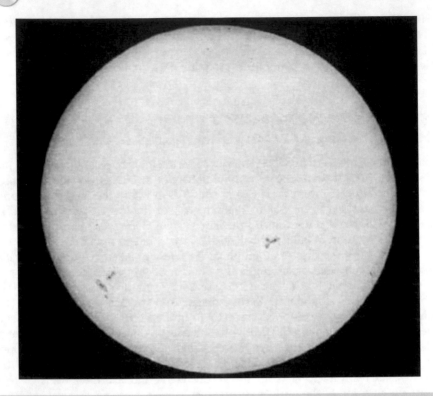

Figure 4.4. First photograph of the Sun (Credit—NASA).

Figure 4.5. Aurora (Credit—John Chumack).

Thursday, April 3

Tonight let's look again and identify the upside down Y of the constellation Cancer. If you can spot M44, the star just south of it is Delta. About three fingerwidths southeast of Delta is Alpha, and we'll begin by exploring this star.

Alpha Cancri (Acubens) which is 130 light-year distant shines at about 4th magnitude, and is also a great double star for a small telescope (Figure 4.6). Its name translates as the "claw" and you will find it clutches a disparate 11.8 magnitude companion star nearby.

Now hop just one fingerwidth west (RA 08 51 18 Dec +11 48 00) for a stunning sight—galactic cluster M67 (Figure 4.7). Hanging out in space some 2500 light-years away and containing more than 500 members, this grand cluster is a rule breaker in age. Believed to be about 10 billion years old, it is one of the oldest star clusters in our galaxy. Its stars have literally "switched off" from the main sequence, and have passed through the red giant stage and are returning back to their blue youth!

In binoculars you will see it as almost galaxy-like in structure, while even small telescopes resolve individual stars. Large telescopes will reveal stars beyond stars, like a globular cluster smeared across the night sky. It is truly one of the most beautiful and mysterious of all open clusters.

Figure 4.6. Alpha Cancri: Acubens (Credit—Palomar Observatory, courtesy of Caltech).

Figure 4.7. M67 (Credit—Palomar Observatory, courtesy of Caltech).

Friday, April 4

We'll return again tonight to Cancer to have a look at some curiosities. The first is about four fingerwidths away from Delta—Zeta Cancri (Figure 4.8). Its name is Tegmeni and it is a handsome double star for the small telescope. Both components are nearly the same magnitude and neatly split for mid-magnification ranges.

About a fingerwidth due east is V Cancri—a Mira-type variable star. While many such variables are difficult to follow with amateur equipment, V Cancri breaks the rules. It changes from magnitude 7.9 to magnitude 12.8 in a period of 125 days. When it swells to its maximum, it reaches a size about the same as the orbit of Mars.

For those of you who use only your eyes to observe—look again at the Beehive and concentrate on Delta to the southeast (Figure 4.9). Known as Asellus Australis, this is a yellow optical double star often called the "southern donkey."

While you're out tonight, be on watch for the Kappa Serpentid meteor shower. Its radiant will be near the "Northern Crown," the constellation known as Corona Borealis. The fall rate is small with an average of 4 or 5 per hour.

Figure 4.8. Zeta Cancri (Credit—Palomar Observatory, courtesy of Caltech).

Figure 4.9. Delta Cancri (Credit—Palomar Observatory, courtesy of Caltech).

Saturday, April 5

The weekend has arrived at last and it's time to do some galaxy hunting in Cancer. You'll find NGC 2672 just about a degree northeast of Delta (RA 08 49 21 Dec +19 04 29) (Figure 4.10).

At magnitude 12.3, it's a galaxy meant for a larger scope. This small elliptical is noteworthy because it also has a companion—NGC 2673—on its eastern edge. Now we're entering the realm of 15th magnitude studies, for very large telescopes only. Just slightly to the southwest (RA 08 50 01 Dec +19 00 34), see if you can spot NGC 2677 as well. This one is extremely challenging.

If you're enjoying this small cluster of galaxies, move no more than a third of a degree west, or let the field drift (RA 08 48 27 Dec +19 01 10). Your reward will be another galactic pair—NGC 2667 (Figure 4.11). These are so close as to be in the same field at high power, and are designated as A and B galaxies. Be aware they require very averted vision and they are among the most difficult of studies! Look for two almost stellar nuclei...

Figure 4.10. NGC 2672/73 (Credit—Palomar Observatory, courtesy of Caltech).

Figure 4.11. NGC 2667 (Credit—Palomar Observatory, courtesy of Caltech).

Sunday, April 6

Ah, yes...tonight is New Moon! With these dark skies, it's time to go Herschel hunting again as we take on spiral galaxy NGC 2775. You'll find it located roughly five degrees southeast of Alpha Cancri (RA 09 10 20 Dec +07 02 18).

At magnitude 11.8, this elongated spiral with a bright core is suited to mid-sized telescopes. Some 60 million light-years away, NGC 2775 is a curious spiral galaxy (Figure 4.12). Its bulging core region and tight pattern of spiral arms have been home to five supernova events within the last 30 years.

For those with very large telescopes, there is a reason why NGC 2775 is an active region. To its northeast is NGC 2777, an amorphous galaxy with a tidally interacting, uncataloged companion—as is the case in the more studied M81/82 system. NGC 2777 is producing a streamer of material flowing toward NGC 2775, yet the "companion" lies between them—possibly a galaxy in formation.

Figure 4.12. NGC 2775 and companions (Credit—Palomar Observatory, courtesy of Caltech).

Northeast is NGC 2773, which is far fainter, but still achievable with a large scope. Even if you only catch NGC 2775, you've not only captured another Herschel, but Caldwell 48 as well!

Monday, April 7

Today in 1991, the Compton Gamma Ray Observatory (CGRO) was deployed (Figure 4.13). While it may sound strange, this observatory sees the sky in gamma ray photons. These photons go off the edge of the ultraviolet—imperceptible to the human eye. Unfortunately, we can't study gamma rays from Earth because our atmosphere blocks them, but the CGRO has shown a universe beyond our direct comprehension.

If there were one constellation we would love to be able to "see" in gamma rays, Cancer would be it. Riddled with quasars, this constellation has got to produce some amazing things! Have a look at a quasar for yourself tonight (Figure 4.14). You'll find 0839+187 about half a degree away from Delta Cancri (Figure 4.15). 0851+202 lies two degrees northeast and 3C215 is five degrees east-southeast. 3C212 and 3C208 are within two degrees north of Alpha, and are less than a degree apart, with radio source 3C208.1 in between them! While they will appear as nothing more

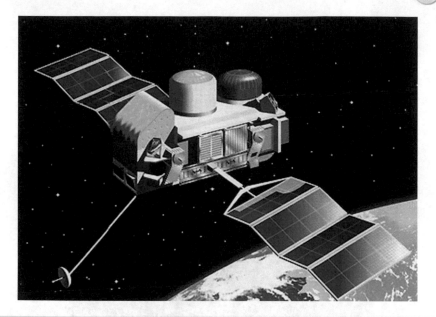

Figure 4.13. Impression of CGRO in flight (Credit—NASA).

than stellar points located in the center of the images, these are quite probably our only visual points of reference for the black holes at their hearts.

While you're out, watch for bright streaks belonging to the Delta Draconid meteor shower. Its radiant is near the Cepheus border. The fall rate is quite low with around five meteors per hour, and with no Moon to interfere tonight, you'll stand a better than average chance of spotting at the "shooting star." Make a wish!

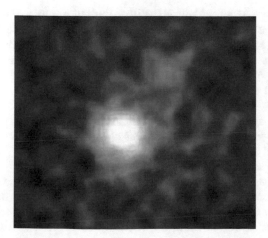

Figure 4.14. Quasar as seen in gamma rays (Credit—CGRO/NASA).

Figure 4.15. Quasar 0839+187 (center of image) (Credit—Palomar Observatory, courtesy of Caltech).

Tuesday, April 8

Did you spot the tender crescent Moon at sunset? With a limited number of days before it prevents deep- sky studies, let's begin in the extended constellation of Hydra. To identify Alpha Hydrae, try extending a full handspan southwest of Regulus. Our next mark is NGC 2811 (Figure 4.16), which resides about a fistwidth southwest of Alpha and just south of a parallelogram of finderscope stars (RA 09 16 11 Dec −16 18 49).

As Herschel object H II.502, you'll find this 13th magnitude galaxy to be just within the grasp of mid-sized aperture. With a larger telescope you will discover it to be fairly direct, and elongated north to south. This galaxy with a bright core and a hint of a spiral arm had a supernova event as recently as 2005—but don't confuse the line of sight stars along its border with supernovae!

Our next study will involve a more difficult star hop. For this one, it is best for those in northern observing locations to wait until the constellation of Corvus is fairly high in the south, since the area we're studying is quite low. The marker star we will need is Xi Hydrae which is around a fistwidth southwest of Alpha

Figure 4.16. NGC 2811 (Credit—Palomar Observatory, courtesy of Caltech).

Figure 4.17. NGC 3621 (Credit—Palomar Observatory, courtesy of Caltech).

Corvi. It will be the southwest corner of a line of five dim stars in this area. NGC 3621 is about four degrees west-southwest and will be just east of a triangle of finderscope stars (RA 11 18 16 Dec –32 48 48)(Figure 4.17).

Fortunately, Herschel H I.241 is a big, bright galaxy—one bright enough to be caught in large finderscopes and decent binoculars. Its elongated structure shows quite well and offers up a bright nucleus for even mid-sized telescopes. For large aperture, expect a nearly face-on spiral galaxy which shows some knots in its loose arms. At a distance of 22 million light-years, astronomers have used some of NGC 3621's brightest stars as standard candles. While these faint beauties are not for everyone, spring is the season of galaxies!

Wednesday, April 9

Tonight our first voyage is to the Moon's surface. Look along the terminator in the southern quadrant for ancient old crater Furnerius (Figure 4.18). Named for the French Jesuit mathematician George Furner, this crater spans approximately 125 kilometers and is a Lunar Club challenge. Power up and look for two interior craters. The smaller is crater A and it spans a little less than 15 kilometers and drops to a depth of over 1000 meters. The larger crater C is about 20 kilometers in diameter, but goes far deeper, to more than 1400 meters. That's about as deep as a coral will grow under the Earth's oceans!

Now let's return to eastern Hydra and pick up another combination Messier/Herschel object. You'll find M48 easily just a little less than a handspan southeast of Procyon (RA 08 13 42 Dec –05 45 00) (Figure 4.19).

Figure 4.18.
Furnerius as imaged by Apollo 15 (Credit—NASA).

Figure 4.19. M48 (Credit—Palomar Observatory, courtesy of Caltech).

Often called the "missing Messier," Charles discovered this one in 1711, but cataloged its position incorrectly. Even the smallest of binoculars will enjoy this rich galactic cluster filled with more than 50 members, including some yellow giants. Look for a slight triangle shape with a conspicuous chain of stars across its center. Larger telescopes should use lowest power since this will fill the field of view and resolve splendidly. Be sure to mark your notes for both a Messier object and a Herschel catalog H VI.22!

Thursday, April 10

It's a clear night and we've got Moon! No matter, what we really want to do is check out a changeable, sometimes transient, and eventually bright feature on the lunar surface—crater Proclus (Figure 4.20). At around 28 kilometers in diameter and 2400 meters deep, Proclus will appear on the terminator on the west mountainous border of Mare Crisium. For many viewers tonight, it will seem to be about two-thirds black, but one-third of the exposed crater will be exceptionally brilliant—and with good reason. Proclus has an albedo, or surface reflectivity, of about 16%—an unusually high value for a lunar feature. Watch this area over the next few nights as two rays from the crater will widen and

Mare Crisium

Figure 4.20. Proclus (Credit—NASA).

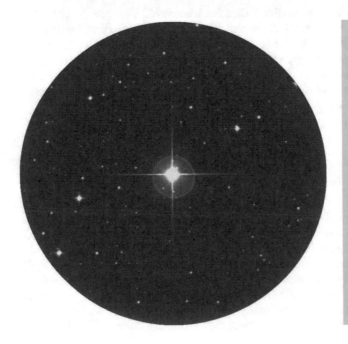

Figure 4.21.
C Hydrae
(Credit—Palomar
Observatory, courtesy
of Caltech).

lengthen, extending approximately 322 kilometers to both the north and the south. Congratulations on another Lunar Club challenge!

Now let's check out a dandy little group of stars about a fistwidth southeast of Procyon and just slightly more than a fingerwidth northeast of M48 (RA 08 25 39 Dec –02 54 23). Called C Hydrae (Figure 4.21), this group isn't truly gravitationally bound, but is a real pleasure to large binoculars and telescopes of all sizes. While they share similar spectral types, this mixed magnitude collection will be sure to delight you!

Friday, April 11

Today is the birthday of William Wallace Campbell (Figure 4.22). Born in 1862, Campbell went on to become the pioneer observer of stellar motions and radial velocities. He was the director of Lick Observatory from 1901 to 1930 and he also served as president of both the University of California and the National Academy of Sciences. Also born on this day, but in 1901, was Donald H. Menzel— an assistant astronomer at Lick Observatory, Director of Harvard Observatory, expert on the Sun's coronosphere, and genuine UFO buff. Today in 1960, the first radio search for extraterrestrial civilizations, Project Ozma, was started by Frank Drake. In 1986, Halley's Comet was also closest to the Earth for its latest apparition and was at a distance of 65 million kilometers.

Tonight we're heading toward the lunar surface to view a very fine old crater on the northwest shore of Mare Nectaris—Theophilus (Figure 4.23). Slightly south of midpoint on the terminator, this crater contains an unusually large multiple-peaked central mountain which can be spotted in binoculars. Theophilus is an odd crater: it's shaped like a parabola—with no area on the floor being flat. It stretches across a distance of 100 kilometers and dives down 440 meters below

Figure 4.22. William Campbell (widely used public image)

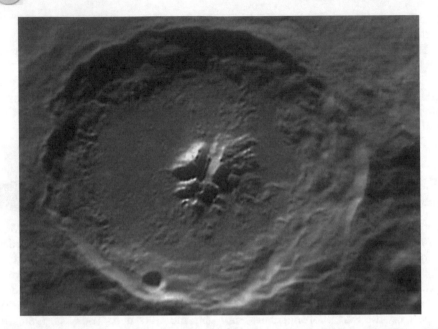

Figure 4.23. Theophilus (Credit—Wes Higgins).

the surface. Tonight it will appear dark, shadowed by its massive west wall, but look for sunrise on its 1400 meter summit!

Saturday, April 12

Today in 1961, Yuri Gagarin became the first man in space as he made one full orbit of the Earth aboard Vostok 1. And, in 1981, Columbia became the first Space Shuttle to be launched. Today also celebrates in 1851 the birth of Edward Maunder—a bank-teller turned assistant Royal Astronomer (Figure 4.24). Assigned to photographing and cataloging sunspots, Maunder was the first to discover solar minimum times and equate these with climate change. Maunder was also the first to suggest that Mars had no "canals," only delicate changes in surface features. Smart man!

Tonight Mars will play a very important role in observing as it will be slightly more than a degree away from the Moon's limb. For many observers this will be a very grand and well placed occultation event, so be sure to check IOTA for precise times in your location!

While we're waiting for the Red Planet to appear again, let's have a look at the lunar surface. An outstanding feature will be crater Maurolycus (Figure 4.25) just southwest of the three rings of Theophilus, Cyrillus, and Catharina. This Lunar Club challenge spans 114 kilometers and goes below the lunar surface by 4730 meters. Be sure to look for Gemma Frisius just to its north.

Figure 4.24. Edward Maunder (widely used public image)

Figure 4.25. G. B. Hodierna (archival image)

Sunday, April 13

Born today (1597) in Sicily was an astronomer who has a familiar name, but whom few of us know much about—Giovanni Battista Hodierna (Figure 4.26). By the age of 21, the self-educated Hodierna had observed three comets in less than a year with a primitive 20 power Galilean telescope. Indeed, his hero was Galileo and before the age of 30 this Roman Catholic priest had returned to his home town to teach mathematics and astronomy.

Figure 4.26. Maurolycus (Credit—Wes Higgins).

Figure 4.27. Rima Sulpicius Gallus (Credit—Wes Higgins).

Although his treatises never became as popular as those of some of his contemporaries, Hodierna was the first to study light seen through a prism, and he developed an early microscope. He also drew Saturn's ring system, and observed the eclipses and transits of Jupiter's moons—making the first ephemerides for these. He gave detailed descriptions of the lunar surface and sunspots, and calculated eclipse times. Not enough? Then know that by the time he had reached the age of 47, he had already proposed a theory about the origin of comets and nebulae! Along the way, Hodierna also cataloged 40 deep sky objects—some of which he was able to resolve, classify, and sketch. Even more than 3.5 centuries later, his drawing of the Orion Nebula still exists!

Before we explore deeper space tonight, let's have a look at the Moon as challenge craters Cassini and Cassini A have now come into view just south of the black slash of the Alpine Valley. For more advanced lunar observers, head a bit further south to the Haemus Mountains to look for the bright punctuation of a small crater. You'll find it right on the southwest shore of Mare Serenitatis! Now power up and look for a curious feature with an even more curious name...Rima Sulpicius Gallus (Figure 4.27). It is nothing more than a lunar wrinkle which accompanies the crater of the same name—a long-gone Roman counselor. Can you trace its 90 kilometer length?

Now let's have a look at 140 light-year distant Epsilon Hydrae (Figure 4.28)— the northernmost star in the small circlet east of Procyon. While Epsilon and Rho will make a beautiful visual double for binoculars, Epsilon itself is a multiple

Figure 4.28. Epsilon Hydrae (Credit—Palomar Observatory, courtesy of Caltech).

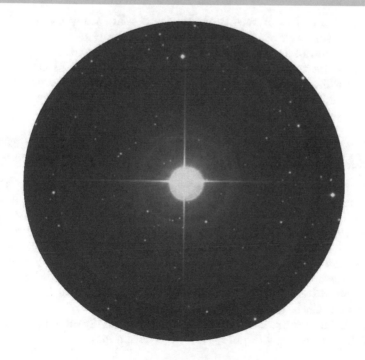

Figure 4.29. Iota 1 Cancri (Credit—Palomar Observatory, courtesy of Caltech).

system. Its A and B components are tough to split in any scope, but the 8th magnitude C star is easier. The D component is a dwarf star.

Another beautiful double star for a bright night is Iota 1 Cancri (Figure 4.29), located at the northern tip of the upside-down Y (RA 09 46 41 Dec +28 45 35). Telescopes will find this one especially interesting because the primary star is orange in hue, making a nice contrast with its greenish secondary.

Monday, April 14

Today is the birthday of Christian Huygens (Figure 4.30). Born in 1629, the Dutch scientist went on to become one of the leaders in his field during the 17th century. Among his achievements were developing a theory of light and patenting the pendulum clock. And Huygens was the first to discover Saturn's rings (Figure 4.31) and its largest satellite—Titan. While you're out enjoying the night, be sure to at least take a glance at Saturn and pay your respects to a man who was chasing photons some 380 years before us—and with equipment far more primitive. As you look at Titan, remind yourself of how much we've achieved since Chris' time!

As we look at the lunar surface, the terminator is silently moving west revealing old craters in a new light (Figure 4.32). Let's have a look.

Figure 4.30.
Christian Huygens
(widely used public
image).

Figure 4.31. Saturn (Credit—Wes Higgins).

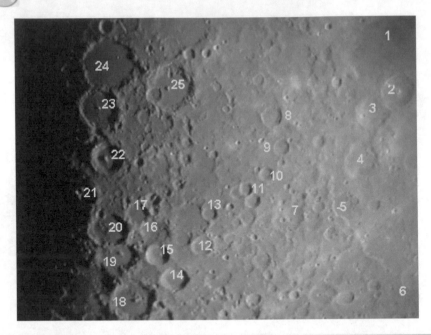

Figure 4.32. Ptolemaeus area (Credit—Greg Konkel, annotation by Tammy Plotner). (1) Sinus Asperitatis, (2) Theophilus, (3) Cyrillus, (4) Catharina, (5)Rupes Altai,(6) Piccolomini, (7) Sacrobosco, (8) Abulfeda, (9) Almanon, (10) Taylor, (11) Abenezra, (12) Apianus, (13) Playfair, (14) Aliacensis, (15) Werner, (16) Blanchinus, (17) Lacaille, (18) Walter, (19) Regiomontanus, (20) Purbach, (21) Thebit, (22) Arzachel, (23) Alphonsus, (24) Ptolemaeus, (25) Albategnius.

Tuesday, April 15

Tonight bright Regulus waltzes within a degree of the lunar limb. Coming this close could mean an occultation event in your area, so check IOTA resources and be sure to note Saturn isn't much more than a fingerwidth away either!

On the lunar surface, we can enjoy a strange, thin feature. If you use last night's map, you would be well acquainted with this area! Look toward the lunar south where you will note the prominent rings of craters Ptolemaeus, Alphonsus, Arzachel, Purbach, and Walter descending from north to south. Just west of them, you'll see the emerging Mare Nubium. Between Purbach and Walter you will see the small, bright ring of Thebit with a crater caught on its edge. Look further west and you will see a long, thin, dark feature cutting across the mare. Its name? Rupes Recta (Figure 4.33)—better known as "The Straight Wall," or sometimes Rima Birt. It is one of the steepest known lunar slopes, rising about 366 meters from the surface at a 41° angle. Be sure to mark your lunar challenge notes and we'll visit this feature again!

While outside, keep a watch for the "April Fireballs." This unusual name has been given to what may be a branch of the complex Virginid stream which began earlier in the week. The absolute radiant is unclear, but keep your eyes on the

Figure 4.33. Rupes Recta: The Straight Wall (Credit—Greg Konkel).

ecliptic near Virgo. These bright bolides can possibly arrive in a flurry depending on how much Jupiter's gravity has perturbed the meteoroid stream. Even if you only see one tonight, keep watching in the days ahead. The April Fireballs will last for two more weeks!

Wednesday, April 16

Tonight we'll use what we learned last month to locate another unusual feature—Montes Recti or the "Straight Range" (Figure 4.34). You'll find this curiosity tucked between Plato and Sinus Iridum on the north shore of Mare Imbrium.

To binoculars or small scopes at low power, this isolated strip of mountains will appear as a white line drawn across the gray mare. It is believed this feature may be all that is left of a crater wall from the Imbrium impact. It runs for a distance of around 90 kilometers, and is approximately 15 kilometers wide. Some of its peaks reach as high as 2072 meters! Although this doesn't sound particularly impressive, that's over twice as tall as the Vosges Mountains in west–central Europe, and on the average very comparable to the Appalachian Mountains in the eastern United States.

Now have a look at 27 Hydrae (Figure 4.35) about a fingerwidth southwest of Alpha. It's an easy double for any equipment with its slightly yellow 5th magnitude primary and distant, white, 7th magnitude secondary. Although it is wide, the pair is a true binary system.

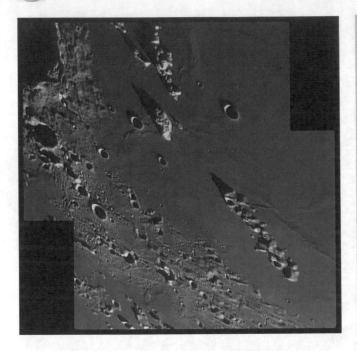

Figure 4.34. Montes Recti: The Straight Range (Credit—Wes Higgins).

Figure 4.35. 27 Hydrae (Credit—Palomar Observatory, courtesy of Caltech).

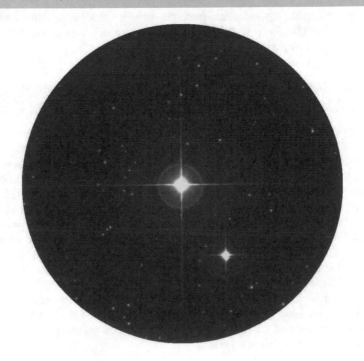

Thursday, April 17

Today in 1976, the joint German and NASA probe Helios 2 came closer to the Sun than any other spacecraft so far (Figure 4.36). One of its most important contributions was helping us to understand the nature of gamma ray bursts.

Tonight you are on your own without a map. Lunar features are easy when you become acquainted with them! Return to the Moon and explore with binoculars or telescopes the area to the south around another easy and delightful lunar feature, the crater Gassendi (Figure 4.37). At around 110 kilometers in diameter and 2010 meters deep, this ancient crater contains a triple mountain peak in its center. As one of the most "perfect circles" on the Moon, the south wall of Gassendi has been eroded by lava flows over a 48 kilometer expanse and offers a great amount of detail to telescopic observers on its ridge- and rille-covered floor. For those observing with binoculars, Gassendi's bright ring stands on the north shore of Mare Humorum—an area about the size of the state of Arkansas!

Now, are you ready for even more meteors? Tonight is the peak of the Sigma Leonids. The radiant is located at the Leo/Virgo border, but has migrated to Virgo in recent years. Thanks to Jupiter's gravity, this shower may

Figure 4.36. Helios 2 (widely used public image)

Figure 4.37. Gassendi (Credit—Wes Higgins).

eventually become part of the Virginid Complex as well. The fall rate is very low at around 1–2 per hour.

Friday, April 18

Return to the Moon tonight to have a look on the terminator near the southern cusp for two outstanding features. The easiest is crater Schickard (Figure 4.38)—a class V mountain-walled plain spanning 227 kilometers. Named for German astronomer Wilhelm Schickard, this beautiful old crater with subtle interior details has another crater caught on its northern wall which is named Lehmann.

Look further south for one of the Moon's most incredible features—Wargentin. Among the many strange things on the lunar surface, Wargentin is unique. Once upon a time, it was a very normal crater and had been so for hundreds of millions of years, then it happened: either a fissure opened in its interior or the meteoric impact which formed it caused molten lava to begin to rise. Oddly enough, Wargentin's walls did not have large enough breaks to allow the lava to escape, and it continued to fill the crater to the rim. Often referred to as "the Cheese," enjoy Wargentin tonight for its unusual appearance…and be sure to note Nasmyth and Phocylides as well!

While we're out, have a look at R Hydrae (Figure 4.39) about a fingerwidth east of Gamma—which is itself a little more than fistwidth south of Spica. R is

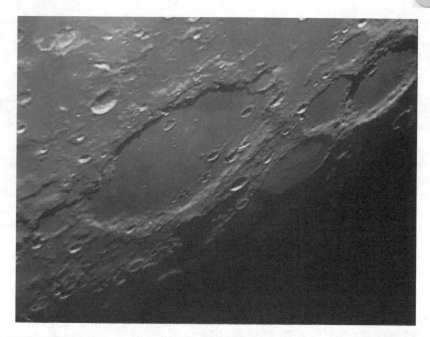

Figure 4.38. Schickard and Wargentin (Credit—Roger Warner).

Figure 4.39. R Hydrae (Credit—Palomar Observatory, courtesy of Caltech).

a beautiful, red, long-term variable first observed by Hevelius in 1662. Located about 325 light-years from us, it's approaching—but not so very fast. Be sure to look for a visual companion star as well!

Saturday, April 19

Today in 1971, the world's first space station was launched—the Soviet research vessel Salyut 1 (Figure 4.40). Six weeks later, Soyuz 11 and its crew of three docked with the station, but a mechanism failed denying them entry. The crew carried out their experiments, but were sadly lost when their re-entry module separated from the return spacecraft and depressurized. Although this tragedy cast an early shadow over Salyut 1, the mission continued to enjoy success through the early 1980s and it paved the way for Mir.

It's big. It's bright. It's nearly full (Figure 4.41). What else can it be besides the Moon? While the light can be almost overpowering in a telescope, try switching to binoculars and see how many features you remember which now appear as bright points. Do you recognize Proclus on the edge of Mare Crisium? How about Furnerius along the southeastern limb? Look at how things can change just by the amount of light reflected from the Sun!

Despite all the light, you may have noticed brilliant blue–white Spica very near the Moon tonight (Figure 4.42). Take the time to look at this glorious helium star, which shines 2300 times brighter than the Sun which lights tonight's

Figure 4.40. Salyut 1 (Credit—NASA).

Figure 4.41. Nearly Full Moon (Credit—Roger Warner).

Figure 4.42. Spica (Credit—Palomar Observatory, courtesy of Caltech).

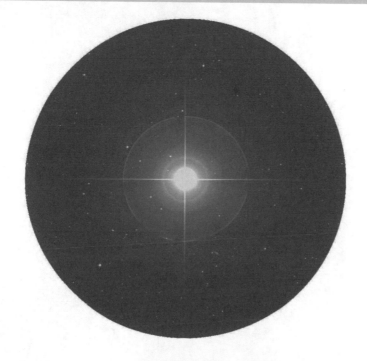

Moon. Roughly 275 light-years away, Alpha Virginis is a spectroscopic binary. The secondary star is about half the size of the primary and orbits it about every 4 days from its position of about 18 million kilometers from center to center... That's less than one-third the distance at which Mercury orbits the Sun! The two stars can actually graze during an eclipse. Oddly enough, Spica is also a pulsating variable and the very closeness of this pair make for fine viewing—even without a telescope!

Sunday, April 20

Tonight's Full Moon is often referred to as the "Pink Moon" of April. As strange as the name may sound, it actually comes from the herb moss pink or wild ground phlox. April is the time of blossoming and the "pink" is one of the earliest widespread flowers of the spring season. As always, it is known by other names as well, such as the Full Sprouting Grass Moon, the Egg Moon, and the coastal tribes referred to it as the Full Fish Moon. Why? Because spring was the season the fish swam upstream to spawn!

While skies are bright, let's take this opportunity to have a look at Alpha Canis Minoris (Figure 4.43), now heading west. If you're unsure of which bright star it is, you'll find it in the center of the diamond shape grouping in the southwest area of the early evening sky in the Northern Hemisphere. Known to the ancients as Procyon, "The Little Dog Star," it's the eighth brightest star in the night sky

Figure 4.43. Alpha Canis Minoris: Procyon (Credit—Palomar Observatory, courtesy of Caltech).

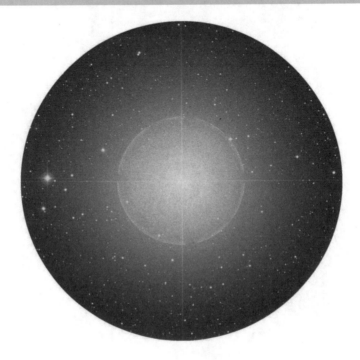

and the fifth nearest to our solar system. For over 100 years astronomers have known this brilliant star was not alone—it had a companion, and a very unusual one. About 15,000 times fainter than the parent star, Procyon B is an example of a white dwarf whose diameter is only about twice that of Earth. But its density exceeds 2 tons per square inch! (Or, a third of a metric ton per cubic centimeter!) While only very large telescopes can resolve this second closest of the white dwarf stars, even the moonlight can't dim its beauty!

Monday, April 21

Tonight for binoculars, try dropping about a handspan south of brilliant Regulus for a beautiful collection of stars around 4th magnitude Lambda Hydrae (RA 10 10 35 Dec –12 21 14). While this loose association is not a genuine cluster, it makes a nice showing for low power. Although we cannot see Lambda's spectroscopic companion, the star and this collection holds a place in ancient history. According to ancient Chinese legend, Lambda belonged to the stellar mansion of "Tschang"—and the star's presence meant rule over the kitchens, feasts, and hunting. If the stars of this small cluster couldn't be seen, it was bad news for the Emperor. For large telescopes, have a look just southwest of Lambda for faint NGC 3145 (Figure 4.44). This small spiral with a bright core is also known as Herschel III.516!

Figure 4.44. Lambda Hydrae and NGC 3145 (Credit—Palomar Observatory, courtesy of Caltech).

Figure 4.45. NGC 2610 (Credit—Palomar Observatory, courtesy of Caltech).

For those working on planetary nebula studies, here's a good one—NGC 2610 (RA 08 33 23 Dec–16 08 58) near the Hydra/Puppis/Pyxis border (Figure 4.45). At 13th magnitude, it's not for the beginner, but a worthy study for seasoned veterans. Its location near two 7th magnitude stars will help reveal its position at low power. Magnify to catch a slightly elliptical shell, a stellar point on its northeast edge and a wink of a central star. This planetary is also cataloged as Herschel IV. 65 and, though it's not easy, it is a prime study object on many observing lists.

Tuesday, April 22

Today celebrates the birthday of Sir Harold Jeffreys (Figure 4.46), who was born in 1891. Jeffreys was an astrogeophysicist and the first person to envision Earth's fluid core. He also helped in our understanding of tidal friction, general planetary structure, and the origins of our solar system.

Start your Moon-filled morning off before dawn with a chance to view the peak of the Lyrid meteor shower. Since the radiant is near Vega, you will improve your chances of spotting them when the constellation of Lyra is as high as possible. This stream's parent is Comet Thatcher, and it produces around 15 bright, long-lasting meteors per hour. Placing an obstruction between the Moon's glare and your observing point will help greatly!

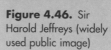

Figure 4.46. Sir Harold Jeffreys (widely used public image)

Tonight we continue our planetary nebula studies, and our mark lies due west of Alpha Crateris. It can be found more easily by hopping northeast to Nu Hydrae, west past Phi 1 and 2, then center on Mu and drop south less than two degrees (RA 10 24 46 Dec –18 38 32).

Suitable for even small telescopes and larger binoculars, this handsome planetary nebula—NGC 3242—is often referred to as the "Ghost of Jupiter" (Figure 4.47). For aperture at high power, this nebula comes alive with an inner star revealed and an eye-like ring structure. With high surface brightness, it will

Figure 4.47. NGC 3242: The Ghost of Jupiter (Credit—Palomar Observatory, courtesy of Caltech).

take all the magnification sky conditions allow, and still return structure. Be sure to use averted vision to pick up many faint details in this incredible planetary nebula, which is also known as Herschel H IV.27 and Caldwell 59!

Wednesday, April 23

Pioneer quantum physicist Max Planck was born on this day in 1858 (Figure 4.48). In 1900, Max developed the Planck equation to explain the shape of blackbody spectra (a function of temperature and wavelength of emission). A "blackbody" is any object which absorbs all incident radiation—regardless of wavelength. For example, heated metal has blackbody properties because the energy it radiates is thermal. The blackbody spectrum's shape remains constant, and the peak and height of an emitter can be measured against it—be it cosmic background radiation or our own bodies.

Now let's put this knowledge into action. Stars themselves approximate blackbody radiators, because their temperature directly controls the color we see. A prime example of a "hot" star is Alpha Virginis, better known as Spica. Compare its color to the cooler Arcturus... What colors do you see? There are other astronomical delights which radiate like blackbodies over some or all parts of the spectrum. You can observe a prime example in a nebula such as M42, in Orion. By examining the radio portion of the spectrum, we find its temperature properly matches that of electrons fluorescing. Much like in a common household fixture, this process is what produces the visible light we see.

For our friends in the Southern Hemisphere, why not have a look tonight at the awesome cluster IC 2391.

Located just slightly north-northwest of Delta Velorum (RA 08 40 36 Dec –53 02 00), IC 2391 is often called the Omicron Velorum Cluster (Figure 4.49). First

Figure 4.48. Max Planck (widely used public image)

Figure 4.49. IC 2391 (Credit—Palomar Observatory, courtesy of Caltech).

recorded by Al Sufi in 964 ad, this beauty was re-cataloged by Lacaille as object II.5 on February 11, 1752. Consisting of about 30 stars, this sparkling gemstone holds an average visual magnitude of 2. Not bad for a 36 million year old! Take the time to really study this cluster, for it is a textbook in stellar evolution. Containing low mass stars, brown dwarfs, a supersaturated rotator star, and even an area similar to the Trapezium, IC 2391 continues to be a focus of research. Enjoy this fantastic cluster for its visual beauty—with or without optical aid!

Skywatchers both north and south should wait for the Moon to rise, because Antares is less than half a degree away. Check IOTA, this could be an occultation!

Thursday, April 24

Today in 1970, China launched its first satellite. Named Shi Jian 1, it was a successful technological and research craft. This achievement made China the fifth country to send a vessel to space.

With tonight's dark skies, this would be a perfect time for larger telescopes to discover an unusual galaxy grouping in Hydra about five degrees due west of the Xi pairing (RA 10 36 35 Dec −27 31 03).

Centralmost are two fairly easy ellipticals, NGC 3309 and NGC 3311, accompanied by spiral NGC 3322. Far fainter are other group members, such as NGC

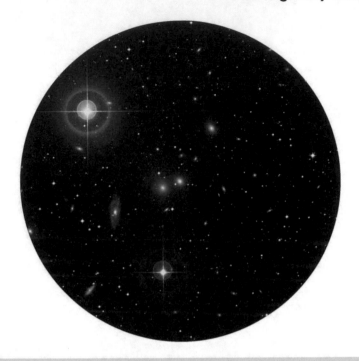

Figure 4.50. NGC 3308 Region (Credit—Palomar Observatory, courtesy of Caltech).

Figure 4.51. M83 (Credit—Brad Moore).

3316 and NGC 3314 to the east of the 7th magnitude star, NGC 3305 to the north, and NGC 3308 due west of the 5th magnitude star (Figure 4.50). While such galaxy clusters are not for everyone, studying those very faint fuzzies is a rewarding experience for those with large aperture telescopes.

Before binocular observers begin to feel we have deserted them, let's drop in on a galaxy suitable for binoculars or very small telescopes, which resides roughly a handspan below Spica—M83 (Figure 4.51). Starhop instructions are not easy for this one, but look for a pair of twin stars just west of the easily recognized "box" of Corvus—Gamma and R Hydrae. You'll find M83 about four fingerwidths further south of R.

As one of the brightest galaxies around, the "Southern Pinwheel" was discovered by Lacaille in 1752. Roughly 10 million light-years distant, M83 has been home to a large number of supernova events—one of which was even detected by an amateur observer. To binoculars it will appear as a fairly large, soft, round glow with a bright core set in a delightful stellar field. As aperture increases, so do details—revealing three well-defined spiral arms, a dense nucleus, and knots of stars. It is truly a beauty and will become an observing favorite!

Friday, April 25

Today marks the 18th anniversary of the deployment of Hubble Space Telescope. While everyone in the astronomical community is well aware of what this magnificent telescope "sees," did you know you can see it with just your eyes? The HST is a satellite which can be tracked and observed. Visit www.heavens-above.com and enter your location. This page will provide you with a list of visible passes for your area. Although you can't see details of the scope itself, it's great fun to track with binoculars or see the Sun glinting off its surface in a scope.

It's Friday night. You've got dark skies ahead and an itch to see something out of the ordinary with your telescope. If this is the case, then look no further than two degrees west of Upsilon Hydrae (RA 09 45 42 Dec –14 19 35) for a unique galactic pair—NGC 2992/93 (Figure 4.52).

At magnitude 13, these two small galaxies will show their core regions cleanly to mid-aperture telescopes, but come to life in larger ones. NGC 2992 is a Seyfert type-2 galaxy which has a wisp extending from its galactic plane—divided by a dustlane. It is surmised that an incredibly powerful stellar wind drives from the core of this galaxy, perhaps producing massive starbursts.

NGC 2992 is interacting with starburst galaxy NGC 2993 and they share mutual tidal forces. When they began to approach each other, a "bridge" was created between the galactic nuclei in the form of ejected stars and gas—triggering violent star formation. Eventually this action will destroy their disc structures and they will merge to create a single elliptical galaxy.

While you're out, keep an eye turned toward the sky as the Mu Virginid meteor shower reaches its peak at 7–10 per hour. With dark skies early tonight, you might catch one of these medium-speed meteors radiating from a point near the constellation of Libra.

Figure 4.52. NGC 2992/93 (Credit—Palomar Observatory, courtesy of Caltech).

Saturday, April 26

On this date in 1920, the Shapely-Curtis debate raged in Washington on the nature of (and distance to) spiral nebulae. Shapely claimed they were part of one huge galaxy to which we all belonged, while Curtis maintained they were distant galaxies of their own. Thirteen years later on the same date, Arno Penzias was born (Figure 4.53). He went on to become a Nobel Prize winner for his part in the discovery of the cosmic microwave background radiation, through searching for the source of the "noise" coming from a simple horn antenna. His discovery helped further our understanding of cosmology in ways Shapely and Curtis could never have dreamed of.

Tonight we're off to study another Herschel object (H II.506) in Hydra that's a seven degree drop south of Alpha—NGC 2907 (RA 09 31 42.1 Dec −16 44 04) (Figure 4.54).

While it will require at least a mid-aperture telescope to reveal, this edge-on galaxy is quite worth the trouble. It's highly prized because of research on its dust extinction properties, which in some ways greatly resemble those of our

Figure 4.53. Arno Penzias (widely used public image)

Figure 4.54. NGC 2907 (Credit—Palomar Observatory, courtesy of Caltech).

own Milky Way galaxy. For larger telescopes, averted vision will call up a hint of a dark dustlane across a bright core. While it is neither particularly huge, nor particularly bright, it will present an interesting challenge for those with larger scopes looking for something a bit out of the ordinary.

Sunday, April 27

While we don't have an early evening Moon to contend with, let's head out in search of an object which is one royal navigation pain for the Northern Hemisphere, but which makes up for it in beauty. Start with the southernmost star in Crater—Beta. If you have difficulty identifying it, it's the brightest star east of the Corvus rectangle. Now hop a little more than a fistwidth southeast to reddish Alpha Antilae. Less than a fistwidth below, you will see a dim 6th magnitude star which may require binoculars in the high north. Another binocular field further southwest and about four degrees northwest of Q Velorum is our object—NGC

Figure 4.55. NGC 3132 (Credit—Palomar Observatory, courtesy of Caltech).

3132 (RA 10 07 01 Dec −40 26 11) (Figure 4.55). If you still have no luck, try waiting until Regulus has reached your meridian and head 52 degrees south.

More commonly known as the "Southern Ring" or the "Eight Burst Planetary," this gem is brighter than the northern Ring (M57), and definitely shows more details. Capturable in even small instruments, larger ones will reveal a series of overlapping shells, giving this unusual nebula its name.

While you're out, be sure to do a little sky watching as well. Tonight the dance of the ecliptic becomes wonderfully apparent as Mars and Pollux share the sky around five degrees apart, and Saturn is about two degrees from Regulus. If you're still awake when the Moon rises, you'll be treated to a close apparition of Jupiter near Luna as well. Only your eyes are needed to see these sky gems!

Monday, April 28

Today we fondly remember Eugene Shoemaker on the date of his birth (1928). The famous astronomer and geologist devoted much of his life to the United States Geological Survey, which enabled him to study features on the Moon and examine meteorite impacts on the Earth (Figure 4.56). This keen interest led to

Figure 4.56. Eugene Shoemaker (Credit—NASA).

Figure 4.57. Jan Oort (widely used public image).

Figure 4.58. Bart Jan Bok (widely used public image).

the study of the asteroids which may have formed these features—which in turn led to his discovery of several asteroids. Added to this list are 32 comets which bear the name Shoemaker. Gene will live forever as "the father of the science of near-Earth objects."

Also on this day in 1774, Francis Baily was born. He went on to revise several star catalogs and explain the phenomenon seen at the beginning and ending of a total solar eclipse which we now know as "Baily's Beads." The year 1900 saw the birth of Jan Hendrick Oort (Figure 4.57), who quantified the Milky Way's rotational characteristics and envisioned the vast, spherical area of comets at the edge of our solar system which we now call the Oort Cloud. Last, but not least, was the birth of Bart Jan Bok (1906) (Figure 4.58), who studied the structure and dynamics of the Milky Way.

For skywatchers, no equipment is necessary to enjoy the "leftovers" of a traveler from the Oort Cloud. Tonight the Alpha Boötid meteor shower will reach its maximum. Pull up a comfortable seat and face orange Arcturus as it climbs the sky in the east. These slow meteors have a fall rate of 6–10 per hour and leave very fine trails, making an evening of quiet contemplation most enjoyable.

Figure 4.59. Arcturus (Credit—John Chumack).

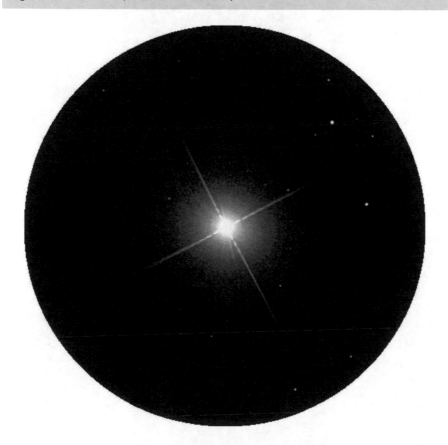

While we're waiting on a silver streak to crease the sky, let's talk about Arcturus (Figure 4.59). Its proper name is Alpha Boötis and it is the fourth brightest star in the night. It was known in mythology as the "watcher of the bear," and the Greeks called it Arktos—the root word for the English "arctic." Can you see how it looks a little orange? That means it is a K-type star—cooler than our own Sun, yet shining twice as brightly. It's around 37 light-years away and the light you see tonight left in 1971. Arcturus is a very ancient star, belonging to the oldest population in our Galaxy. It is even believed it may have come from a smaller galaxy which merged with the Milky Way billions of years ago!

Tuesday, April 29

Tonight let's start our starhop by identifying the lopsided rectangle of the constellation of Corvus ("The Crow") southwest of bright Spica. Now identify its Beta star at the southeast corner. About 110 light-years away, this G-type star is heading

Figure 4.60. Beta Corvi (Credit—Palomar Observatory, courtesy of Caltech).

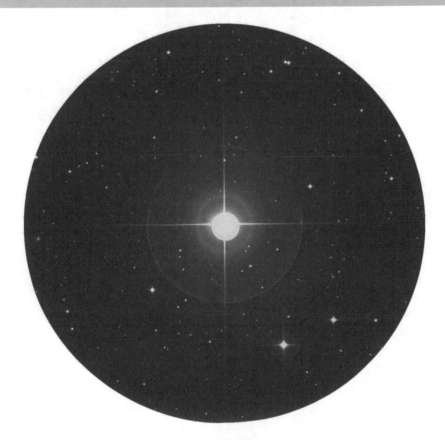

toward us at a speed of a little over 7 kilometers per second—which should put it in our neighborhood in say...roughly a quadrillion years? In other words, there's no need to worry about Beta catching up to us any time soon, but have a look even in binoculars because it has a nice, unrelated companion star as well.

Now let's try picking up a globular cluster in Hydra which is located about three fingerwidths southeast of Beta Corvi (Figure 4.60) and just a breath northeast of double star A8612—M68 (RA 12 39 28 Dec –26 44 34) (Figure 4.61).

This class X globular was discovered in 1780 by Charles Messier and first resolved into individual stars by William Herschel in 1786. At a distance of approximately 33,000 light-years, it contains at least 2000 stars, including 250 giants and 42 variables. It will show as a faint, round glow in binoculars, and small telescopes will perceive individual members. Large telescopes will fully resolve this small globular to the core!

Figure 4.61. M68 (Credit—Palomar Observatory, courtesy of Caltech).

Wednesday, April 30

Karl Frederich Gauss was born on this day in 1777 (Figure 4.62). Known as the "Prince of Mathematics," Gauss contributed to the field of astronomy in many ways, from computing asteroid orbits to inventing the heliotrope. Among

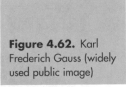

Figure 4.62. Karl Frederich Gauss (widely used public image)

Figure 4.63. Sunspot (Credit—NASA).

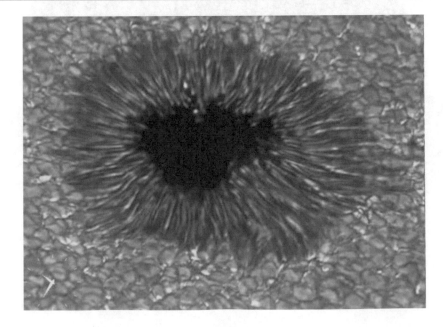

Gauss' many endeavors, he is most recognized for his work in magnetism. We understand the term "gauss" as a magnetic unit—a refrigerator magnet carries about 100 gauss while an average sunspot might go up to 4000. On the most extreme ends of the magnetic scale, the Earth produces about 0.5 gauss at its poles, while a magnetar can produce as much as 10^{15} gauss!

While we cannot directly observe a magnetar, those living in the Southern Hemisphere can view a region of the sky where magnetars are known to exist—the Large Magellanic Cloud—or you can use the projection method to view a sunspot (Figure 4.63)! Or, if you have a proper solar filter to avoid eye damage, you may be able to see how magnetism distorts sunspots as they near the limb—this is called the "Wilson Effect."

Figure 4.64. Sirsalis Rille (Credit—Alan Chu

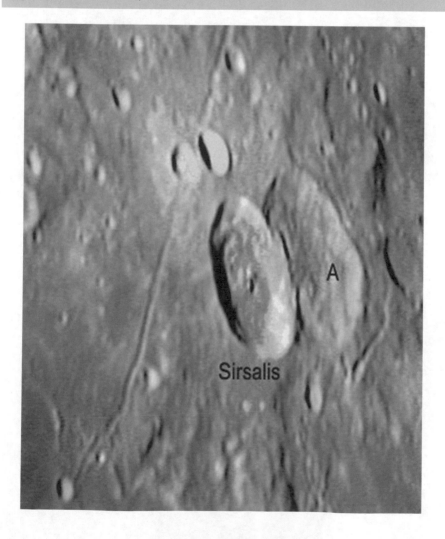

While both magnetars and sunspots are areas of awesome magnetic energy, what happens when you find magnetism in a very unlikely place? I encourage you to be out before dawn to have a look at the lunar surface just a little southeast of the gray oval of Grimaldi. The area we are looking for is called the Sirsalis Rille (Figure 4.64) and on an orb devoid of magnetic fields—it's magnetic! Like a dry river bed, this ancient "crack" on the surface runs 480 kilometers along the surface and branches in many areas. Be sure to look for Spica nearby!

CHAPTER FIVE

May, 2008

On this day in 1949 Gerard Kuiper discovered Nereid, a satellite of Neptune. If you're game, you can find Neptune in Capricorn about two degrees north of Deneb Algedi or about 30 degrees east of Jupiter about an hour before sunrise. While it can be seen in binoculars as a bluish "star," it takes around a 6″ telescope and some magnification to resolve its disc. Today's imaging technology can even reveal its moons!

While you're out this morning, keep an eye on the sky for the peak of the Phi Boötid meteor shower, whose radiant is near the constellation of Hercules. The best time to view this meteor shower is around 2 AM local time, and you will have best success watching for these meteors when the radiant is at its highest. The average fall rate is about six per hour.

For amateurs all over the world, Astronomy Week is about to begin and tonight would be a great time to celebrate with a serious galactic study. For the large telescope and seasoned observer, your challenge for this evening will be 5.5 degrees south of Beta Virginis and one half degree west (RA 11 46 45 Dec –03 50 53). Classified as Arp 248, and more commonly known as Wild's Triplet, these three very small interacting galaxies are a real treat (Figure 5.1)! Best with a 9 millimeter eyepiece, use wide aversion and try to keep the star just north of the trio at the edge of the field in order to cut glare. Be sure to mark your Arp Galaxy challenge list!

Figure 5.1. Arp 248: Wild's Triplet (Credit—Palomar Observatory, courtesy of Caltech).

Friday, May 2

With plenty of dark sky tonight, we're heading for the galaxy fields of Virgo about four fingerwidths east-southeast of Beta Leonis. As part of Markarian's Chain, this set of galaxies can all be fitted within the same field of view with a 32 millimeter eyepiece and a 12.5". scope, but not everyone has the same equipment. Set your sights toward M84 and M86 (RA 12 25 03 Dec +12 53 13) and let's discover!

Good binoculars and small telescopes reveal this pair with ease as a matched set of ellipticals (Figure 5.2). Mid-sized telescopes will note the western member of the pair—M84—is slightly brighter and visibly smaller. To the east and slightly north is larger M86—whose nucleus is broader and less intensely brilliant. In a larger scope, we see the galaxies literally leap out of the eyepiece at even the most modest magnifications. Strangely, though, additional structure fails to be seen.

As aperture increases, one of the most fascinating features of this area becomes apparent. While studying the bright galactic forms of M84/86 with direct vision,

Figure 5.2. M84/86 region (Credit—Palomar Observatory, courtesy of Caltech).

aversion begins to welcome many other mysterious strangers into view. Forming an easy triangle with the two Messiers, and located about 20 arcminutes south, lies NGC 4388. At magnitude 11.0, this edge-on spiral has a dim star-like core to mid-sized scopes, but a classic edge-on structure in larger ones.

At magnitude 12, NGC 4387 is located in the center of a triangle formed by the two Messiers and NGC 4388. 4387 is a dim galaxy—hinting at a stellar nucleus to smaller scopes, while larger ones will see a very small face-on spiral with a brighter nucleus. Just a breath north of M86 is an even dimmer patch of nebulosity—NGC 4402—which needs higher magnification to be detected in smaller scopes. Large apertures at high power reveal a noticeable dustlane. The central structure forms a curved "bar" of light. Luminosity appears evenly distributed end-to-end, while the dustlane cleanly separates the central bulge of the core.

East of M86 are two brighter NGC galaxies—4435 and 4438. Through average scopes, NGC 4435 is easily picked out at low power with a simple star-like core and wispy, round body structure. NGC 4438 is dim, but even with large apertures elliptical galaxies seem a bit boring. The beauty of NGC 4435 and NGC 4438 is simply their proximity to each other. 4435 shows true elliptical structure, evenly illuminated, with a sense of fading toward the edges... But 4438 is quite a different story! This elliptical is much more elongated. A highly conspicuous wisp of galactic material can be seen stretching back toward the brighter, nearby galaxy pair M84/86.

Happy hunting!

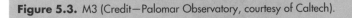

Saturday, May 3

Tonight let's use our binoculars and telescopes to hunt down one of the best globular clusters of the Northern Hemisphere—M3 (Figure 5.3). You will discover this ancient beauty about halfway between the pair of Arcturus and Cor Caroli—just east of Beta Comae (RA 13 42 11 Dec +28 22 31). The more aperture you use, the more stars you will resolve. Discovered by Charles Messier on this day in 1764, this ball of approximately a half million stars is one of the oldest formations in our galaxy. At around 40,000 light-years away, this awesome cluster spans about 220 light-years and is believed to be as much as 10 billion years old. To get a grip on this concept, our own Sun is less than half that age!

Let's further our understanding of distance and how it affects what we see. As you know, light travels at an amazing speed of about 300,000 kilometers per

Figure 5.3. M3 (Credit—Palomar Observatory, courtesy of Caltech).

second. To get a feel for this, how many seconds are there in a minute? An hour? A week? A month? How about a year? Ah, you're beginning to see the light! For every second, 300,000 kilometers. M3 is 40,000 years away traveling at the speed of light. In terms of kilometers, that's far more zeros than most of us can possibly understand—yet amazingly we can still see this great globular cluster.

Now let's locate M53 near Alpha Comae (Figure 5.4). Aim your binoculars or telescopes there and you will find M53 about a degree northeast (RA 13 12 55 Dec +18 10 09). This very rich, magnitude 8.7 globular cluster is almost identical to M3, but look at what a difference an additional 25,000 light-years can make to how we see it! Binoculars can pick up a small round fuzzy, while larger telescopes will enjoy the compact bright core as well as resolution at the cluster's outer edges. As a bonus for scopes, look one degree to the southeast for the peculiar round cluster, NGC 5053. Classed as a very loose globular, this magnitude 10.5 grouping is one of the least luminous objects of its type, due to its small stellar population and the wide separation between members—yet its distance is almost the same as that of M3.

Figure 5.4. M53 (Credit—Palomar Observatory, courtesy of Caltech).

Sunday, May 4

Tonight let's head for another trio of galaxies best suited for mid-to-large aperture telescopes. Begin by heading west about a fistwidth from Regulus and identify 52 Leonis. Our mark is 1.5 degrees south. At lower power you will see a triangle of galaxies.

The largest and brightest is M105 (Figure 5.5), discovered by Méchain on March 24, 1781. This dense elliptical galaxy would appear to be evenly distributed, but the Hubble Space Telescope revealed a huge area within its core to contain about 50 million solar masses. The companion elliptical to the northeast—NGC 3384— will reveal a bright nucleus as well as an elongated form. The faintest of this group—NGC 3389—is a receding spiral and larger scopes will reveal patchiness in structure.

Continue another degree south and enjoy another galactic pair. The widely spaced M96 and M95 are part of the galaxy grouping known as Leo I. The dusty spiral M96 will appear as a silver oval, whose nucleus is much sharper than its faint spiral arms (Figure 5.6). M96 hosted a supernova as recently as 1998. To its west, you will discover one very beautiful barred spiral—M95 (Figure 5.7). While both of these were discovered by Méchain only four days earlier than M105, it wasn't until recent years that they became a prime target for the Hubble Space Telescope. We enjoy M95 for its unique ring-like arms and unmistakable

Figure 5.5. M105 (Credit—Palomar Observatory, courtesy of Caltech).

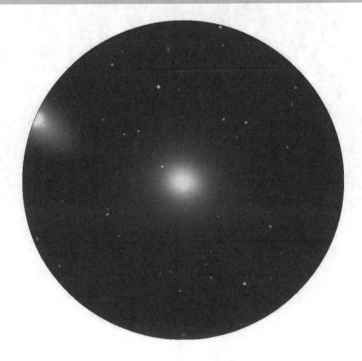

Figure 5.6. M96 (Credit—Palomar Observatory, courtesy of Caltech).

Figure 5.7. M95 (Credit—Palomar Observatory, courtesy of Caltech).

barred core, but the HST was looking for Cepheid variables to help determine the Hubble constant. While we don't need a space telescope to view this group of galaxies, we can now appreciate knowing we can see 38 million light-years away from our own backyard!

Monday, May 5

In 1961 Alan Shepard became the first American in "space" (as we now refer to the region above the sky), taking a 15 minute suborbital ride aboard the Mercury craft Freedom 7.

Tonight is New Moon, and it's time to hunt! Let's start about five degrees north of Eta Virginis for M61 (RA 12 21 55 Dec +04 28 28). This 9.7 magnitude galaxy's 1779 discovery was credited to Barnabus Oriani during the fateful year of 1779 when Messier was so avid about chasing a comet that he mistook it for one. While Charles had seen it on this same night, it took him two days to

Figure 5.8. M61 (Credit—Palomar Observatory, courtesy of Caltech).

figure out it wasn't moving...and four more before he cataloged it. (Way to go, Chuck!) Fortunately, seven years later Herschel assigned it his own number of H I.139, even though he wasn't fond of assigning his own number to Messier catalog objects.

M61 is one of the largest galaxies of the Virgo Cluster, and small telescopes will make out a faint, round glow with a brighter nucleus, while larger aperture will see the core as more stellar with notable spiral structure (Figure 5.8). Four supernova events have been observed in M61, as recently as 1999, and surprisingly two of them were exactly 35 years apart... But don't confuse such an event with foreground stars!

There's no confusing the stars of another discovery made tonight as well. In 1702, Gottfried Kirch was also at the eyepiece looking just southwest of Arcturus and a breath northwest of 5 Serpentis (RA 15 18 33 Dec +02 04 57). Both he and his wife Maria had been observing a comet when they stumbled across a magnificent globular cluster. It was later recovered independently by Messier and cataloged as M5 (Figure 5.9).

Located 24,500 light-years away, this class V globular is easily seen in small binoculars and shows incredible detail and distinct ellipticity to the larger telescope. Believed to be as much as 13 billion years old, this fine object is also one of the largest of its kind in the Milky Way. Enjoy both these "discoveries" tonight!

Figure 5.9. M5 (Credit—Palomar Observatory, courtesy of Caltech).

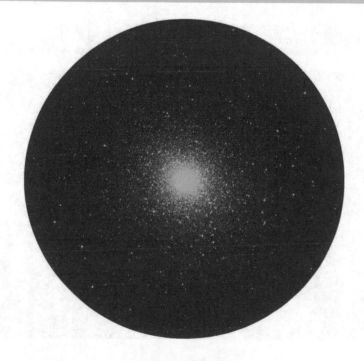

Tuesday, May 6

Tonight let's head for a galaxy pair which is relatively easy for larger binoculars and smaller telescopes. You'll find them almost perfectly mid-way between Theta and Iota (RA 11 18 55 Dec +13 05 32) and their names are M65 and M66 (Figures 5.10 and 5.11).

Discovered by Méchain in March 1780, apparently Messier didn't notice the bright pair when a comet passed between them in 1773. At around 35 million light-years away, you will find M66 to be slightly brighter than its 200,000 light-year distant western neighbor M65. While both are classed as Sb spirals, the two couldn't appear more different. M65 has a bright nucleus and a smooth spiral structure with a dark dustlane at its eastern edge. M66 has a more stellar core region with thick, bright arms which show knots to larger scopes—as well as a wonderful extension from the southern edge.

If you are viewing with a larger scope, you may notice to the north of this famous pair yet another galaxy! NGC 3628 is a similar magnitude edge-on beauty with a great dissecting dark dustlane (Figure 5.12). This pencil-slim, low surface brightness galaxy is a bit of a challenge for smaller scopes, but larger ones will find its warped central disc well worth for high power study.

Congratulations on spotting the "Leo Trio," another member of Arp's Peculiar Galaxy Catalog!

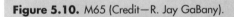

Figure 5.10. M65 (Credit—R. Jay GaBany).

Figure 5.11. M66 (Credit—R. Jay GaBany).

Figure 5.12. NGC 3628 (Credit—R. Jay GaBany).

Wednesday, May 7

Tonight the slender crescent Moon will make a very brief appearance at dusk along the western skyline. If your atmosphere is very steady, why not set your telescope on it and look for some very unusual features which will soon wash out as the Sun overtakes the moonscape. Almost in the center of the eastern limb, look for Mare Smythii and Mare Marginis to its north. Between them you will see the long oval crater Neper bordered by Jansky at the very outer edge.

Once the Moon has set and the sky has become fully dark, let's go study. For those who like curiosities, our target for tonight will be 1.4 degrees northwest of 59 Leonis, which is itself about a degree southwest of Xi. While this type of observation may not be for everyone, what we are looking for is a very special star—a red dwarf named Wolf 359 (RA 10 56 28 Dec +07 00 52). Although it is very faint at approximately 13th magnitude, you will find it precisely at the center of the highly accurate half degree field in Figure 5.13.

Figure 5.13. Wolf 359 (Credit—Palomar Observatory, courtesy of Caltech).

Discovered photographically by Max Wolf in 1959, charts from that time period will no longer be accurate because of the star's large proper motion. It is one of the least luminous stars known, and we probably wouldn't even know it was there except for the fact that it is the third closest star to our solar system. Located only 7.5 light-years away, this miniature star is about 8% the size of our Sun—making it roughly the size of Jupiter. Oddly enough, it is also a "flare star"—capable of jumping another magnitude brighter at random intervals.

It might be faint and difficult to spot in mid-sized scopes, but Wolf 359 is definitely one of the most unusual things you will ever observe!

Thursday, May 8

Tonight the Moon is a little bit older and brilliantly lit with earthshine. Power up and let's go look for a crater named for historian and theologian Denis Pétau, who is better known as Petavius! (Figure 5.14)

Figure 5.14. (1) Petavius, (2) Snellius, (3) Vallis Snellius, (4) Stevinus (Credit—Alan Chu).

Located almost centrally along the terminator in the southeast quadrant, a lot will depend tonight on your viewing time and the age of the Moon itself. Perhaps when you look, you'll see 177 kilometer diameter Petavius cut in half by the terminator. If so, this is a great time to take a close look at the small range of mountain peaks contained in its center, as well as a deep rima which runs for 80 kilometers across its otherwise fairly smooth surface. To the east lies a long furrow in the landscape. This deep runnel is Palitzsch and its Valles. While the primary crater which forms this deep gash is only 41 kilometers wide, the valley itself stretches for 110 kilometers. Look for crater Haas on Petavius' southern edge with Snellius to the southwest and Wrottesley along its northwest wall.

Once the Moon has quit the sky, let's take a look at an object which can be viewed unaided from a dark location and is splendid in binoculars. Just northeast of Beta Leonis, look for a hazy patch of stars known as Melotte 111. Often called the "Queen's Hair," this five degree span of 5th to 10th magnitude stars is wonderfully rich and colorful. As legend has it, Queen Berenice offered her beautiful long tresses to the gods for the King's safe return from battle. Touched by her love, the gods took Berenice's sacrifice and immortalized it in the stars.

The cluster is best in binoculars because of its sheer size, but you'll find other things of interest there as well. Residing about 260 light-years away, this collection is one of the nearest of all star clusters, including the Pleiades and the Ursa Major moving group. Although Melotte 111 is more than 400 million years old, it contains no giant stars, but its brightest members have just begun their evolution. Unlike the Pleiades, the Queen's Hair has no red dwarfs and a low stellar concentration which leads astronomers to believe it is slowly dispersing. Like many clusters, it contains double stars—most of which are spectroscopic. For binoculars, it is possible to split star 17, but it will require very steady hands.

Friday, May 9

Today in 1962, the first Earth-based laser was aimed at crater Albategnius. Although it isn't visible tonight, let's take a look at what is visible just 1.5 light seconds away! First is a Lunar Club challenge which won't prove difficult because you'll be working with a map. Relax! This will be much easier than you think. Starting at Mare Crisium, move along the terminator to the north following the chain of craters until you identify a featureless oval which looks similar to Plato seen on a curve. This is Endymion...and if you can't spot it, don't worry. Let's take a look at some features which will point you to it!

Most prominent of all will be two craters to the north named Atlas and Hercules. The easternmost Atlas was named for the mythical figure who bore the weight of the world on his shoulders, and the crater spans 87 kilometers and contains a vivid Y-shaped rima in the interior basin. Western Hercules is considerably smaller at 69 kilometers in diameter, and shows a deep interior crater called G. Power up and look for the tiny E crater which marks the southern

Figure 5.15. Atlas region (Credit—Greg Konkel, annotation by Tammy Plotner). (1) Mare Humboldtianum, (2) Endymion, (3) Atlas, (4) Hercules, (5) Chevalier, (6) Shuckburgh, (7) Hooke, (8) Cepheus, (9) Franklin, (10) Berzelius, (11) Maury, (12) Lacus Somniorum, (13) Daniel, (14) Grove, (15) Williams, (16) Mason, (17) Plana, (18) Burg, (19) Lacus Mortis, (20) Baily, (21) Atlas E, (22) Keldysh, (23) Mare Frigoris, (24) Democritus, (25) Gartner, (26) Schwabe, (27) Thales, (28) Strabo, (29) de la Rue, (30) Hayn.

crater rim. North of both is another unusual feature which many observers miss. It is a much more eroded and far older crater which only shows a basic outline and is only known as Atlas E (Figure 5.15).

Since we're here, let's take a crater walk and see how many features we can identify... Good luck, and clear skies!

Saturday, May 10

Something wonderful is happening in the sky! Somewhere out there, the Moon is silently occulting Mars, and the Red Planet will be hauntingly close to the limb as the skies darken... Tonight let's journey to the Moon as we look at a beautiful series of craters—Fabricius, Metius, and Rheita (Figure 5.16). Bordered on the south by shallow Jannsen, Lunar Club challenge Fabricius is a 78 kilometer diameter crater highlighted by two small interior mountain ranges. To its northeast is Metius, which is slightly larger with a diameter of 88 kilometers. Look carefully at the two. Metius has much steeper walls, while Fabricius shows differing levels and heights. Metius' smooth floor also contains

Figure 5.16. Fabricius, Metius, and Rheita (Credit—Alan Chu).

a very prominent B crater on the inside of its southeast crater wall. Further northeast is the lovely Rheita Valley which stretches almost 500 kilometers and appears more like a series of confluent craters than a fault line. Crater Rheita, which is 70 kilometers in diameter, is far younger than this formation because it intrudes upon it. Look for a bright point inside the crater which is its central peak.

Since tonight will be our last chance to galaxy hunt for a while when the Moon has westered, let's take a look at one of the brightest members of the Virgo Cluster—M49 (Figure 5.17). Located about eight degrees northwest of Delta Virginis almost directly between a pair of 6th magnitude stars (RA 12 29 46 Dec +07 59 59), the giant elliptical M49 holds the distinction of being the first galaxy in the Virgo cluster to be discovered—and only the second beyond our local group. At magnitude 8.5, this type E4 galaxy will appear as an evenly illuminated egg shape in almost all scopes, and as a faint patch in binoculars.

Figure 5.17. M49 (Credit—Palomar Observatory, courtesy of Caltech).

While a possible supernova event occurred in 1969, don't confuse the foreground star noted by Herschel with something new!

Although most telescopes won't be able to pick this region apart—especially with the Moon so near—there are also many fainter companions near M49, including NGC 4470. But a sharp-eyed observer named Halton Arp noticed them and listed them as Peculiar Galaxy 134—one with "fragments!"

Sunday, May 11

Tonight no two lunar features in the north will be more prominent than Aristoteles and Eudoxus. Viewable even in small binoculars, let's take a closer look at larger Aristoteles to the north (Figure 5.18).

Figure 5.18. Aristoteles (Credit—Wes Higgins).

As a Class I crater, this ancient old beauty has some of the most massive walls of any lunar feature. Named for the great philosopher, it stretches across 87 kilometers of lunar landscape and drops below the average surface level to a depth of 366 meters—a distance which is similar to Earth's tallest waterfall, the Silver Cord Cascade. While it has a few scattered interior peaks, the crater floor remains almost unscarred. As a telescopic Lunar Club challenge, be sure to look for a much older crater sitting on Aristoteles' eastern edge. Tiny Mitchell is extremely shallow by comparison and only spans 30 kilometers. Look carefully at this formation, for although Aristoteles overlaps Mitchell, the smaller crater is actually part of the vast system of ridges which supports the larger one.

When you're done, let's have a look at another delightful pair that's joined together—Gamma Virginis (Figure 5.19).

Better known as Porrima, this is one cool binary; whose members are nearly equal in spectral type and brightness. Discovered by Bradley and Pound in 1718, John Herschel was the first to predict this pair's orbit in 1833 and state that one day they would become inseparable to all but the very largest of telescopes— and he was right. In 1920 the A and B stars had reached their maximum separation, but during 2007 they were as close together as they will ever be in our lifetimes. Observed as a single star in 1836 by William Herschel, its 171-year periastron now puts Porrima at almost the same position as it was when Sir William saw it!

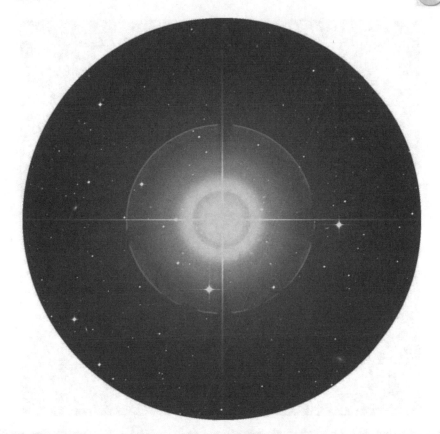

Figure 5.19. Gamma Virginis: Porrima (Credit—Palomar Observatory, courtesy of Caltech).

Monday, May 12

Tonight Regulus will be less than a fingerwidth away from the lunar limb. For some this could be an occultation event! While you're waiting, are you ready to explore some more history? Then tonight have a look at the Moon and identify Alphonsus—it's the centermost in a line of rings which looks much like the Theophilus, Cyrillus, and Catharina trio.

Alphonsus is a very old, Class V crater which spans 118 kilometers in diameter and drops below the surface by about 2730 meters, and it contains a small central peak (Figure 5.20). Partially flooded, Eugene Shoemaker made a study of this crater's formation and found dark haloes on the floor. He attributed this to volcanism—in fact Shoemaker believed these craters were maar volcanoes, and that the haloes were dark ash. Power up and look closely at the central peak (Figure 5.21), for not only did Ranger 9 hard land just northeast of there, but

Figure 5.20. Ranger 9 image of Alphonsus taken 3 minutes before impact (Credit—NASA).

this is the only area on the Moon where an astronomer has observed a change and was able to back up the observation with photographic proof...

On November 2, 1958 Nikolai Kozyrev's long and arduous study of Alphonsus was about to be rewarded. Some 2 years earlier, Dinsmore Alter had taken a series of photographs from Mt. Wilson using 60" reflector showing hazy patches in this area which could not be accounted for. Night after night, Kozyrev continued to study at the Crimean Observatory, but with no success. During the process of guiding the scope for a spectrogram the unbelievable happened—a cloud of gas containing carbon molecules had been captured! Selected as the last target for the Ranger photographic mission series, Alphonsus delivered 5814 spectacular high-resolution images of this mysterious region before Ranger 9 splattered nearby. Capture it yourself tonight!

Now let's have a look at telescopic star W Virginis located about 3.5 degrees southwest of Zeta (RA 13 26 01 Dec –03 22 43) (Figure 5.22). This 11,000 light-year distant Cepheid variable is, oddly enough, a Population II star residing outside the galactic plane. This expanding and contracting star goes through its changes

Figure 5.21. Alphonsus central peak taken 54 seconds before Ranger 9 impact (Credit—NASA).

in a little over 17 days and will vary between 8th and 9th magnitude. Although it is undeniably a Cepheid, it breaks the rules by being both out of place in the cosmic scheme and displaying abnormal spectral qualities.

Tuesday, May 13

Tonight is all about lunar studies. While the Moon moves quietly toward Virgo, our first challenge for the evening will be a telescopic one on the lunar surface known as the Hadley Rille (Figure 5.23). Using our past knowledge of Mare Serenitatis, look for the break along its western shoreline dividing the Caucasus and Apennine mountain ranges. Just south of this break is the bright peak of Mons Hadley. You'll find this area of highest interest for several reasons, so power up as much as possible.

Figure 5.22. W Virginis (Credit—Palomar Observatory, courtesy of Caltech).

Impressive Mons Hadley measures about 24 by 48 kilometers at its base and reaches up an incredible 4572 meters. If this mountain was indeed caused by volcanic activity on the lunar surface, this would make it comparable to some of the very highest volcanically created peaks on Earth, such as Mt. Shasta or Mt. Rainer. To its south is the secondary peak Mons Hadley Delta—the home of the Apollo 15 landing site (Figure 5.24). It lies just a breath north of where Hadley Delta extends into the cove created by Palus Putredinus.

Along this ridgeline and smooth floor, look for a major fault line known as the Hadley Rille, winding its way across 120 kilometers of lunar surface. In places, the rille spans 1500 meters in width and drops to a depth of 300 meters below the surface. Believed to have been formed by volcanic activity some 3.3 billion years ago, we can see the impact the lower lunar gravity has had on this type of formation, since earthly lava channels are usually less than 10 kilometers long and only around 100 meters wide. During the Apollo 15 mission, Hadley Rille was visited at a point where it is only 1.6 kilometers wide—still a considerable

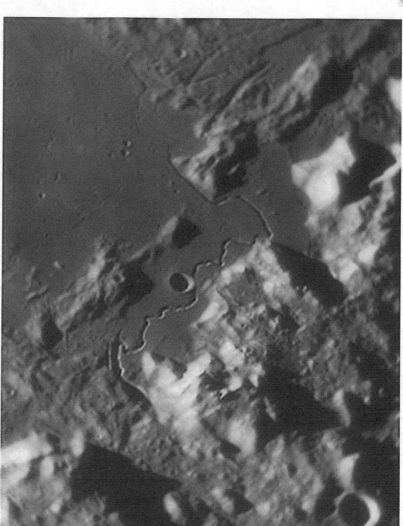

Figure 5.23. Hadley Rille (Credit—Wes Higgins).

distance as seen in comparison to astronaut James Irwin and the lunar rover. Over a period of time, lava may have continued to flow through this area, yet it remains forever buried beneath years of regolith.

Now let's go south for another challenging feature and a crater which conjoins it—Stofler and Faraday (Figure 5.25).

Located along the terminator, crater Stofler was named for Dutch mathematician and astronomer Johan Stofler. Consuming lunar landscape with an immense diameter of 126 kilometers and dropping 2760 meters below the surface,

Figure 5.24. Apollo 15 at Hadley Rille (Credit—NASA).

Stofler is a wonderland of small details in eroded surroundings. Breaking its wall on the north is Fernelius, but sharing the southeast boundary is Faraday. Named for English physicist and chemist Michael Faraday, it is more complex and deeper at 4090 meters, but far smaller at 70 kilometers in diameter. Look for myriad smaller strikes which bind the two together!

Wednesday, May 14

As we begin the evening, let's have a look at awesome crater Clavius (Figure 5.26). As a huge mountain-walled plain, Clavius will appear near the terminator tonight in the lunar Southern Hemisphere, rivaled only in sheer size by similarly structured Deslandres and Baily. Rising 1646 meters above the surface, the interior wall slopes gently downward for a distance of almost 24 kilometers and a span of 225 kilometers. Its crater-strewn walls are over 56 kilometers thick!

Figure 5.25. Stofler and Faraday (Credit—Wes Higgins).

Clavius is punctuated by many pockmarks and craters; the largest on the southeast wall is named Rutherford. Its twin, Porter, lies to the northeast. Long noted as a test of optics, Clavius crater can offer up to thirteen such small craters on a steady night at high power. How many can you see?

If you want to continue with tests of resolution, why not visit nearby Theta Virginis (RA 13 09 56 Dec –05 32 20)? It might be close to the Moon, but it is 415 light-years away from Earth (Figure 5.27)! The primary star is a white A-type subgiant, but it's also a spectroscopic binary consisting of two stars which orbit each other about every 14 years. In turn, the visual secondary is orbited by a 9th

Figure 5.26. Clavius (Credit—Wes Higgins).

Figure 5.27. Theta Virginis (Credit—Palomar Observatory, courtesy of Caltech).

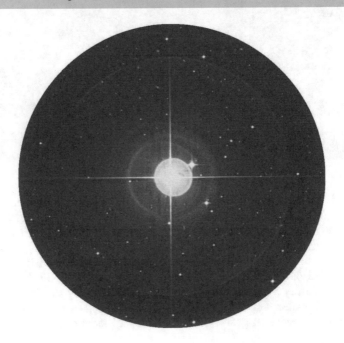

magnitude F-type star which is a close 7.1 arc seconds away from the primary. Look for the fourth member of the Theta Virginis system well away some 70 arc seconds, but shining at a feeble magnitude of 10.4.

Thursday, May 15

As the Sun sets, you get a clear western horizon and see if you can spot faint Mercury about a handspan above the orange glow. It reached its greatest elongation within the last 24 hours.

Tonight as we look at the Moon, crater Copernicus will try to steal the scene. Remember it and head further south to capture another Lunar Club challenge—Bullialdus (Figure 5.28). Even binoculars can make out this crater with ease near the center of Mare Nubium. If you're scoping—power up—this one is fun! A very similar crater to Copernicus, note Bullialdus' thick, terraced walls and central peak. If you examine the area around it carefully, you can note it is a much newer crater than shallow Lubiniezsky to its north and almost non-existent Kies to the south. On Bullialdus' southern flank, it's easy to make out its A and B craters, as well as the interesting little Koenig to the southwest.

Now let's head about four fingerwidths northwest of Beta Virginis for another unusual star—Omega (RA 11 38 27 Dec +08 08 03). Classed as an M-type red giant, this 480 light-year distant beauty is also an irregular variable (Figure 5.29).

Figure 5.28. Bullialdus (Credit—Wes Higgins).

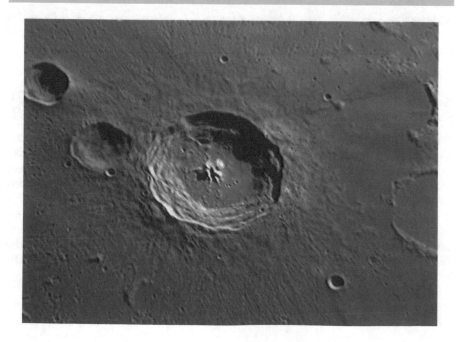

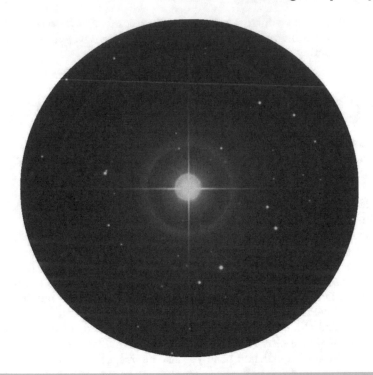

Figure 5.29. Omega Virginis (Credit—Palomar Observatory, courtesy of Caltech).

Although you won't notice much change in this 5th magnitude star (it cycles over about half a magnitude), it has a very pretty red coloration and is worth the time to view.

Friday, May 16

Tonight would be a wonderful opportunity for Moongazers to return to the surface and have a look at the peaceful Sinus Iridum area. If you've been clouded out before, be sure to have a look for the telescopic Lunar Club challenges, Promontoriums Heraclides and LaPlace (Figure 5.30). What other craters can you discover in the area?

If you're up for a bit more of a challenge, then let's head about 59 light-years away for star 70, in Virgo. You'll find it located about six degrees northeast of Eta (RA 13 28 25 Dec +13 46 43) and right in the corner of the Coma-Boötes-Virgo border. So what's so special about this G-type, very normal-looking, 5th magnitude star?

It's a star that has a planet.

Long believed to be a spectroscopic binary because of its 117 day shift in color, closer inspection has revealed that 70 Virginis actually has a companion planet

Figure 5.30. Detail view of Sinus Iridum (Credit—Greg Konkel, annotation by Tammy Plotner).

Figure 5.30. Detail view of Sinus Iridum (Credit—Greg Konkel, annotation by Tammy Plotner).

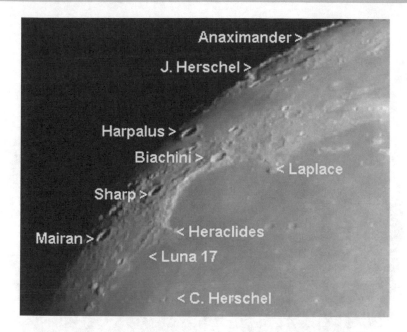

(Figure 5.31). Roughly 7 times larger than Jupiter and orbiting no further away than Mercury from its cooler-than-Sol parent star, 70 Virginis B just might well be a planet cool enough to support water in its liquid form.

How "cool" is that? Try, about 85°C...

Saturday, May 17

Today in 1835, J. Norman Lockyer was born (Figure 5.32). While that name might not stand out, Lockyer was the first to note previously unknown absorption lines while making visual spectroscopic studies of the Sun in 1868. Little did he know at the time, he had correctly identified the second most abundant element in our universe—helium—an element not discovered on Earth until 1891! Also known as the "Father of Archeoastronomy," Sir Lockyer was one of the first to make the astronomical connection with ancient structures such as Stonehenge and the Egyptian pyramids. (As a curious note, 14 years after Lockyer's notation of helium, a Sun-grazing comet made its appearance in photographs of the solar corona taken during a total eclipse in 1882... It hasn't been seen since.)

If you would like to see a helium-rich star, look no further tonight than Alpha Virginis—Spica (Figure 5.33). You can't miss it because it's so near the Moon! As the sixteenth brightest star in the sky, this brilliant blue–white "youngster" appears to be about 275 light-years away and is about 2300 times

Figure 5.31. 70 Virginis (Credit—Palomar Observatory, courtesy of Caltech).

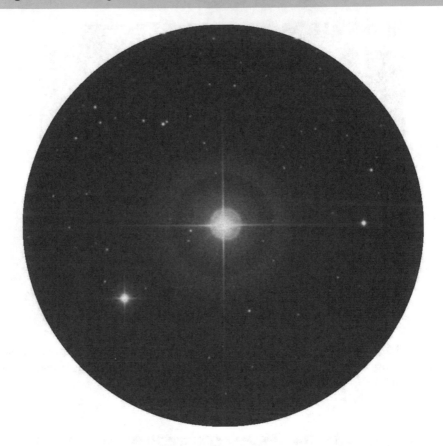

Figure 5.32. Norman Lockyer (widely used public image).

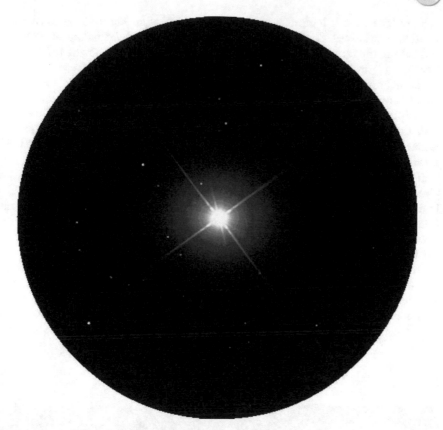

Figure 5.33. Alpha Virginis: Spica (Credit—John Chumack).

brighter than our own Sun. Although we cannot see it visually, Spica is a double star. Its spectroscopic companion is roughly half its size and is also rich in helium.

Sunday, May 18

On this day in 1910, Comet Halley transited the Sun, but could not be detected visually. Since the beginning of astronomical observation, transits, eclipses, and occultations have provided science with some very accurate determinations of size. Since Halley could not be spotted against the solar surface, we knew almost a century ago that the nucleus had to be smaller than about 100 kilometers. To get a rough idea of this size, take a look at crater Copernicus about midway along the western hemisphere of the Moon. What's its diameter? Oh, about the same size as a certain comet!

Now let's have a look at a very bright and changeable lunar feature which is often overlooked. Starting with the great gray oval of Grimaldi, let your eyes slide along the terminator toward the south until you encounter the bright crater Byrgius (Figure 5.34).

Named for Joost Bürgi, who made a sextant for Tycho Brahe, this "seen on the curve" crater is really quite large with a diameter of 87 kilometers. Perhaps its most interesting feature is the high-albedo Byrgius A, which sits along its eastern wall line and produces a wonderfully bright ray system. While it's noted as a Lunar Club II challenge, it's also a great crater to help add to your knowledge of selenography!

It's time to add to our double star list as we hunt down Zeta Boötes located about seven degrees southeast of Arcturus (RA 14 41 08 Dec +13 43 42). This is a delightful multiple star system for even small telescopes—but not an easy one (Figure 5.35). The Zeta pairing has an extremely elliptical orbit: the distance between the stars varies from as little as the Earth–Sun distance to as much as 1.5 times the radius of Pluto's orbit!

Another great target for a bright night is Delta Corvi (RA 12 29 51 Dec – 16 30 55). It is 125 light-years away and displays a yellowish-colored primary and a slightly blue secondary which is an easily split pair in any telescope and a nice visual double with Eta in binoculars (Figure 5.36). Use low power and see if you can frame this bright grouping of stars in the same eyepiece field.

Figure 5.34. Byrgius (Credit—Alan Chu).

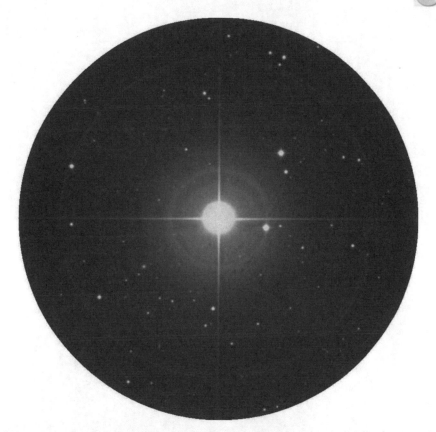

Figure 5.35. Zeta Boötes (Credit—Palomar Observatory, courtesy of Caltech).

Monday, May 19

While tonight the Moon will appear about as full as it gets to some observers, the date won't be "official" until tomorrow. Even as it is, Luna is an awesome sight for those who have never seen it through a telescope. Invite someone to visit with you, or offer to take your telescope to a public area. Power up on bright features like Tycho's rays, it's inspiring!

Although the glare will make it difficult to do many things, we can still have a look at a few other bright objects. Let's start tonight by going just north of Zeta Boötes for Pi (RA 14 40 43 Dec +16 25 03). With a wider separation, this pair of whites will easily resolve to the smaller telescope (Figure 5.37).

Now skip up northeast about a degree for Omicron Boötes (RA 14 45 14 Dec +16 57 51) (Figure 5.38). While this is not a multiple system, it makes for a nice visual pairing and a binocular challenge. For telescopes, the southeastern star holds interest as a small asterism.

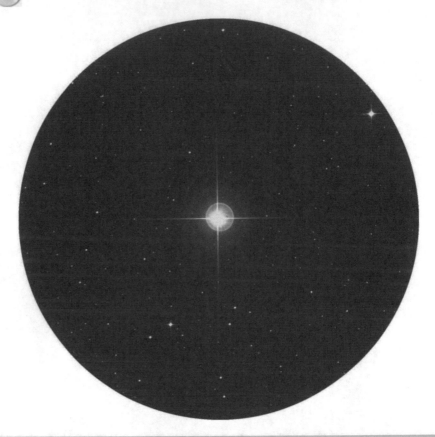

Figure 5.36. Delta Corvi (Credit—Palomar Observatory, courtesy of Caltech).

Continue northeast another two degrees to discover Xi Boötes (RA 14 51 23 Dec +19 06 01)(Figure 5.39). This one is a genuine multiple star system with magnitude 5 and 7 companions. Not only will you enjoy this G-type sun for its duplicity, but for the fine field of stars in which it resides!

Tuesday, May 20

Tonight is Full Moon (Figure 5.40). By May in most northern areas, flowers are everywhere, so it's not hard to imagine how this became to be known as the "Full Flower Moon." Since the Earth is awakening again, agriculture has re-emerged and so it is sometimes known as the "Full Corn Planting Moon" or the "Milk Moon." No matter what you call it, it's still majestic to watch it rise!

To participate in another Lunar Club challenge and do some outreach work, you can demonstrate the "Moon Illusion" to someone. While we know it's purely psychological and not physical, the fact remains: the Moon seems larger on the horizon. Using a small coin held at arm's length, compare it to Luna as it rises,

Figure 5.37. Pi Boötes (Credit—Palomar Observatory, courtesy of Caltech).

Figure 5.38. Omicron Boötes (Credit—Palomar Observatory, courtesy of Caltech).

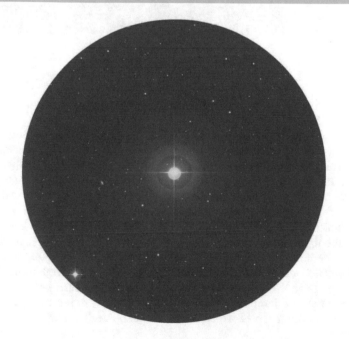

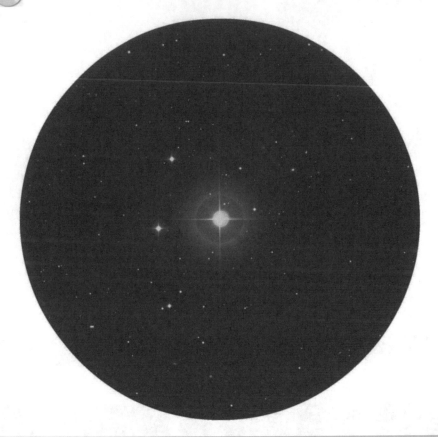

Figure 5.39. Xi Boötes (Credit—Palomar Observatory, courtesy of Caltech).

and then again as it seems to "shrink" while moving higher! You've now qualified for extra credit...

Even though the Moon is very bright when full, try using colored or Moon filters to have a look at the many surface features which throw amazing patterns across its surface. If you have no filters, a pair of sunglasses will suffice. Look for things you might not ordinarily notice—such as the huge streak which emanates from crater Menelaus. Look at the pattern projected from Proclus—or the intense little dot of little-known Pytheas north of Copernicus. It's hard to miss the blinding beacon of Aristarchus! Check the southeastern limb where the edge of Furnerius lights up the landscape—or how a nothing crater like Censorinus shines on the southeast shore of Tranquillitatis, while Dionysus echoes it on the southwest. Could you believe Manlius just north of central could be such a perfect ring—or that Anaxagoras would look like a northern polar cap?

While it might be tempting to curse the Moon for hiding the stars when it's full, there is no other world we can view in such detail... Even if you just look with your eyes!

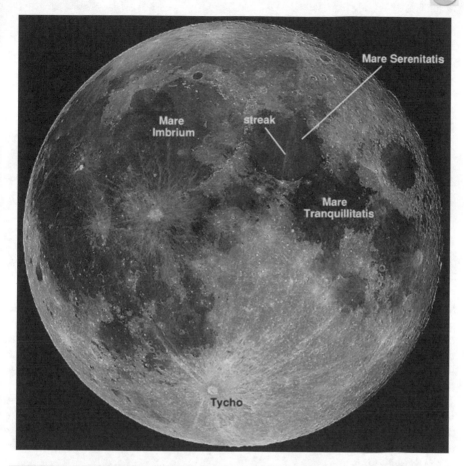

Figure 5.40. Full Moon (Credit—NASA).

Wednesday, May 21

In 1961, United States President John F. Kennedy launched the country on a journey to the Moon as he made one of his most famous speeches to Congress:

I believe this nation should commit itself to achieving the goal, before this decade is out, of landing a man on the Moon and returning him safely to Earth. No single space project in this period will be more impressive to mankind, or more important for the long-range exploration of space.

With just a little bit of time before the Moon rises, let's take a look at the constellation of Leo and its brightest stars (Figure 5.41). For our first destination we'll travel 85 light-years to learn about "The Little King"—Regulus.

Ranked as the 21st brightest star in the night sky, 1.35 magnitude Alpha Leonis is a helium star about 5 times larger and 160 times brighter than our own Sun.

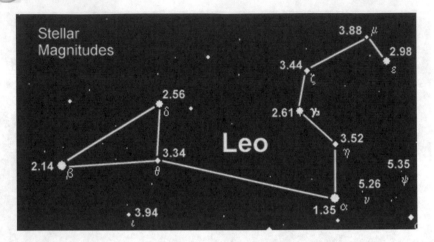

Figure 5.41. Stellar Magnitudes in Leo (Credit—NASA).

Speeding away from us at 3.7 kilometers per second, Regulus is also a multiple system whose 8th magnitude companion is easily seen in small telescopes. The companion is itself a double at around magnitude 13 and is a dwarf of an uncertain type. There is also a 13th magnitude fourth star in this grouping, but it is believed not to be associated with Regulus, since the Little King is moving toward it, and will be about 14 arc seconds away in 785 years.

Northeast of Regulus by about a fistwidth is 2.61 magnitude Gamma Leonis—also known as Algieba. This is one of the finest double stars in the sky, but a little difficult at low power since the pair is both bright and close. Separated by about twice the diameter of our own solar system, this 90 light-year distant pair is slowly widening.

Another two fingerwidths north is 3.44 magnitude Zeta—named Aldhafera. Located about 130 light-years away, this excellent star has an optical companion which is viewable in binoculars—35 Leonis. Remember this pair, because it will lead you to galaxies! Before we leave, let's have a look east at 3.34 magnitude Theta. Also known as Chort, mark this one in your memory, as well as 3.94 magnitude Iota to the south, as future markers for galaxy hops. Last is easternmost 2.14 magnitude Beta. Called Denebola, it is the "Lion's Tail" and has several faint optical companions. Now behold the intensely close pairing of Antares and the Moon! For some observers this could be an occultation event, so be sure to check IOTA for precise times and locations.

Thursday, May 22

Tonight start by locating 5th magnitude 6 Comae Berenices about three finger-widths east of Beta Leonis. Remember this star! We are going on a galaxy hop to a Méchain discovery less than one degree west of 6 Comae—its designation is M98 (RA 12 13 48 Dec +14 53 58) (Figure 5.42).

At magnitude 10, this beautiful galaxy is a telescope-only challenge and a bit on the difficult side for small aperture. Long considered to be part of the Virgo Cluster, M98 is approaching us at a different rate from other cluster members, giving rise to speculation it is merely in our line of sight. Quite simply put, it has a blue shift instead of red! But all these galaxies (and far fainter ones we can't see) are in close proximity, and this leads some researchers to believe M98 is a true member because of the extreme tidal forces which must exist in the area—pushing it toward us at this point in time, rather than away.

In a small telescope, M98 will appear like a slim line with a slightly brighter nucleus—a characteristic of an edge-on galaxy. To large aperture, its galactic disk is hazy and contains patchiness in structure. These are regions of newly forming stars and vast regions of dust—yet the nucleus remains a prominent feature. It's a very large galaxy, so be sure to use a minimum of magnification and plenty of aversion to make out small details in this fine Messier object!

Figure 5.42. M98 (Credit—Palomar Observatory, courtesy of Caltech).

Friday, May 23

While our destination tonight isn't quite, what you'd call romantic, I think you'll enjoy getting a "Blackeye." You'll find it located just one degree east-northeast of 35 Coma Berenices and it is most often called M64 (RA 12 56 43 Dec +21 41 00).

Originally discovered by Bode about a year before Messier cataloged it, M64 is about 25 million light-years away and holds the distinction of being one of the more massive and luminous of spiral galaxies (Figure 5.43). It has a very unusual structure and is classified as an Sa spiral in some catalogs and as an Sb in others. Overall, its arms are very smooth and show no real resolution to any scope—yet its bright nucleus has an incredible dark dustlane consuming the northern and eastern regions around its core, giving rise to its nickname, the Blackeye Galaxy.

In binoculars, this 8.5 magnitude galaxy can be perceived as a small oval with a slightly brighter center. Small telescope users will pick out the nucleus more easily, but will require both magnification and careful attention to dark adaptation to catch the dustlane. In larger telescopes, the structure is

Figure 5.43. M64 (Credit—Palomar Observatory, courtesy of Caltech).

easily apparent and you may catch the outer wisps of arms on nights of exceptional seeing.

No matter what you use to view it, this is one compact and bright little galaxy!

Saturday, May 24

Tonight we'll return once again to 6 Coma Berenices and head no more than a half degree southwest for another awesome galaxy—M99 (RA 12 18 49 Dec +14 25 00).

Discovered by Pierre Méchain on the same night he found M98, M99 is one of the largest and brightest of the spiral galaxies in the Virgo Cluster (Figure 5.44). Recognized second after M51 for its structure, Lord Rosse proclaimed it to be "a bright spiral with a star above." It is an Sc class and, unlike its similarly structured neighbors, it rotates clockwise. Receding from us at 2324 kilometers per second, its speedy retreat through the galaxy fields and its close encounter with the approaching M98 may be the reason it is asymmetrical—with a wide arm extending to the southwest. Three documented supernovae have been recorded in M99—in 1967, 1972, and 1986.

Figure 5.44. M99 (Credit—Palomar Observatory, courtesy of Caltech).

Possible in large binoculars with excellent conditions, this roughly 9th magnitude object is of low surface brightness and requires clean skies to see details. For a small telescope, you will see this one as fairly large, round, and wispy, and with a bright nucleus. But unleash aperture if you have it! For large scopes, the spiral pattern is very prominent and the western arm shows well. Areas within the structure are patchworked with bright knots of stars and thin dustlanes which surround the concentrated core region. During steady seeing, a bright, pinpoint stellar nucleus will come out of hiding. A worthy study!

Sunday, May 25

Before we leave Leo to softly exit west, there is another galaxy very worthy your time, and even binoculars can spot it. You'll need to identify slightly fainter Lambda Leonis to the southwest of Epsilon and head south about one fingerwidth for NGC 2903 (RA 09 32 09 Dec +21 30 02).

Figure 5.45. NGC 2903 (Credit—Palomar Observatory, courtesy of Caltech).

This awesome oblique spiral galaxy was discovered by William Herschel in 1784 (Figure 5.45). At a little brighter than magnitude 9, it is easily in range of most binoculars. It is odd that Messier missed this one considering its brightness… And three of the comets he discovered passed by it! Perhaps it was cloudy when Messier was looking, but we can thank Herschel for cataloging NGC 2903 as H I.56.

While small optics will only perceive this 25 million light-year distant beauty as a misty oval with a slightly brighter core region, larger aperture will light this baby up. Soft suggestions of its spiral arms and concentrations will begin to appear. One such knot is the star cloud NGC 2905—a detail in a distant galaxy so prominent it received its own New General Catalog designation.

NGC 2903 is roughly the same size as our own Milky Way, and includes a central bar—yet the nucleus of our distant cousin has "hot spots" which were studied by the Hubble Telescope and extensively by the Arecibo telescope. While our own galactic halo is filled with ancient globular clusters, this galaxy sports brand new ones!

Be sure to mark your notes with your observations, because many different organizations consider this to be on their "Best of" lists.

Monday, May 26

Tonight we'll return again to 6 Comae and our hunt will be for the last of the three galaxies discovered by Méchain on the same wonderful night in 1781. You'll find it just a fingerwidth northeast of star 6 (RA 12 22 54 Dec +15 49 19). After our recent studies of M98 and M99, you won't be surprised to learn its name is M100 (Figure 5.46)!

This spiral is one of the brightest members of the Virgo Cluster—and its design is much like that of our own galaxy. From our point of view, we see M100 "face on," and even Lord Rosse in 1850 was able to detect a spiral form. Thanks to its proximity to other cluster members, it has two grand arms in which recently formed, young, hot, and massive stars reside. Regardless of what seems to be its perfect form, in the nucleus younger stars have formed more toward the south side than the north. Perhaps interaction with its dwarf neighbors?

Achievable in binoculars as a soft round glow, and about the same in a small telescope, extensive photography has shown M100 to be far larger than previously believed—with a substantial portion of its mass contained in faint outer regions. The Hubble Telescope discovered over 20 Cepheid variables and one nova in our spiral friend, and was able to accurately determine its distance as 6 million light-years. In addition, NASA's Ultraviolet Imaging Telescope has shown starburst and star formation activity at the edges of M100's inner spiral arms.

Larger telescopes will see this galaxy's intense core region as slightly elliptical and will sometimes reveal patchiness in the structure. With good sky conditions, even smaller scopes can reveal a spiral pattern, and this improves significantly with aperture. Be sure to look carefully because five supernova events have been observed in this hot galaxy—one as recently as February 2006!

Figure 5.46. M100 (Credit—Palomar Observatory, courtesy of Caltech).

Tuesday, May 27

No galactic tour through Coma Berenices would be complete without visiting one of the most incredible "things Messier missed". You'll find NGC 4565 located less than two degrees east of 17 Comae (RA 12 35 21 Dec +25 59 13)...

Residing at a distance of around 30 million light-years, this large 10th magnitude galaxy is probably one of the finest edge-on structures you will ever see (Figure 5.47). Perfectly suited for smaller scopes, this ultra-slender galaxy with the bright core has earned its nickname of "The Needle." Although photographs sometimes show more than what can be observed visually, mid-to-large aperture can easily trace out NGC 4565's full photographic diameter.

Although Lord Rosse in 1855 saw the nucleus of the Needle as stellar, most telescopes will resolve a bulging core region with a much sharper point in the center and a dark dustlane upon aversion. The core itself has been extensively studied because of its cold gas and emission lines, indicating it has a barred

Figure 5.47. NGC 4565 (Credit—R. Jay GaBany).

structure. This is how the Milky Way would look if viewed from the same angle! It, too, shines with the light of 30 billion stars...

Chances are NGC 4565 is an outlying member of the Virgo Cluster, but its sheer size indicates it is probably closer to us than any of the others. If we were to gauge it at a distance of 30 million years as is accepted, its diameter would be larger than any galaxy yet known!

Get acquainted with it tonight...

Wednesday, May 28

Rudolf Leo Bernhard Minkowski was born on this date in 1895 (Figure 5.48). Although the name might not seem familiar, it should. Minkowski was a research assistant to Walter Baade. Together they broadened the field of spectroscopy with their research into the Orion Nebula, supernova remnants, and comets—not to mention their discovery of the small central star in the Crab Nebula! Minkowski and Baade found optical counterparts to several newly discovered radio sources, such as Cygnus A. When he discovered the galaxy with the highest

Figure 5.48. Rudolf Minkowski (widely used public image).

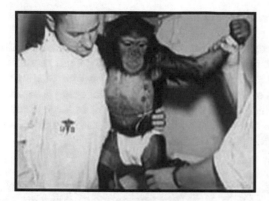

Figure 5.49. Abel (Credit—NASA).

redshift then known (in 1960), he was reputed to have joined other astronomers in the library of the 200" telescope's dome with a bottle of whisky—the night was then declared "overcast." You can thank Minkowski for something else too...the Palomar Digital Sky Survey's POSS images, which you see throughout this book!

Also on this day, in 1959, the first primates made it to space. Abel (a rhesus monkey) and Baker (a squirrel monkey) lifted off in the nosecone of an Army Jupiter missile and were carried to sub-orbital flight. Recovered unharmed, Abel died just 3 days later from the effects of anesthesia during an electrode removal (Figure 5.49), but Baker lived on to a ripe old age of 27.

Now shake your fist at Spica...because that's all it takes to find the awesome M104, 11 degrees due west (RA 12 39 59 Dec –37 23). (If you do have trouble finding M104, don't worry. Try this trick! Look for the upper left hand star in the rectangle of Corvus—Delta. Between Spica and Delta is a diamond-shaped

Figure 5.50. M104: The Sombrero (Credit—R. Jay GaBany).

pattern of 5th magnitude stars. Aim your scope or binoculars just above the one furthest south.)

Also known as the "Sombrero" (Figure 5.50), this gorgeous 8th magnitude galaxy was discovered by Pierre Méchain in 1781, added by hand to Messier's catalog, and observed independently by Herschel as H I.43—who was probably the first to note its dark inclusion. The Sombrero's rich central bulge is comprised of several hundred globular clusters and can just be hinted at in large binoculars and small telescopes. Large aperture will revel this galaxy's "see through" qualities and bold, dark dustlane—making it a seasonal favorite!

Thursday, May 29

Today in 1919, a total eclipse of the Sun occurred, and stellar measurements taken along the limb agreed with predictions based on Einstein's General Relativity theory—the first such confirmation. Although we call it gravity, spacetime

Figure 5.51. NGC 4038/39 (Credit—Palomar Observatory, courtesy of Caltech).

curvature deflects the light of stars near the limb, causing their apparent positions to differ slightly. Unlike today, at that time you could only observe stars near the Sun's limb (within less than an arc second) during an eclipse. It's interesting to note that even Newton had his own theories on light and gravitation which predicted some deflection!

Now let's have a look at what gravitation can do on a mass scale as we head slightly less than four degrees west-southwest of Gamma Corvi (RA 12 01 52 Dec –18 52 02) for one of the coolest sights in the night—the "Ring-Tail Galaxy."

Designated NGC 4038 and 4039, this gravitationally interacting pair comes in at magnitude 11 and is best suited for the intermediate scope (Figure 5.51). Discovered by William Herschel on February 7, 1785, the master classified the pair as a planetary nebula (H IV.28). His son, John, also cataloged them—but under separate numbers. As you study, you can see why—the northern portion is far brighter than the southern extension. It takes a fine night or larger aperture to see much more than a heart-shaped, glowing structure.

Gloriously revealed to the world by the Hubble Telescope and dubbed the "Antennae" for its strange extensions, this colliding system shows tremendous tidal filaments in long exposure photography. Does the Antennae broadcast? You bet. In 1957, the first radio radiation was detected from this strange and beautiful galaxy! Be sure to mark your notes for another "peculiar" study!

Friday, May 30

In an early night sky filled with thousands of faint galaxies, is there anything a very small telescope or a pair of binoculars can find to do on a dark night? The answer is yes. Your mission is to have a look at a very neat cluster of stars cataloged as Upgren 1 (RA 12 35 01 Dec +36 22 18). You'll find it easily about one binocular field (five degrees) southwest of Cor Caroli (Alpha CnV).

This bright collection of F-type stars lies in the general direction of the north galactic pole. Although there are only about 10 bright stars, Upgren 1 is believed to be one of the most ancient of all open clusters (Figure 5.52). At around 380 light-years away from our solar system, it has taken billions of years for its stars to slowly drift apart. According to research done by Upgren and Rubin (1965), there is a possibility that this collection may be remnants of a much older cluster. While their stellar distances are about the same, their velocities are different—suggesting this bright gathering belongs to two radically different stellar groups.

Regardless of its pedigree, it's an easy and handsome object for all optics—even under urban lighting. Hunt it down tonight!

Figure 5.52. Upgren 1 (Credit—Palomar Observatory, courtesy of Caltech).

Saturday, May 31

Tonight we'll take a closer look at the work of Abbé Nicholas Louis de la Caille (or de Lacaille). Born in 1731, the French astronomer and mapmaker was the first to demonstrate Earth's bulge at its equator. From 1751 to 1753, he had the great fortune to observe the southern skies and, putting his cartographic skills to use, he mapped them and established the 14 constellations which remain in use to this day—including Musca. Even though Lacaille was best known for the constellation names, he and his productive 0.5 inch telescope also cataloged 9766 stars during his two-year observing period. Of these, one stands out for good reason—Lacaille 8760.

Its designation is also AX Microscopii (RA 21 17 15 Dec –38 52 02), and it is a dwarf red flare star which resides only 12.9 light-years from us (Figure 5.53). While this star might not seem so important, it is one target for interferometric studies searching for planets which may have formed in "habitable zones" around

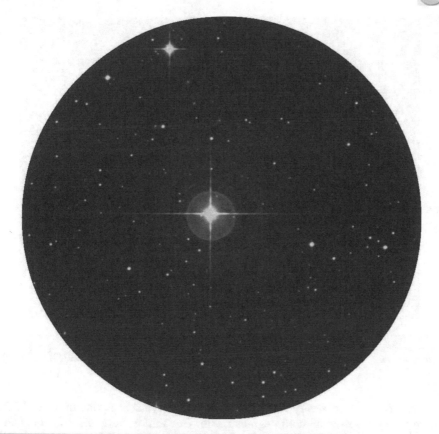

Figure 5.53. AX Microscopii/Lacaille 8760 (Credit—Palomar Observatory, courtesy of Caltech).

stars similar to our own. Even though AX is slightly smaller than Sol, this particular cool main sequence star might in fact be inhospitable due to its daily flare activity.

Since it will be awhile before the constellation of Microscopium rises high enough for southern observers to capture this star, let's have a look at an object from Lacaille's catalog known as I.5.

Located less than two handspans south of Spica (RA 13 26 45 Dec −47 28 26), most of us know this globular cluster best as NGC 5139—or Omega Centauri (Figure 5.54). As the most luminous of all globular clusters, Lacaille reported it as a "nebula in Centaurus; with simple view, it looks like a star of 3rd magnitude viewed through light mist and through the telescope like a big comet badly bounded." Yet, through even the most modest of today's telescopes, Omega Centauri will explode into a fury of stars. Located about 17,000 light-years away, it took around 2 million years to form, and it may be the remnant of another galaxy's core which was captured by our own. With more than one million members, it is the size of a small galaxy in itself!

Figure 5.54. NGC 5139: Omega Centauri (Credit—Mike Sidonio).

While this object is very low for northern observers, it is not impossible for those who live lower than 40 degrees north. Our atmosphere will rob this giant of a cluster of some of its beauty, but I encourage you to try! It's a sight you will never forget...

CHAPTER SIX

June, 2008

So far we've studied many Herschel objects in disguise as Messier catalog items—but we haven't really focused on some mighty fine galaxies within the reach of the intermediate-to-large telescope. Tonight let's take a serious skywalk as we head to 6 Comae and drop two degrees south.

At magnitude 10.9, Herschel catalog object H I.35 is also known by its New General Catalog number of 4216. This splendid edge-on galaxy has a bright nucleus and will walk right out in larger telescopes with no aversion required. But the most fascinating part about studying anything in the Virgo cluster is about to be revealed.

While studying structure in NGC 4216, averted vision picks up 12th magnitude NGC 4206 to the south. This is also a Herschel object—H II.135. While it is smaller and fainter, the nucleus will be the first thing to catch your attention—and then you'll notice it is also an edge-on galaxy! As if this weren't distracting enough, while re-centering NGC 4216, sometimes the movement is just enough to allow the viewer to catch yet another edge-on galaxy to the north—NGC 4222. At magnitude 14, you can only expect to be able to see it in larger scopes, but what a treat this trio is (Figure 6.1)!

Is there a connection between certain types of galaxy structures within the Virgo cluster? Scientists certainly seem to think so. While studies involving these low-metallicity galaxies are ongoing, research into the evolution of galaxy

Figure 6.1. NGC 4222 (top), NGC 4216 (center), and NGC 4206 (bottom) (Credit—Palomar Observatory, courtesy of Caltech).

clusters themselves continues to make new strides forward in our understanding of the universe.

Capture them tonight!

Monday, June 2

Tonight while we're out, let's have a look at a Herschel discovery made on this date in 1788. Triangulating with Theta and Iota Draconis to the north is a 5th magnitude star which will lead you to NGC 6015 to its east. This near 12th magnitude spiral galaxy with the bright core is fairly large, but soft-spoken in the eyepiece. Studied extensively for its rotation rate, research has shown we are looking at the combined light of 28 million solar masses...but a peculiar lot. NGC 6015 is not only isolated and ringed, but it is also warped. Its spiral arms formed in a way unlike other galaxies of its type: the inner and outer arms rotate at different rates. But don't expect to see details in this one! With low surface brightness, you'll do best staying at minimum magnification.

Figure 6.2. NGC 4281 and NGC 4273 (Credit—Palomar Observatory, courtesy of Caltech).

For an added Herschel treat tonight for larger scopes, hop back to star 17 and head about one half degree due west for the close galactic pair NGC 4281 (H II.573) and NGC 4273 (H II.569) (Figure 6.2). Here is a study of two galaxies similar in magnitude (12) and size—but of very different structure. Northeastern NGC 4281 is an elliptical, and by virtue of its central concentration will appear slightly larger and brighter—while southwestern NGC 4273 is an irregular spiral which will appear brighter in the middle but more elongated and faded along its frontiers. Sharp-eyed observers may also note fainter (13th magnitude) NGC 4270 north of this pairing.

Tuesday, June 3

If you're up early, why not keep a watch out for the peak of the Tau Herculids meteor shower? These are the offspring of comet Schwassman-Wachmann 3, which broke up in 2006. The radiant is near Corona Borealis and we'll be in this stream for about a month. At its best when the parent comet has passed perihelion, you'll catch about 15 per hour maximum. Most are quite faint, but with no Moon to interfere, sharp-eyed observers will enjoy it!

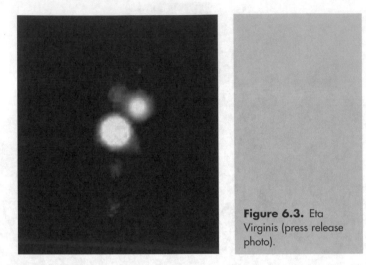

Figure 6.3. Eta Virginis (press release photo).

Figure 6.4. Rho Virginis and NGC 4596 (Credit—Palomar Observatory, courtesy of Caltech).

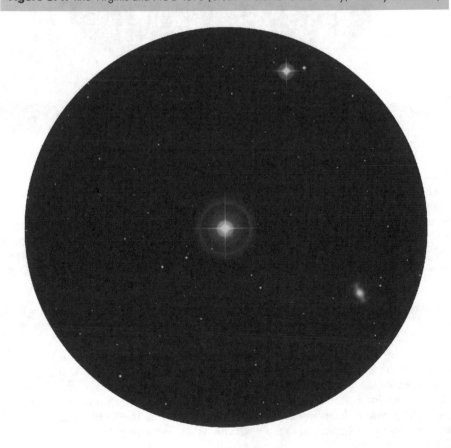

Tonight is officially New Moon, and it's time to try a visual double for the unaided eye—Eta Virginis (Figure 6.3). Can you distinguish between a 4th and 6th magnitude star?

The brighter of the two is Zaniah (Eta), which through occultation had been discovered to be a triple star. In 2002, Zaniah became the first star imaged by combining multiple telescopes, using the Navy Prototype Optical Interferometer. This was the first time the three were split visually. Two of them are so close together that they orbit at less than half the distance between the Earth and the Sun!

Binocular users should take a look at the visual double Rho Virginis about a fistwidth west-southwest of Epsilon. This pair is far closer and will require optical aid to separate. The brighter of this pair—Rho—is a white main sequence dwarf with a secret... It's a variable! Known as a Delta Scuti type, this odd star can vary slightly in magnitude in anywhere from 0.5 to 2.5 hours as it pulsates.

For mid-to-large telescopes, Rho offers just a little bit more. The visual companion star has a visual companion as well! Less than a half degree southwest of Rho is a small, faint spiral galaxy—NGC 4608. At 12th magnitude, it's hard to see because of Rho's brightness...but it's not alone. Look for a small, but curiously shaped galaxy labeled NGC 4596 (Figure 6.4). Its resemblance to the planet Saturn makes it well worthwhile!

Wednesday, June 4

Tonight we'll use Rho Virginis as a stepping stone to yet more galaxies. Get on your mark and move 1.5 degrees north for M59 (Figure 6.5).

Discovered in 1779 by J. G. Koehler while studying a comet, this 11th magnitude elliptical galaxy was observed and labeled by Messier who was just a bit behind him. Much denser than our own galaxy, M59 is only about one-fourth the size of the Milky Way. In a smaller telescope, it will appear as a faint oval, while larger telescopes will make out a more concentrated core region.

Now shift a half degree east for brighter and larger M60 (Figure 6.6). Also caught first by Koehler on the same night as M59, it was "discovered" a day later by yet another astronomer who had missed M59! It took Charles Messier another four days until this 10th magnitude galaxy interfered with his comet studies and was cataloged.

At around 60 million light-years away, M59 is one of the largest ellipticals known and has five times more mass than our galaxy. As a study object of the Hubble Telescope, this giant has shown a concentrated core containing over two billion solar masses. Photographed and studied by large terrestrial telescopes, M59 may contain as many as 5100 globular clusters in its halo.

While our backyard equipment essentially reveals only M59's core, there is a curiosity here. It shares "space" with spiral galaxy NGC 4647. Telescopes of even modest aperture will pick up the nucleus and faint structure of this small face-on galaxy. Harlow Shapely found the pair odd because—while they are relatively close in astronomical terms—they are very different in age and development. Halton Arp also studied this combination of an elliptical galaxy affecting a spiral and cataloged it as Peculiar Galaxy 116. Be sure to mark your notes!

Figure 6.5. M59 (Credit—Palomar Observatory, courtesy of Caltech).

Figure 6.6. M60 and NGC 4647 (Credit—Palomar Observatory, courtesy of Caltech).

Thursday, June 5

Tonight we'll go back to Rho Virginis once again and find about a finger-width northwest for yet another bright spiral galaxy, M58—which *was* actually discovered by Messier in 1779!

As one of the brightest galaxies in the Virgo cluster, M58 is one of only four having a barred structure (Figure 6.7). It was cataloged by Lord Rosse as a spiral in 1850. In binoculars, it will look much like our previously studied ellipticals, but a small telescope under good conditions will pick up the bright nucleus and a faint halo of structure—while larger ones will see the central concentration of the bar across the core. Chalk up another Messier study for both binoculars and telescopes and let's get on to something really cool!

Around a half degree southwest are NGC 4567 and NGC 4568 (Figure 6.8). L. S. Copeland dubbed them the "Siamese Twins," but this galaxy pair is also considered part of the Virgo cluster. While seen from our viewpoint as touching galaxies, no evidence exists of tidal filaments or distortions in structure, making

Figure 6.7. M58 (Credit—Palomar Observatory, courtesy of Caltech).

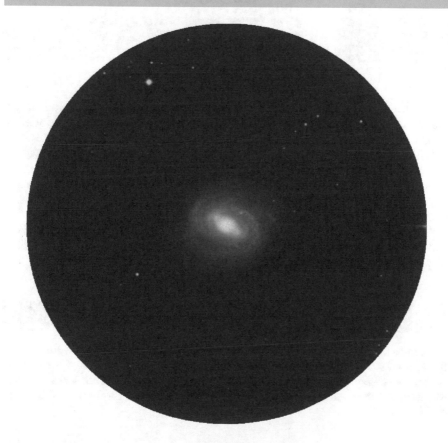

Figure 6.8. NGC 4567/68: The Siamese Twins (Credit—Palomar Observatory, courtesy of Caltech).

them a line of sight phenomenon and not interacting members. While this might take little of the excitement away from the "Twins," a supernova was spotted in NGC 4568 as recently as 2004.

While visible in smaller scopes as a duo, with soft twin nuclei, intermediate and larger scopes will see an almost V-shaped or heart-shaped pattern where the structures overlap. If you're doing double galaxy studies, this is a fine, bright one! If you see a faint galaxy in the field as well, be sure to add NGC 4564 to your notes.

For all you stargazers, keep watch for the Scorpid meteor shower. Its radiant will be near the constellation of Ophiuchus, and the average fall rate will be about 20 per hour with some fireballs!

Friday, June 6

Be sure to be outside as the Sun sets this evening to catch the very beginnings of the Moon as it joins bright star Pollux. This conjunction puts one of the Gemini "twins" less than three fingerwidths north. For the next 2 months, skywatchers will have some wonderful opportunities to view some of the best conjunctions of the year!

While I'm sure unaided-eye viewers and binocular users are tired of the galaxy hunt, be sure to take the time to look at the many old favorites now in view. To the eye, one of the most splendid signs of the changing seasons is the Ursa Major Moving Group, which sits above Polaris for Northern Hemisphere observers. For the Southern Hemisphere, the return of Crux serves the same purpose.

Old favorites have now begun to appear again, such as Hercules, Cygnus, and Scorpius...and as the night deepens and the hour grows late, a wealth of starry clusters and nebulae will come into view. Before we leave Virgo for the year, there is one last object which is seldom explored, and is such a worthy target we must visit it before we go. Its name is NGC 5634 (Figure 6.9) and you'll find it halfway between Iota and Mu Virginis (RA 14 29.37 Dec –05 58.35). Discovered by Sir William Herschel on March 5, 1785 and cataloged as H I.70, this small, magnitude 9.5 globular cluster isn't for everyone, but thanks to an 11th magnitude line-of-sight star on its eastern edge, it sure is interesting. At class IV, it's more concentrated than many globular clusters, although its 19th magnitude members make it nearly impossible to resolve with backyard equipment.

Located a bit more than 82,000 light-years from our solar system and about 69,000 light-years from the galactic center, you'll truly enjoy this globular for the randomly scattered stellar field which accompanies it. In the finderscope, an 8th magnitude star will lead the way—not truly a member of the cluster, but one lying between us. Capturable in scopes as small as 4.5", look for a concentrated

Figure 6.9. NGC 5634 (Credit—Palomar Observatory, courtesy of Caltech).

central area surrounded by a haze of stellar members—a huge number of which are recently discovered variables. While you look at this globular, keep this in mind...

Based on observations with the Italian Telescopio Nazionale Galileo, it is now surmised that the NGC 5634 globular cluster has the same position and radial velocity as does the Sagittarius dwarf spheroidal galaxy. Because of the dwarf galaxy's metal-poor population of stars, it is believed NGC 5634 may in fact have once been part of the dwarf galaxy—and been pulled away by our own tidal field to become part of the Sagittarius stream!

Saturday, June 7

If you're up before dawn the next two days or out just after sunset, enjoy the peak of the June Arietid meteors—the year's strongest daylight shower—with up to 30 visible per hour.

If you'd like to try your ear at radio astronomy with the offspring of sungrazing asteroid Icarus, tune an FM radio to the lowest frequency not receiving a clear signal. An outdoor antenna pointed at the zenith increases your chances, but even a car radio can pick up strong bursts! Simply turn up the static and listen. Those hums, whistles, beeps, bongs, and occasional snatches of signals are our own radio signals being reflected off the meteor's ion trail!

Figure 6.10. Detail view of Crisium (Credit—Greg Konkel, annotation by Tammy Plotner).

Take the time to view the lunar surface for a couple of telescopic challenges which are easy to catch—all you have to know is Mare Crisium (Figure 6.10)! On the southeastern shoreline is a peninsula which reaches into Crisium's dark basin. This is Promontorium Agarum. On the western shore, bright Proclus lights the banks, but look into the interior for the two dark pock marks of Pierce to the north and Picard to the south. Be sure to mark them on your notes!

Last but not least in Virgo, let's study a radio galaxy which is so bright that it can be seen through binoculars—8.6 magnitude M87 (Figure 6.11), about two fingerwidths northwest of Rho Virginis. This giant elliptical was discovered by Charles Messier in 1781. Spanning 120,000 light-years, it's an incredibly luminous galaxy containing far more mass and stars than the Milky Way—gravitationally distorting its four dwarf satellites galaxies. M87 is known to contain many thousands of globular clusters—perhaps up to 150,000—and far more than our own galaxy's mere 200.

In 1918, H. D. Curtis of Lick Observatory discovered something else—M87 has a jet of gaseous material extending from its core and pushing out for several thousand light-years into space (Figure 6.12). This highly perturbed jet exhibits the same polarization as synchrotron radiation—the kind emitted by neutron stars. It contains a series of small knots and clouds as observed by Halton Arp at Palomar in 1977, and he also discovered a second jet in 1966 erupting in the opposite direction. Thanks to these two properties, M87 made Arp's Catalog of Peculiar Galaxies as number 152.

Figure 6.11. M87 and satellites (Credit—Palomar Observatory, courtesy of Caltech).

Figure 6.12. Jet from M87's nucleus (Credit—NASA).

In 1954 Walter Baade and R. Minkowski identified M87 with radio source Virgo A, discovering a weaker halo in 1956. Its position over an X-ray cloud extending through the Virgo cluster makes M87 a source of an incredible amount of X-rays. Because of its many strange properties, M87 remains a prime target for scientific investigation. The Hubble has shown a violent nucleus surrounded by a fast-rotating accretion disc, whose gaseous make-up may be part of a huge system of interstellar matter. As of today, only one supernova event has been recorded—yet M87 remains one of the most active and highly prized study galaxies of all. Capture it tonight!

Sunday, June 8

Born on this date in 1625 was Giovanni Cassini (Figure 6.13)—the most notable observer following Galileo. As head of the Paris Observatory for many years, he was the first to observe seasonal changes on Mars and measure its parallax (and so, its distance). This set the scale of the solar system for the first time. Cassini was the first to describe Jovian features, and studied the Galilean moons' orbits. He also discovered four moons of Saturn, but he is best remembered for being the first to see the namesake division between the rings of planes A and B.

Figure 6.13. Giovanni Cassini (widely used public image).

Tonight let's honor Cassini by taking a look at two of these planets—beginning with the westering Saturn. To the unaided eye, this creamy yellow "star" outshines most stars in the region and holds competition with Regulus in Leo. To binoculars, it reveals itself as a planet—one with ears! While great detail cannot be seen, even the slightest optical aid makes it a joy.

To the small telescope, Saturn's ring system becomes very clear, and bright Titan can easily be seen (Figure 6.14). To the mid-sized telescope, the "Lord of the Rings" easily shows the Cassini division and other details, and reveals some of the many smaller moons dancing along the rings' edges. For the large telescope, Saturn continues to be one of the most fascinating of planets. Several ring divisions are easily apparent and subtle shading details on the planet's surface are easily discerned. Titan shines very brightly and under good conditions will display a certain amount of limb darkening, making it perceivable as an orb. Tethys, Rhea, and Dionne are easily visible, and the three-dimensional feel of Saturn revealed through shadow-play is incredible.

Now have a look at Mars. Although it is much closer to us than Saturn, it's also very close in the sky to the Moon! So close, in fact, this could be an occultation event of the finest kind. Be sure to check out IOTA for information and then check out the surface. What details can you see? Are the polar caps visible? Dark markings? While we wait on the occultation to end and Jupiter to rise, let's see why we love the Moon and collect study craters Messier and Messier A (Figure 6.15).

Located in Mare Fecunditatis about one-third its width from west to east, this pair of twin craters will be difficult in binoculars, but not hard for even a small telescope and intermediate power. Indeed named for famed French

Figure 6.14. Saturn and Moons (Credit—R. Jay GaBany).

Figure 6.15. Mare Fecunditatis/Messier/Messier A (Credit—Greg Konkel).

astronomer Charles Messier, the easternmost crater is somewhat oval in shape with dimensions of 9 by 11 kilometers. At high power, Messier A to the west appears to have overlapped a smaller crater during its formation and it is slightly larger at 11 by 13 kilometers. Although it is not on the challenge list, you'll find another point of interest to the northwest. Rima Messier is a long surface crack which runs diagonally across Mare Fecunditatis' northwestern flank and reaches a length of 100 kilometers.

Monday, June 9

Today is the birthday of Johann Gottfried Galle (Figure 6.16). Born in Germany in 1812, Galle was the first observer to locate Neptune. He is also known for being Encke's assistant—and he's one of the few astronomers ever to have observed Halley's Comet twice. Unfortunately, he died two months after the comet passed perihelion in 1910, but at a ripe old age of 98! I wonder if he knew Mark Twain?

Somewhere out in space, something very wonderful and very quiet is happening... The Sun, Mercury, and Venus are aligning. No, this isn't the dawning of the age of Aquarius, or even an apocalypse—just a perfectly natural and perfectly normal event in our solar system. Just like clockwork our orbits pass each other in the twinkling of a Cosmic Eye. Before anyone can cry "doom," be sure to remind them of how many times a day the hands of a clock align in hours, minutes, and seconds. The watch doesn't fall apart, and neither will our solar system!

Also, be sure to watch tonight as the Moon and Saturn make a splendid union in tonight's sky. If you haven't had the chance to pick up crater Posidonius for

Figure 6.16. Johann Galle (widely used public image).

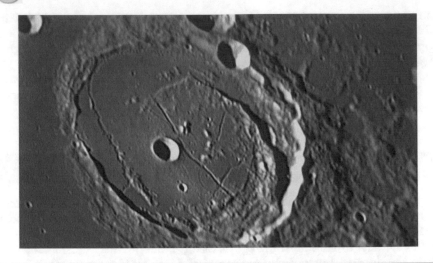

Figure 6.17. Posidonius (Credit—Wes Higgins).

your lunar studies, now would be a great time to power up and really study this ancient mountain-walled plain (Figure 6.17). Standing on the northeast shore of Mare Serenitatis near the terminator, this melted-down and lava-filled area looks very flat, with gently stepped walls. Still, it stretches an admirable 95 kilometers in diameter and drops to a depth of 2300 meters at the floor. During this sunrise phase, it is possible to spot some of the huge systems of rimae which line its floor, even using binoculars. This extended system of cracks gives mute testimony to Posidonius' volcanic history, and several strikes—such as the A crater near the center—help this crater to show its age well!

Tuesday, June 10

If you observe the Moon tonight, be sure to note both Gemma Frisius and Maurolycus on your lunar challenge list. While at Gemma Frisius, let's look for a little more "off the beaten path" crater about halfway toward Catharina. Under tonight's lighting at low power, you'll see it first as a sunken oval, but power up and let's explore Sacrobosco (Figure 6.18).

Named for the English mathematician "John of Holywood" (Johannes Sacrobuschus), this class III crater spans 98 kilometers and drops down to a floor level of 2800 meters—making those crater walls about as high as the West Ridge of Mt. Everest. On its troubled floor you will see the evidence of three far newer impacts: Crater C to the north which spans 13 kilometers and drops down 2630 meters; Crater A to the west which is 18 kilometers in diameter and 1830 meters deep; and Crater B to the east at 15 kilometers wide and 1210 meters deep. While these strikes are fascinating...look again: Sacrobosco itself is imprinted over the top of a far older crater!

Figure 6.18. Sacrobosco (Credit—Wes Higgins).

Figure 6.19. R Coronae Borealis (Credit—Palomar Observatory, courtesy of Caltech).

When you're finished, point your binoculars or telescopes toward Corona Borealis and about three fingerwidths northwest of Alpha for variable star R (RA 15 48 34 Dec +28 09 24). This star is a total enigma. Discovered in 1795, most of the time R carries a magnitude near 6, but can drop to magnitude 14 in a matter of weeks—only to unexpectedly brighten again! Scientists believe R occasionally emits a carbon cloud which blocks its light (Figure 6.19). When studied at a minimum, the light curve resembles a "reverse nova," and has a peculiar spectrum. It is very possible this ancient Population II star has used all of its hydrogen fuel and is now fusing helium to carbon. It's so very odd that we can't even accurately determine its distance!

Wednesday, June 11

Our lunar challenge feature for this evening is prominent enough to be spotted in binoculars, but is well worth the time to power up with the telescope and explore. Starting with the recognizable slash of the Alpine Valley, follow the mountain trail south to the double strike of crater Cassini (Figure 6.20).

Named for Giovanni Cassini, this smashing old Class V crater rises above the lunar topography by 1067 meters, making its shallow walls alone as tall as the Catskill Mountains. It covers about 57 kilometers of lunar landscape in its rough diameter, and the crater floor is 1240 kilometers below the surface. At one time Cassini may very well have had a central peak, but something obliterated it when Cassini A was formed. This double-stepped feature is 57 kilometers in diameter and drops down to an additional 2830 meters. While both Cassini and Cassini A are Lunar Club challenges, look carefully for yet another interior crater. Small crater B is often referred to as the "Washbowl" for its almost perfect concave structure.

Figure 6.20. Cassini and Cassini A (Credit—Wes Higgins).

Figure 6.21. Delta Serpentis (Credit—Palomar Observatory, courtesy of Caltech).

While we're out, let's have a look at Delta Serpentis (RA 15 34 48 Dec +10 32 19) (Figure 6.21). To the eye and binoculars, 4th magnitude Delta is a widely separated visual double star... But power up in the telescope to have a look at a wonderfully difficult binary. Divided by no more than four arc seconds, 210 light-year distant Delta and its 5th magnitude companion could be as old as 800 million years, and on the verge of becoming evolved giants. Separated by about nine times the distance of Pluto from our Sun, the white primary is a Delta Scuti–type variable which changes subtly in less than four hours. Although it takes the pair 3200 years to orbit each other, you'll find Delta Serpentis to be an excellent challenge for your optics.

Thursday, June 12

Tonight look at the wasted southern highlands area of the Moon with new eyes... Many of these craters you see were caused by impacts—some are volcanic formations—and all are beautiful (Figure 6.22)! Are you ready to learn a few more?

Now let's turn binoculars or telescopes toward magnitude 2.7 Alpha Librae— the second brightest star in the celestial "Scales" (RA 14 50 52 Dec −16 02 30) (Figure 6.23). Its proper name is Zuben El Genubi, and even as *Star Wars* as that name sounds, the "Southern Claw" is actually quite close to home at a distance of only 65 light-years.

Figure 6.22. Southern Highlands (Credit—Greg Konkel, annotation by Tammy Plotner).
(1) Goodacre, (2) Gemma Frisius, (3) Maurolycus, (4) Barocius, (5) Clairaut, (6) Baco,
(7) Mutus, (8) Manzinus, (9) Lilius (10) Zach, (11) Curtius, (12) Moretus, (13) Deluc,
(14) Maginus, (15) Proctor, (16) Saussure, (17) Miller, (18) Stofler, (19) Lexell, (20) Walter.

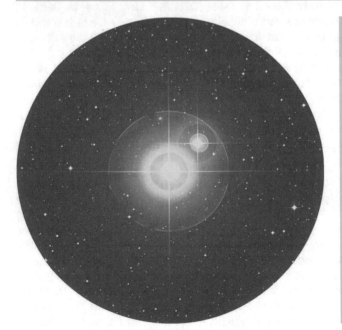

Figure 6.23. Alpha
Librae (Credit—Palomar
Observatory, courtesy
of Caltech).

No matter what size optics you are using, you'll easily see Alpha's 5th magnitude companion, which is widely spaced and shares the same proper motion. Alpha itself is a spectroscopic binary which was verified during an occultation, and its inseparable companion is only a half magnitude dimmer according to the light curves. Enjoy this easy pair tonight!

Friday, June 13

Today in 1983, Pioneer 10 became the first manmade object to leave the solar system (Figure 6.24). What wonders would it see? Are there other galaxies out there like our own? Will there be life like ours? While we can't see through Pioneer's "eyes," tonight let's take an historic journey to the Moon, as we look at the northeast shore of Mare Cognitum and the Apollo 14 mission landing site—Fra Mauro.

As craters go, 3.9 billion year old Fra Mauro is on the shallow side and spans 95 kilometers (Figure 6.25). At some 730 meters deep, standing at the foot of one of its walls would be like standing at the bottom of the Grand Canyon... Yet, time has so eroded this crater that its west wall is completely missing and its floor is covered with fissures.

Even though ruined Fra Mauro seems like a forbidding place to land a manned mission, it remained high on the priority list because it is geologically rich (Figure 6.26). Ill-fated Apollo 13 was to land in a formation north of the crater which was formed by ejecta belonging to the Imbrium Basin—material which had already been mapped telescopically. By returning samples of this material

Figure 6.24. Pioneer 10 "Greetings" (Credit—NASA).

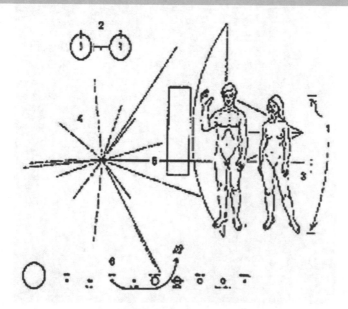

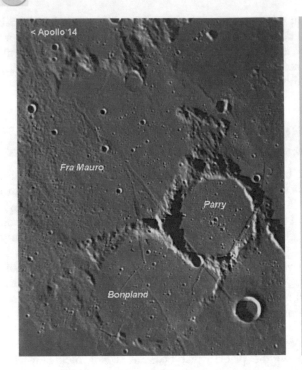

Figure 6.25. Detail view of Fra Mauro (Credit—Wes Higgins, annotation by Tammy Plotner).

Figure 6.26. Apollo 14: Alan Shepard at Fra Mauro (Credit—NASA).

from deep within the Moon's crust, scientists would have been able to determine the exact time these changes came about.

As you view Fra Mauro tonight, picture yourself in a lunar rover traversing this barren landscape and viewing the rocks thrown out from a long-ago impact. How willing would you be to take on the vision of others and travel to another world?

Saturday, June 14

As the day begins and you wait on dawn, keep watch for the peak of the Ophiuchid meteor shower with its radiant near Scorpius. The fall rate is poor with only 3 per hour, but fast-moving bolides are common. Today is about the midpoint—and the activity peak—of this 25 day long stream.

Tonight let's venture toward the south shore of Palus Epidemiarum to have a high power look at crater Capuanus (Figure 6.27). Named for Italian astronomer Francesco Capuano di Manfredonia, this 60 kilometer wide crater boasts a still-tall southwest wall, but the northeast one was destroyed by lava flow. At its highest, it reaches around 1900 meters above the lunar surface, yet drops to no more than 300 meters at the lowest. Look for several strikes along the crater walls as well as more evidence of a strong geological history. To its north is the Hesiodus Rima...a huge fault line extending 300 kilometers across the surface!

To the east, Jupiter is now rising... But give it some time to clear the atmospheric distortion! By far brighter than neighboring stars to the unaided

Figure 6.27. Capuanus (Credit—Wes Higgins).

eye, giant Jupiter will move slowly along the ecliptic plane over the course of the evening (Figure 6.28). To smaller binoculars it is easily observed as an orb with two gray bands across the middle. To larger binoculars, the equatorial belts become much clearer and the four Galilean moons are easily seen with steady hands. To the small telescope, no planet offers greater details. Even at very low magnifying power, the north, south, and central equatorial zones are easily observable and all four moons are clear and steady.

For most observers, tonight will show Callisto, Ganymede, Europa, and Io grouped to the east of the Mighty Jove, but as time progresses, so do their positions! Try observing over a period of several hours and watch just how quickly these four bright moons shuttle around... You might even catch a possible transit of Io!

To the mid-sized telescope, far greater details begin to appear—such as temperate belts on the planet's surface and the soft appearance of the Great Red Spot. Finer details are visible during steady seeing, and small things like being able to see which satellite is closer to—or further away from—our vantage point become very easy. Simple things, like watching a moon transit the surface and the resulting shadow on the planet are much easier. With a large telescope, seeing details on Jupiter depends more on seeing conditions. While more aperture allows finer views, conditions are everything when it comes to the Mighty Jove!

Figure 6.28. Jupiter (Credit—Wes Higgins).

Sunday, June 15

As we wait on the sky to darken tonight, let's start our adventures by taking a close look at crater Kepler (Figure 6.29). Situated just north of central along tonight's terminator, this great crater named for Johannes Kepler only spans 32 kilometers, but drops to a deep 2750 meters below the surface. This class I crater is a geological hotspot!

As the very first to be mapped by the US Geological Survey, the area around Kepler contains many smooth lava domes reaching no more than 30 meters above the plains. According to records, in 1963 a glowing red area was spotted near Kepler and extensively photographed. Normally one of the brightest regions of the Moon, the brightness value at the time nearly doubled! Although it was rather exciting, scientists later determined the phenomenon was caused by high-energy particles from a solar flare reflecting from Kepler's high-albedo surface. In the days ahead all details around Kepler will be lost, so take this opportunity to have a good look at one awesome small crater!

When skies are dark, it's time to have a look at the 250 light-year distant silicon star Iota Librae (RA 15 12 13 Dec −19 47 28) (Figure 6.30). This is a real challenge for binoculars—but not because the components are so close. In Iota's case, the near 5th magnitude primary simply overshadows its 9th magnitude companion! In 1782, Sir William Herschel measured them and determined them to be a true physical pair. Yet, in 1940 Librae A was determined to have an equal magnitude

Figure 6.29. Kepler (Credit—Wes Higgins).

Figure 6.30. Iota Librae (Credit—Palomar Observatory, courtesy of Caltech).

companion only 0.2 arcseconds away... And the secondary was proved to have a companion of its own which echoes the primary. A four star system!

Monday, June 16

No matter if you stayed up late chasing deep sky, or got up early, right now is the time to catch the peak of the June Lyrids meteor shower. Although the Moon will make observing difficult, it's still an opportunity for those wishing to log their meteor observations. Look for the radiant near bright Vega—you may see up to 15 faint blue meteors per hour from this branch of the May Lyrid meteor stream.

Today in 1963, Valentina Tereshkova, aboard the Soviet Vostok 6, became the first woman ever to go into space. Her solo flight is still unique. Twenty years later, on the 18th, Sally Ride became the first American woman in orbit, aboard the Space Shuttle.

Are you ready to study the Moon again tonight? Be sure to look for the "Cow Jumping over the Moon"—but power up with a telescope to study some very wild looking features—lunar lava domes (Figure 6.31).

North of Aristarchus, west of Promontorium Heraclides, and near the termi- nator is Rumker—the largest of the lunar lava domes. Only visible when near

Figure 6.31. Marian, Rumker, and the Megadome (Credit—Wes Higgins).

Figure 6.32. Upsilon Librae (Credit—Palomar Observatory, courtesy of Caltech).

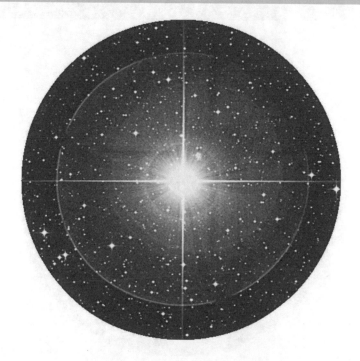

the terminator, this roughly 77 kilometer diameter "soft hill" ranges anywhere from 60 to 760 meters tall. Although it is not much more than a bump on the lunar surface, it does contain a few summit craters at its highest points. What we are looking at is really an important part of the geology which shaped the Moon's surface. In all likelihood, Rumker is a shield volcano...in an area which has many!

When you're finished with your lunar observations, tonight let's try a challenging double star—Upsilon Librae (RA 15 37 01 Dec –28 08 06). This beautiful red star is right at the limit for a small telescope, but quite worthy, since the pair is a widely disparate double (Figure 6.32). Look for the 11.5 magnitude companion to the south in a very nice field of stars!

Tuesday, June 17

Now let's go deep south and have look at an area which once held something almost half as bright as tonight's Moon and over four times brighter than Venus. Only one thing could light up the skies like that—a supernova.

According to historical records from Europe, China, Egypt, Arabia, and Japan, 1002 years ago one of the very first supernovae events was noted. Appearing in the constellation of Lupus, it was at first believed to be a comet by the Egyptians, yet the Arabs saw it as an illuminating "star."

Figure 6.33. Field of SN 1006 (Credit—Palomar Observatory, courtesy of Caltech).

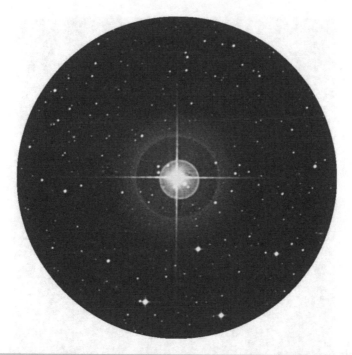

Figure 6.34. Mu Librae (Credit—Palomar Observatory, courtesy of Caltech).

Located less than a fingerwidth northeast of Beta Lupi (RA 15 02 48.40 Dec −41 54 42.0) and a half degree east of Kappa Centauri, no visible trace is left of a once grand event which could be seen five months beginning in May and lasting until it dropped below the horizon in September, 1006 (Figure 6.33). It is believed most of its mass was converted to energy and very little of the star remains. In the area, a magnitude 17 star shows a tiny gas ring, and radio source 1459-41 remains our best candidate for pinpointing this incredible event.

Still up for more? Then try your hand at a super-challenging double—Mu Librae (RA 14 49 19 Dec −14 08 56) (Figure 6.34). This pair is only a magnitude apart in brightness and right at the limit for a small telescope. Up the power slowly and look for the companion just to the southwest of the primary. Good luck, and mark your notes because Mu's blues are on many observing lists!

Wednesday, June 18

Tonight the Moon is full (Figure 6.35). Often referred to as the Full Strawberry Moon, this name was a constant to every Algonquin tribe in North America. But our friends in Europe referred to it as the Rose Moon. The North American version came about because the short season for harvesting strawberries comes

Figure 6.35. Nearside of the Moon as imaged by Apollo 11 (Credit—NASA).

each year during the month of June—so the full Moon occurring during this month was named for this tasty red fruit!

Tonight before it rises and the light commands the sky, let's have a look at a tasty red star—R Hydrae (Figure 6.36). You'll find it about a fistwidth south of Spica or about a fingerwidth west of Gamma Hydrae.

R was the third long-term variable star to be discovered, and it is credited to Maraldi in 1704. While it had been observed by Hevelius some 42 years earlier, it was not recognized immediately because its changes happen over more than a year. At maximum, R reaches near 4th magnitude—but drops well below perception to the human eye to magnitude 10. During Maraldi's and Hevelius' time, this incredible star took over 500 days to cycle, but it has speeded up to around 390 days in the present century.

Why such a wide range? Science isn't really sure. R Hydrae is a pulsing M-type giant whose evolution may be progressing more rapidly than expected due to changes in structure. We do know it is about 325 light-years away, and it is approaching us at around 10 kilometers per second.

In the telescope, R will have a pronounced red coloration which deepens near minima. Nearby is 12th magnitude visual companion star Ho 381, which was first

Figure 6.36. R Hydrae (Credit—Palomar Observatory, courtesy of Caltech).

measured for position angle and distance in 1891. Since that time no changes in separation have been noted, which leads us to believe the pair may be a true binary.

Thursday, June 19

Tonight let's return to R Hydrae. While observing a variable star with either the unaided eye, binoculars or a telescope can be very rewarding, it's often quite difficult to catch changes in a long-term variable, because there are times when its constellation is not visible. While R Hydrae is unique in color, let's drop about half a degree to the southeast to visit another variable star—SS Hydrae (Figure 6.37).

SS is a quick change artist—the Algol type. While you will need binoculars or a telescope to see this normally 7.7 magnitude star, at least its fluctuations are far more rapid, with a period of only 8.2 days. With R Hydrae we have a star which expands and contracts causing the changes in brightness—but SS is an eclipsing binary. While less than a half magnitude is not a noteworthy amount, you will notice a difference if you view it over a period of time. Be sure to note: this is actually a triple star system, because there is also a 13th magnitude companion

Figure 6.37. SS Hydrae (Credit—Palomar Observatory, courtesy of Caltech).

Figure 6.38. Crater Einstein (Credit—Alan Chu).

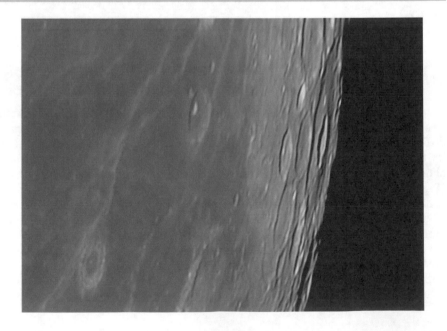

located 13 arcseconds from the primary. Watch it as often as possible in the next few weeks and see if you can detect any changes!

When the Moon rises well tonight, take a look at the northwestern limb about half the distance between Grimaldi and Sinus Iridum. Our search is for an "on the edge" crater known as Einstein (Figure 6.38). Use the prominent crater Kraft to help guide you to this extreme edge feature!

Friday, June 20

Although we will have Moon to contend with in the predawn hours, we welcome the "shooting stars" as we pass through another portion of the Ophiuchid meteor stream. The radiant for this pass will be nearer Sagittarius and the fall rate varies from 8 to 20, but it can sometimes produce unexpectedly more.

For variable star fans, let's return again to Corona Borealis and focus our attention on S—located just west of Theta—the westernmost star in the constellation's arc formation (Figure 6.39). At magnitude 5.3, this long-term variable takes almost a year to go through its changes. It usually far outshines the 7th magnitude star to its northeast—but will drop to a barely visible magnitude 14 at minimum. Compare it to the eclipsing binary U Coronae Borealis about a degree northwest. In slightly over three days this Algol-type star will range by a full magnitude as its companions draw together.

Figure 6.39. Theta Coronae Borealis (Credit—Palomar Observatory, courtesy of Caltech).

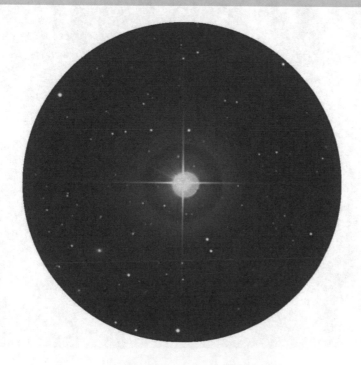

Saturday, June 21

Summer Solstice occurs today at the zero hour. So what exactly is it? Solstice is nothing more than an astronomical term for the moment when one hemisphere of the Earth is tilted the most toward the Sun (Figure 6.40). Today, the Sun is about 24 degrees above the celestial equator—its highest point of the year. The day of summer solstice also has the longest period of daylight...and the shortest of night, this occurs around six months from now for the Southern Hemisphere.

Tonight let's look forward to the coming summer as we hop a fingerwidth northeast of Beta Ophiuchi (RA 17 46 18 Dec +05 43 00) to a celebration in starlight known as IC 4665 (Figure 6.41). Very well suited to binoculars or even the smallest optics at low power, this magnificent open cluster is even visible to the unaided eye as a hazy patch.

Hanging out in space far from the galactic plane, IC 4665 is anywhere from 30 to 40 million years old—relatively young in astronomical terms! This places the cluster somewhere between the age of the Hyades and the Pleiades. At one time the cluster was believed to have been home to an unusually large number of spectroscopic binaries. While this has been disproved, scopists will enjoy powering up on the approximate 50 members of this association to search for true multiple stars. Enjoy it tonight!

Figure 6.40. Solstice and Equinox (Credit—NASA).

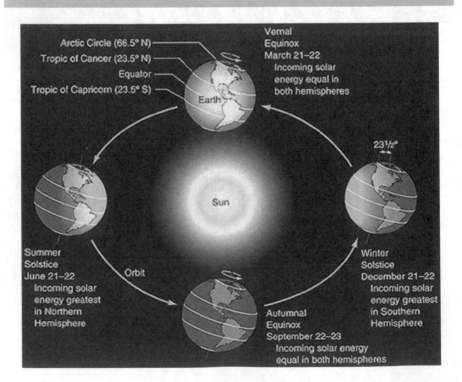

Figure 6.41. IC 4665 (Credit—Palomar Observatory, courtesy of Caltech).

Sunday, June 22

Today celebrates the founding of the Royal Greenwich Observatory in 1675. That's 333 years of astronomy! Also on this date in history, in 1978, James Christy of the US Naval Observatory in Flagstaff, Arizona, discovered Pluto's satellite Charon (Figure 6.42).

While observing Pluto is quite possible with a mid-sized (8″) telescope, careful work is needed to separate and identify it from field stars. Just a few days ago, Pluto reached opposition, meaning it is viewable all night. Since it will take several nights of observation for confirmation, right now would be an excellent time to begin your Pluto quest. With a little research you'll find plenty of online locator charts to help guide you on your way!

For observers of all skill levels, it's simply time to stop and have a look at a seasonal favorite which is now nearly overhead—M13 (Figure 6.43). You'll find this massive globular cluster quite easy to locate on the western side of the Hercules "keystone" about a third of the way between the northern and the southern stars—Eta and Zeta.

At a little brighter than magnitude 6, this 25,100 light-year distant globular cluster can be seen unaided from a dark-sky location. First noted by Edmond Halley in 1714, the "Great Hercules Cluster" was cataloged by Messier on June 1,

Figure 6.42. Pluto and Charon (Credit—NASA).

Figure 6.43. M13: The Great Hercules Cluster (Credit—R. Jay GaBany).

1764. Filled with hundreds of thousands of stars, yet with only one young blue star, M13 could be as much as 14 billion years old.

Thirty four years ago, the Great Hercules Cluster was chosen by the Arecibo Observatory as the target for the first radio message delivered into space, yet it will be a message which won't be received for over 25 centuries. Look at it with wonder tonight... For the light which left it as you are viewing tonight did so at a time when the Earth was coming out of the Ice Age. Our early ancestors were living in caves and learning to use rudimentary tools. How evolved would our civilization be if we ever received an answer to our call?!

Monday, June 23

As the sky darkens, let's discover the wonderful world of low power. Start by re-locating the magnificent M13 and then move about three degrees northwest. What you will find is a splendid loose open cluster of stars known as Dolidze/Dzimselejsvili (DoDz) 5—and it looks much like a miniature of the constellation of Hercules. Just slightly more than four degrees to its east and just about a degree south of Eta Hercules is DoDz 6, which contains a perfect diamond pattern and an asterism of brighter stars which resembles the constellation of Sagitta.

Figure 6.44. Abell 4065: The Corona Borealis Galaxy Cluster (Credit—Palomar Observatory, courtesy of Caltech).

Now we're going to move across the constellation of Hercules toward Lyra. East of the "keystone" you will see a tight configuration of three stars—Omicron, Nu, and Xi. At about the same distance which separates these stars, and to the northeast, you will find DoDz 9. Using minimal magnification, you'll see a pretty open cluster of around two-dozen mixed-magnitude stars that are quite attractive. Now look again at the "keystone" and identify Lambda and Delta to its south. About midway between them and slightly to the southeast you will discover the stellar field of DoDz 8. The last is easy—all you need to do is know the beautiful red and green double, Ras Algethi (Alpha). Move about one degree to the northwest to discover the star-studded open cluster DoDz 7. These great open clusters are very much off the beaten path and will add a new dimension to your large binocular or low power telescoping experiences.

While you're out tonight, take a look at the skies for a circlet of seven stars residing about halfway between orange Arcturus and brilliant blue–white Vega. This quiet constellation is named Corona Borealis, the "Northern Crown" (Figure 6.44). Just northwest of its brightest star is a huge concentration of over 400 galaxies located over a billion light-years from us. This area is so small, from our point of view, that we could cover it with a thumbnail held at arm's length!

Tuesday, June 24

On this day in 1881, Sir William Huggins made the first photographic spectrogram of a comet (1881 III), and discovered cyanogen (CN) emission at violet wavelengths (Figure 6.45). Unfortunately, his discovery caused public panic 29 years later when Earth passed through the tail of Halley's Comet. What a shame the public didn't realize cyanogens are also released organically! More than fearing what is in a comet's tail, they should have been thinking about what might happen should a comet strike the Earth.

Tonight Sir William Herschel was busy at the eyepiece. After 2 months of incredible galaxy research, the master himself had located an open cluster which would survive the years as one of the sky's greatest mysteries. Although it will be awhile before it reaches prime position, let's go have a look about three degrees west of Gamma Sagittarius at NGC 6451 (RA 17 50 42 Dec –30 13 00).

Cataloged on this night in 1784 as H VI.13 and part of the "400" list, sky-savvy observers may know this 8th magnitude stellar collection as the "Tom Thumb Cluster" (Figure 6.46). For a great many years, NGC 6451 was considered to be the second oldest galactic cluster, but Strömgren photometry changed all that. After careful research, scientists proved that not only did it contain a chemically peculiar star, but there was a wide discrepancy in interstellar reddening values as well. With the realization that it contains an A2 type, and that some of the readings were originally taken from a foreground star, its status was changed.

Regardless of the true age of and distance to this open cluster, I'm sure even smaller optics can appreciate this round, loose collection of stars set in the magnificent backdrop of one of our galaxy's spiral arms. For those brave observers who stayed out late to view? Take the time to drop about a half degree south and pick up a 33 Doubles challenge—Stone 37. This 1250 light-year distant

Figure 6.45. Sir William Huggins (widely used public image).

Figure 6.46. NGC 6451 (Credit—Palomar Observatory, courtesy of Caltech).

study is an easy split for most optics and the 6.7 magnitude primary white shows a nice color contrast with the 8.2 magnitude pale blue secondary.

Wednesday, June 25

Today celebrates the birth of Hermann Oberth—who has often been considered the father of modern rocketry (Figure 6.47). Born in Transylvania in 1894, Oberth was a visionary who was convinced space travel would one day be possible. Inspired by the works of Jules Verne, Oberth studied rockets and wrote many books devoted to the possibility of achieving spaceflight. He was the first to conceive of rocket "stages"—allowing vehicles to expend their fuel and lose dead weight.

Our object for tonight will be Herschel II.76—also known as NGC 5970. Begin by identifying Beta and Delta Serpentis Caput, and look for finderscope Chi between them. Less than a degree southwest you will see a similar magnitude double star. Hop about one-third degree northwest and you will find your galaxy mark just a fraction southwest of a 7th magnitude star (RA 15 38 30 Dec – 12 11 10).

Although nearly 11th magnitude, NGC 5970 is not particularly easy for smaller scopes because of its low surface brightness (Figure 6.48). But it could be a distant twin of our own galaxy, so similar is it to the Milky Way in structure. At 105 million light-years away, it's no great surprise we see it as faint—for its light left around the time the dinosaurs ruled the Earth. Stretching across 85,000 light-years of space, this grand spiral's nuclear region, obscuring dust regions, and stellar population have been extensively studied. And—like our own galaxy—it is also part of its own local group.

While smaller telescopes will make out a slightly elongated mist, in mid-to-large aperture NGC 5970 will appear oval-shaped with a bright core and evidence of a central bar. While the edges of the galaxy seem well defined, look closely at the narrower ends where material seems more wispy. While averted in this

Figure 6.47.
Hermann Oberth
(widely used public
image).

Figure 6.48. NGC 5970 (Credit—Palomar Observatory, courtesy of Caltech).

fashion, the nucleus will sometimes take on a stellar appearance—yet lose this property with direct vision. Be sure to mark your Herschel notes on this one!

Thursday, June 26

On this day in 1949, asteroid Icarus was discovered on a 48" Schmidt plate made nine months after the telescope went into operation, and just prior to the beginning of the multi-year National Geographic-Palomar Sky Survey. The asteroid was found to have a highly eccentric orbit and a perihelion distance of just 27 million kilometers, closer to the Sun than Mercury, giving it its unusual name. It was just 6.4 million kilometers from Earth at the time of discovery, and variations in its orbital parameters have been used to determine Mercury's mass and test Einstein's theory of general relativity.

But today is even more special because it is the birthday of none other than Charles Messier, the famed French comet hunter (Figure 6.49). Born in 1730, Messier is best known for cataloging the 100 or so bright nebulae and star clusters now referred to as the Messier objects. The catalog was intended to keep both Messier and others from confusing these stationary objects with possible new comets. Tonight get out your telescopes or binoculars and see how many Messier objects you can capture without using a star chart. Be sure to wish Charles a happy 278th birthday!

Figure 6.49. Charles Messier (widely used public image).

While northern observers chase the Messiers, it's up to the Southern Hemisphere to capture NGC 5128 (RA 13 25 28 Dec –43 01 09). Also known as radio source Centaurus A, this peculiar galaxy was discovered by James Dunlop in 1826 and cataloged as number 482 on his list (Figure 6.50).

Later noted by Arp as number 153, and shining at magnitude 7, NGC 5128 is definitely one of the most outstanding galaxies in the sky. Belonging to the M83

Figure 6.50. NGC 5128: Centaurus A (Credit—Greg Bradley).

group, it seems to be a cross between elliptical and spiral—with a huge dustlane you can't miss. It is possible NGC 5128 was a normal elliptical at one time, but during the last billion years or so it may have cannibalized a spiral. Keep a watch on this great galaxy, because a supernova erupted as recently as 1986. Mark your notes...because this one rates as a "best" on every observing list!

Friday, June 27

As with all astronomical projects, there are sometimes difficult ones needed to complete certain fields of study—such as challenging globular clusters. Tonight we'll take a look at one such cluster needed to complete your list and you'll find it by using M5 as a guide.

Palomar 5 is by no stretch of the imagination easy. For those using GoTo systems and large telescopes, aiming is easy...but for star hoppers a bit of instruction goes a long way. Starting at M5 drop south for the double star 5 Serpentis and again south and slightly west for another, fainter double. Don't confuse it with 6 Serpentis to the east. About half a degree west you'll encounter an 8th magnitude star with 7th magnitude 4 Serpentis a half degree south. Continue south another half degree where you will discover a triangle of 9th magnitude stars with a southern one at the apex. This is home to Palomar 5 (RA 15 16 05 Dec –00 06 41).

Figure 6.51. Palomar 5 (center of image) (Credit—Palomar Observatory, courtesy of Caltech).

Discovered by Walter Baade in 1950, this 11.7 magnitude, Class XII globular is anything but easy (Figure 6.51). At first it was believed to be a dwarf elliptical and possibly a member of our own Local Group of galaxies due to some resolution of its stars. Later studies showed Palomar 5 was indeed a globular cluster—but one in the process of being ripped apart by the tidal forces of the Milky Way.

Away from us by 75,000 light-years and 60,000 light-years from the galactic center, Palomar 5's members are escaping and leaving trails spanning as much as 13,000 light-years...a process which may have been ongoing for several billion years. Although it is of low surface brightness, even telescopes as small as 6" can distinguish just a few individual members northwest of the 9th magnitude marker star—but even telescopes as large as 31" fail to show much more than a faint sheen (under excellent conditions) with a handful of resolvable stars. Even though it may be one of the toughest you'll ever tackle, be sure to take the time to make a quick sketch of the region to complete your studies. Good luck!

While you're out, keep a watch for a handful of meteors originating near the constellation of Corvus. The Corvid meteor shower is not well documented, but you might spot as many as 10 per hour.

Saturday, June 28

Before you start hunting down the faint fuzzies and spend the rest of the night drooling on the Milky Way, let's go globular and hunt up two very nice studies worthy of your time. Starting at Alpha Librae, head five degrees southeast for

Figure 6.52. NGC 5897 (Credit—Palomar Observatory, courtesy of Caltech).

Tau, and yet another degree southeast for the splendid field of NGC 5897 (RA 15 17 24 Dec –21 00 36).

This class XI globular might appear very faint to binoculars, but it definitely makes up for it in size and beauty of field (Figure 6.52). It was first viewed by William Herschel on April 25, 1784 and logged as H VI.8—but with a less than perfect notation of position. When he reviewed it again on March 10, 1785 he logged it correctly and relabeled it as H VI.19. At a distance of a little more than 40,000 light-years, this 8.5 magnitude globular will show some details to the larger telescope, but remain unresolved to smaller ones. As a halo globular cluster, NGC 5897 certainly shows signs of being disrupted, and has a number of blue stragglers, as well as four newly discovered variables of the RR Lyrae type.

Now let's return to Alpha Librae and head about a fistwidth south across the border into Hydra and two degrees east of star 57 for NGC 5694—also in an attractive field (RA 14 39 36 Dec –26 32 18).

Also discovered by Herschel, and cataloged as H II.196, this class VII cluster is far too faint for binoculars at magnitude 10, and barely within reach of smaller scopes (Figure 6.53). As one of the most remote globular clusters in our galaxy, few telescopes can hope to resolve this more than 113,000 light-year distant ball of stars. Its brightest member is only of magnitude 16.5, and it contains no known variables. Traveling at 190 kilometers per second, metal-poor NGC 5694 will not have the same fate as NGC 5897...for this is a globular cluster which is not being pulled apart by our galaxy—but escaping it!

Figure 6.53. NGC 5694 (Credit—Palomar Observatory, courtesy of Caltech).

Sunday, June 29

Today we celebrate the birthday of George Ellery Hale, who was born in 1868 (Figure 6.54). Hale was the founding father of the Mt. Wilson Observatory. Although he had no education beyond his baccalaureate in physics, he became the leading astronomer of his day. He invented the spectroheliograph, coined the word astrophysics, and founded the *Astrophysical Journal* and Yerkes Observatory. At the time, Mt. Wilson dominated the world of astronomy, confirming what galaxies were and verifying the expanding universe cosmology, making Mt. Wilson one of the most productive facilities ever built. When Hale went on to found Palomar Observatory, the 5 meter (200") telescope was named for him, and was dedicated on June 3, 1948. It continues to be the largest telescope in the continental United States.

Tonight, while we have plenty of dark skies to go around, let's go south in Libra and have a look at the galaxy pairing NGC 5903 and NGC 5898 (Figure 6.55). You'll find them about three degrees northeast of Sigma, and just north of a pair of 7th magnitude stars.

While northernmost NGC 5903 seems to be nothing more than a faint elliptical with a brighter concentration toward the center and an almost identical elliptical—NGC 5898—to the southwest, you're probably asking yourself... Why the big deal over two small ellipticals? First off, NGC 5903 is Herschel III.139 and NGC 5898 is Herschel III.138...two more to add to your studies. And second? The Very Large Array has studied this galaxy pair in the spectral lines of neutral hydrogen. The brighter of the pair, NGC 5898, shows evidence of ionized gas which has been collected from outside its galactic realm—while NGC 5903 seems

Figure 6.54. George Ellery Hale (widely used public image).

Figure 6.55. Field of NGC 5903 and NGC 5898 (Credit—Palomar Observatory, courtesy of Caltech).

to be running streamers of material toward its neighbor. A double-galaxy, double-accretion event!

But there's more...

Look to the southeast and you'll double your pleasure and double your fun as you discover two double stars instead of just one! Sometimes we overlook field stars for reasons of study—but don't do it tonight. Even mid-sized telescopes can easily reveal this twin pair of galaxies sharing "their stuff," as well as a pair of double stars in the same low-power field of view. (Psst...slim and dim MCG 043607 and quasar 1514-241 are also here!) Ain't it grand?

Monday, June 30

Out of the blue comes a meteor shower! Keep watch tonight for the June Draconids. The radiant for this shower will be near handle of Big Dipper—Ursa Major. The fall rate varies from 10 to 100 per hour, and lack of lunacy means a great time for the offspring of comet Pons-Winnecke. On a curious note, today in 1908 was when the great Tunguska impact happened in Siberia. A fragment of a comet, perhaps?

Figure 6.56. Hickson Compact Group 68 (Credit—Palomar Observatory, courtesy of Caltech).

Galaxy lovers, are you ready to dance? Then tonight follow me as we journey about a fistwidth east of Cor Caroli (RA 13 53 40 Dec +40 19 07) and make a 100 million light-year jump to look at Hickson Compact Group 68 (Figure 6.56).

Dominated by the 11th magnitude, face-on spiral NGC 5350, this group of interacting galaxies spans an area of 300,000 light-years. Before the intermediate scope users think they are out of the running, think again. Group members such and NGCs 5355, 5358, 5353, and 5354 are within reach of an 8" scope on a dark night with good clarity. Very large scopes may even be able to catch a glimpse of a few of the 45 dwarfs which also are part of this group. It certainly isn't the easiest of targets... But it's one of the most fun!

CHAPTER SEVEN

July, 2008

On this date in 2004, the Cassini mission (a joint effort of NASA/ESA/ASI) became the first spacecraft to orbit Saturn (Figure 7.1). It contained the Huygens probe which landed safely on the surface of Saturn's moon Titan on January 15, 2005. This hugely successful mission went on to discover more about the ringed planet and its satellites than we had ever dreamed possible!

And today in 1917, astronomers at Mt. Wilson were celebrating as the 100" primary mirror arrived (Figure 7.2). Up until that time, the 60" Hale telescope (donated by George Hale's father) was the premier creation of St. Gobrain Glassworks—which was later commissioned to create the blank for the Hooker telescope. Thanks to the funds provided by John D. Hooker (and Carnegie), the dream was realized after years of hard work and ingenuity—to create not only a proper building to house it but the telescope workings as well. It saw "first light" five months later on November 1. As anxious astronomers waited for this groundbreaking moment, the scope was aimed at Jupiter... But the image was horrible! To their dismay, workmen had left the dome open and the Sun had heated the massive mirror! Try as they might to rest until it had cooled, no astronomer slept. Fearful of the worst, sometime around 3:00 in the morning, they returned again long after Jupiter had set. Pointing the massive scope toward a star, this time they achieved a perfect image!

Start your stargazing evening off right by having a look at the western skyline at twilight. Mars and Regulus are waltzing together less than a degree apart.

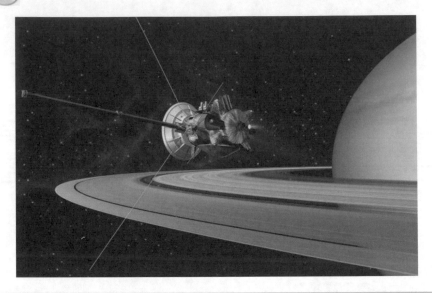

Figure 7.1. Concept of Cassini orbit insertion (Credit—NASA).

Not only is this an eye-appealing conjunction to the unaided observer, but a wonderful opportunity to see both a planet and a star in the same low-power telescope field!

While you're out, take the time to look at lowly Theta Lupi about a fistwidth south-southwest of mighty Antares (Figure 7.3). While this rather ordinary

Figure 7.2. The 100 inch mirror (Credit—Science Museum/Science and Society Library (archival image)

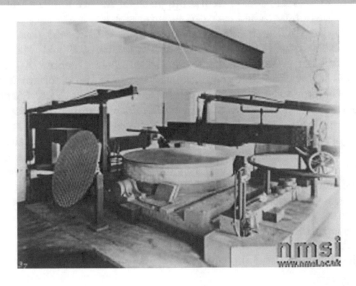

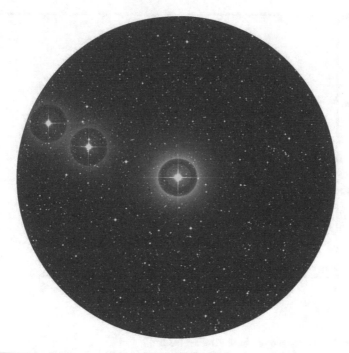

Figure 7.3. Theta Lupi (Credit—Palomar Observatory, courtesy of Caltech).

looking 4th magnitude star appears to be nothing special, there's a lesson to be learned here. So often in our quest to look at the bright and incredible— the distant and impressive—we often forget about the beauty of a single star. When you take the time to seek the path less traveled, you just might find more than you expected. Hiding behind a veil of "the ordinary" are a trio of stars having differing spectral types and magnitudes, all in a diamond-dust field. An underappreciated gem...

Wednesday, July 2

Orbiting in space 41 years ago, the Vela 3 and 4 satellites were quietly keeping watch on Earth's nuclear test ban treaty. The Vela satellites were very successful and far exceeded their life expectancy. Just before the launch of the fifth Vela, scientists were checking the data from July 2, 1967, and found an unusual event recorded by Vela 4—an event strong enough to have also triggered a response from the Vela 3 satellite. While the Velas' pointing accuracy was insufficient to pinpoint the source, the scientists eventually realized they had caught the first known gamma ray burst (Figures 7.4 and 7.5)!

While little is known about these mysterious sources of energy, we do know they occur about once a day with the photon energy of 100 million electric volts.

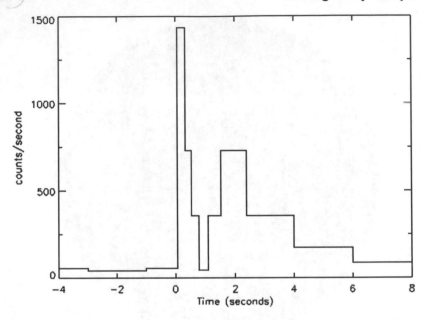

Figure 7.4. First gamma ray burst (Credit—NASA).

Figure 7.5. Gamma ray sky (Credit—NASA).

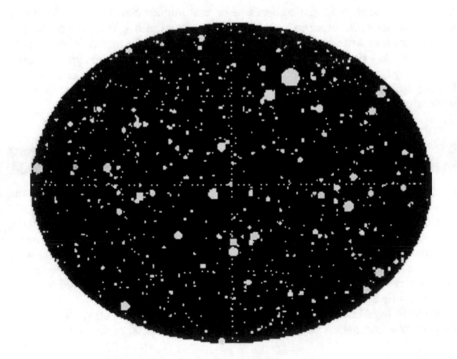

Figure 7.6. IC 4406 (Credit—Palomar Observatory, courtesy of Caltech).

While some of the energy comes from our own Milky Way, many other far more distant sources have also been found—and over 800 have been charted! One such generator of gamma rays is a special type of star known as a Wolf-Rayet—a hot, huge star which is undergoing significant mass loss and exposing its central core.

Tonight for more southern viewers, take the time to look up one such incredible system, IC 4406 (Figure 7.6). You'll find it about five degrees northwest of Alpha Lupi, or just about a fingerwidth northwest of the Tau collection (RA 14 22 26 Dec −44 09 04). This roughly 10th magnitude planetary nebula is sometimes referred to as the "Retina Nebula" for its photographic resemblance to the human retina. This square-appearing patch is a Wolf-Rayet nebula, and color photographs show evidence of gamma rays as green sparkles!

Thursday, July 3

If you're up before dawn, this morning would be a great opportunity to spot the swift inner planet Mercury. It reached its greatest elongation about 48 hours ago.

Tonight is New Moon and what better time to look for some alternate catalog objects? Let's start by Herschel hunting while we continue on our globular cluster studies. Our first stop is to return to brilliant Antares and head a half degree

Figure 7.7. NGC 6144 (Credit—Palomar Observatory, courtesy of Caltech).

northwest for the Bennett list cluster NGC 6144 (RA 16 27 14 Dec –26 01 29) (Figure 7.7).

Originally discovered by Herschel in 1784 and labeled as H VI.10, this 9th magnitude Class II globular is around 8500 light-years away from the galactic core. While it is only about one-third the size of M4, it is also three times more distant from our solar system. If you have trouble spotting it, try high magnification to keep Antares' glare at bay. Situated in the Rho Ophiuchi dust cloud, NGC 6144 has at least one slow variable of the RR Lyrae type.

Now drop a little more than a fistwidth south of Antares for NGC 6139 (RA 16 27 40 Dec –38 50 55) (Figure 7.8). Discovered by James Dunlop in 1820 and cataloged as Dun 536, this 9th magnitude Class II globular is much further from the galactic center at a distance of 11,700 light-years. Within the gravitational pull of this low metallicity cluster, at least s RR Lyrae-type variables still cling to their host.

Now that we've seen two very concentrated globular clusters, let's look at one that isn't even classed. Drop a fingerwidth south of Lambda Scorpii for NGC 6380 (RA 17 34 28 Dec –39 04 09) (Figure 7.9). This 11th magnitude globular is a challenge! Discovered by John Herschel and listed as h 3688, this one is also known as Tonantzintla 1—or Ton 1. It's so vague that it wasn't even classed as a globular cluster until re-examined with a photographic plate. It's very metal rich and contains red giants at its bulging heart... What's left of it!

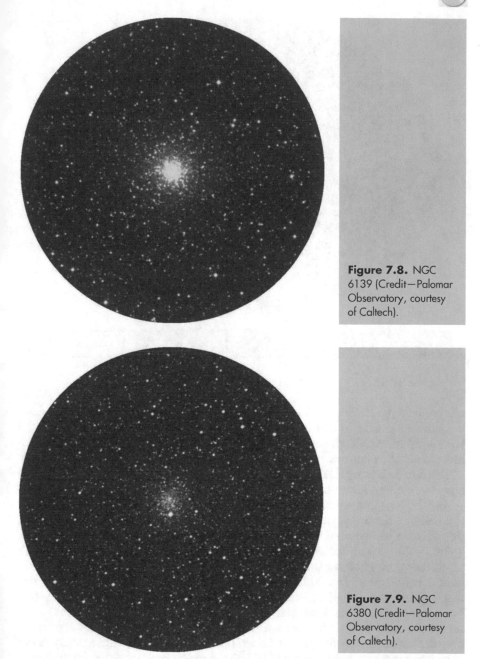

Figure 7.8. NGC 6139 (Credit—Palomar Observatory, courtesy of Caltech).

Figure 7.9. NGC 6380 (Credit—Palomar Observatory, courtesy of Caltech).

Friday, July 4

On this date in 2005, the Deep Impact mission entered the history books as its probe impacted Comet Tempel 1 successfully (Figure 7.10). The spacecraft

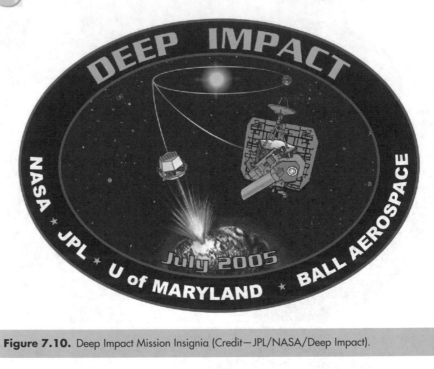

Figure 7.10. Deep Impact Mission Insignia (Credit—JPL/NASA/Deep Impact).

relayed back to Earth a wealth of information about the material released from the surface. Thanks to this incredible mission (a collaboration between JPL, the University of Maryland, and Ball Aerospace Technologies), we've learned much more about the nature of comets and the protosolar nebula in which they formed.

This date in history also marks the 1947 founding of the Astronomical League—a worldwide organization with almost 15,000 members.

And did you know that celestial fireworks occurred in 1054, also on this day? It is believed the bright supernova recorded by Chinese astronomers happened at this point in history, and today we know its remnants as the Crab Nebula (M1).

But could such an event happen again in our own celestial "backyard?" Look no further than HR 8210 (RA 21 26 26 Dec +19 22 32). It may be nothing more than a white dwarf star hiding out in late night Capricornus, but it's a star that's almost run out of fuel (Figure 7.11). This rather ordinary binary system has a companion white dwarf star that's 1.15 times the mass of our Sun. As the companion also expends its fuel, it will add mass to HR 8210 and push it over the Chandrasekhar limit—the point of no return in mass. This will someday result in a supernova event located only 150 light-years away from our solar system...

And that's 50 light-years too close for comfort!

In the Gould Belt, 470 light-years away and roughly 1.5 million years ago, a similarly massive star exploded in the Upper Scorpius association. No longer able to fuel its mass, it unleashed a supernova event which left its evidence as a layer of iron here on Earth, and may have caused a certain amount of biological extinction when its gamma rays directly affected our ozone layer.

Figure 7.11. HR 8210 (Credit—Palomar Observatory, courtesy of Caltech).

Take a long look at Antares tonight—for it is part of that association of stars and is no doubt also a star poised on the edge of extinction. At a safe distance of 500 light-years, you'll find this pulsating red variable equally fascinating to the eye as well as to the telescope. Unlike HD 8210, Alpha Scorpii also has a companion which can be revealed to small telescopes under steady conditions. Discovered on April 13, 1819 during a lunar occultation, this 6.5 magnitude green companion isn't the easiest to split from such a bright primary—but it's certainly fun to try! And the best is yet to come, because Antares will be occulted again in a matter of days...

Saturday, July 5

Tonight the Moon has returned in a position to favor a bit of study. Start by checking IOTA information for a possible visible occultation of Regulus, and also look for Saturn quite nearby as the slender crescent graces the early evening skies.

Although poor position makes study difficult during the first few lunar days, be sure to look for the ancient impact crater Vendelinus just slightly south of central (Figure 7.12). Spanning approximately 150 kilometers in diameter and with walls reaching up to 4400 meters in height, lava flow has long ago eradicated

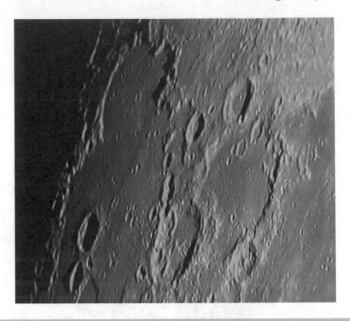

Figure 7.12. Vendelinus (Credit—Alan Chu).

the interior features Its old walls give mute testimony to later impact events, which you can see when viewing crater Holden on the south shore, much larger Lame on the northeast edge, and sharp Lohse northwest. Mark your challenge list!

For all observers, let's take a closer look at the fascinating constellation of Lupus southwest of brilliant Antares. While more northern latitudes will see roughly half of this constellation, it sits well at this time of year for those in the south. So why bother?

Cutting through our Milky Way galaxy at a rough angle of about 18 degrees is a disc-shaped zone called Gould's Belt. Lupus is part of this area whose perimeter contains star-forming regions which came to life about 30 million years ago when a huge molecular cloud of dust and gas was compressed—much like in the Orion area. In Lupus we find Gould's Belt extending above the plane of the Milky Way!

Return again to the beautiful Theta and head around five degrees west for NGC 5986 (RA 15 46 03 Dec –37 47 10), a 7th magnitude globular cluster which can be spotted with binoculars with good conditions (Figure 7.13). While this Class VII cluster is not particularly dense, many of its individual stars can be resolved in a small telescope.

Now sweep the area north of NGC 5986 (RA 17 57 06 Dec –37 05 00) and tell me what you see. That's right! Nothing. This is dark nebula B 288—a cloud of dark, obscuring dust which blocks incoming starlight (Figure 7.14). Look carefully at the stars you *can* see and you'll notice they appear quite red. Thanks to B 288, much of their emitted light is absorbed by this region, providing us with a pretty incredible on-the-edge view of something you can't see—a Barnard dark nebula.

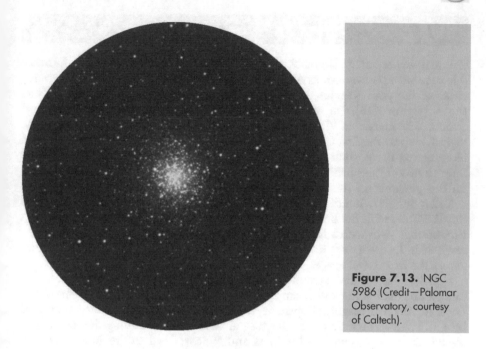

Figure 7.13. NGC 5986 (Credit—Palomar Observatory, courtesy of Caltech).

Figure 7.14. Region of B 288 (Credit—Palomar Observatory, courtesy of Caltech).

Sunday, July 6

Celestial scenery alert! Skywatchers... Mark your calendar and be sure to make this date with the western skyline as sunset marks one of the most picturesque views of the year! Regulus, Mars, and Saturn will all dance with the da Vinci Moon (Figure 7.15). No special equipment is needed to see this event, and thanks to Leonardo da Vinci we can see the ghostly effect on the Moon as quite logical. He was the first to theorize that sunlight was reflecting off the Earth and illuminating the portion of the Moon not lit by the Sun. We more commonly refer to this as "Earthshine"—but no matter how scientific the explanations are for this phenomena, its appearance remains beautiful.

Today in 1687, Isaac Newton's monumental *Principia* was published by the Royal Society with the help of Edmund Halley. Although Newton was indeed a very strange man with a highly checkered history, one of the keys to Newton's work with the theory of gravity was the idea that one body could attract another across the expanse of space.

Now let's have a look at some things gravitationally bound as we start at Eta Lupi, a fine double star which can even be resolved with binoculars (Figure 7.16). Look for the 3rd magnitude primary and 8th magnitude secondary separated by a wide 15". You'll find it by starting at Antares and heading due south two binocular fields to center on bright H and N Scorpii—then one binocular field southwest (RA 16 00 07 Dec –38 23 48).

When you are done, hop another roughly five degrees southeast (RA 16 25 18 Dec –40 39 00) to encounter the fine open cluster NGC 6124 (Figure 7.17). Discovered by Lacaille and known to him as object I.8, this 5th magnitude open

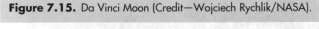

Figure 7.15. Da Vinci Moon (Credit—Wojciech Rychlik/NASA).

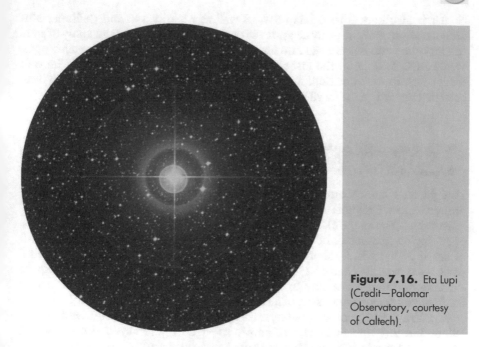

Figure 7.16. Eta Lupi (Credit—Palomar Observatory, courtesy of Caltech).

Figure 7.17. NGC 6124 (Credit—Palomar Observatory, courtesy of Caltech).

cluster is also known as Dunlop 514, as well as Melotte 145 and Collinder 301. Situated about 19 light-years away, it will show as a fine, round, faint spray of stars to binoculars and be resolved into about 100 stellar members to larger telescopes. While NGC 6124 is on the low side for northern observers, it's worth the wait for it to hit its best position. Be sure to mark your notes, because this delightful galactic cluster is a Caldwell object and a southern skies binocular reward!

Monday, July 7

Tonight let's launch our imaginations toward the Moon as we view the area around Mare Crisium and have a look at this month's lunar challenge—Macrobius (Figure 7.18). You'll find it just northwest of the Crisium shore. Spanning 64 kilometers in diameter, this Class I impact crater drops to a depth of nearly 3600 meters—about the same as many of our earthly mines. Its central peak rises back up, and at 1100 meters may be visible as a small speck inside the crater's interior. Be sure to mark you lunar challenges and look for other features you may have missed before!

If you're up to another challenge tonight, wait until the Moon westers and let's go hunting Herschel I.44, also known as NGC 6104 (Figure 7.19). You'll find this 9.5 magnitude globular cluster around two fingerwidths northeast of Theta Ophiuchi and a little more than a degree due east of star 51 (RA 17 38 36 Dec –23 54 31).

Figure 7.18. Macrobius on the edge of Crisium (Credit—Greg Konkel).

Figure 7.19. NGC 6401 (Credit—Palomar Observatory, courtesy of Caltech).

Discovered by William Herschel in 1784 and often classed as "uncertain," today's powerful telescopes have placed this halo object as a Class VIII and given it a rough distance from the galactic center of 8800 light-years. Although neither William nor John could resolve this globular, and listed it originally as a bright nebula, studies in 1977 revealed a nearby suspected planetary nebula named Peterson 1. Thirteen years later, further study revealed this to be a symbiotic star.

Symbiotic stars are a true rarity—not a single star at all—but a binary system. A red giant dumps mass toward a white dwarf from an accretion disc. When the dwarf reaches critical mass, it causes a thermonuclear explosion resulting in a planetary nebula. While no evidence exists that this symbiotic pair is actually located within metal-rich NGC 6401, just being able to see it in the same field makes this journey both unique and exciting!

Tuesday, July 8

With plenty of Moon to explore tonight, why don't we try locating an area where many lunar missions made their mark? Binoculars will easily reveal the fully disclosed areas of Mare Serenitatis and Mare Tranquillitatis, and it is where these two vast lava plains converge that we will set our sights. Telescopically you will

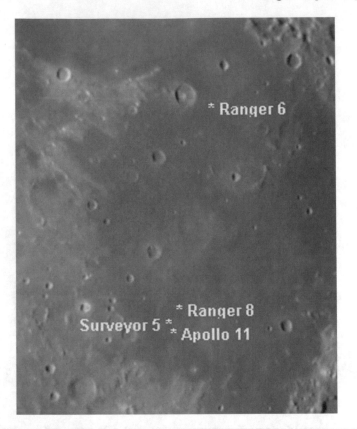

Figure 7.20. Moon shot (Credit—Greg Konkel, annotation by Tammy Plotner).

see a bright "peninsula" westward from where the two conjoin, which extends toward the east: just off that look for bright and small crater Pliny. It is near this rather inconspicuous feature that the remains of Ranger 6 lie forever preserved. It crashed there on February 2, 1964.

Unfortunately, technical errors occurred and Ranger 6 was never able to transmit lunar pictures. Not so Ranger 8! On a very successful mission to the same relative area, this time we received 7137 "postcards from the Moon" in the last 23 minutes before hard landing. On the "softer side," Surveyor 5 safely touched down near this area after two days of malfunctions on September 10, 1967. Incredibly enough, the tiny Surveyor 5 endured temperatures of up to 139°C, but was able to spectrographically analyze the area's soil... And by the way, it also managed to televise an incredible 18,006 frames of "home movies" from its distant lunar locale (Figures 7.20 and 7.21).

Since the moonlight will now begin to interfere with our globular cluster studies, let's waive these for awhile as we take a look at some of the region's most beautiful stars. Tonight your goal is to locate Omicron Ophiuchi (Figure 7.22), about a fingerwidth northeast of Theta (RA 17 18 00 Dec −24 17 12). At a distance

Figure 7.21. Surveyor 5: A foot on the Moon (Credit—NASA).

Figure 7.22. Omicron Ophiuchi (Credit—Palomar Observatory, courtesy of Caltech).

of 360 light-years, this system is easily split by even small telescopes. The primary star is slightly dimmer than magnitude 5 and appears yellow to the eye. The secondary is near 7th magnitude and tends to be more orange in color. This wonderful star is part of many doubles observing lists, so be sure to note it!

Wednesday, July 9

Tonight let's celebrate almost 39 years of space exploration as we walk on the Moon where the first man set foot. For skywatchers, the dark round area you see on the north eastern limb is Mare Crisium and the dark area below that is Mare Fecunditatis. Now look mid-way on the terminator for the dark area that is Mare Tranquillitatis. At its southwest edge, history was made.

In binoculars, trace along the terminator where the Caucasus Mountains stand—and south for the Apennine and Haemus Mountains. As you continue toward the center of the Moon, you will see where the shore of Mare Serenitatis curves east and note the bright ring of Pliny. Continue south along the terminator until you spot the small, bright ring of Dionysius along the edge of Mare Tranquillitatis. Just to the southwest, you may be able to see the soft-spoken rings of Sabine and Ritter. It is near here where the base section of the Apollo 11 landing module—Eagle—lies forever enshrined in "magnificent desolation"...

For telescope users, the time is now to power up! See if you can spot the small craters Armstrong, Aldrin, and Collins just east. But if you cannot, the Apollo

Figure 7.23. Image of Apollo 11 landing area (Credit—Greg Konkel).

Figure 7.24. Jupiter (Credit—Wes Higgins).

11 landing area is about the same distance as Sabine and Ritter are wide to the east-southeast (Figure 7.23). Even if you don't have the opportunity to see it tonight, take the time during the next couple of days to point it out to your children, grandchildren, or even just a friend... The Moon is a spectacular world and we've been there!

Also on this day in 1979, Voyager 2 quietly made its closest approach to Jupiter (Figure 7.24). How about if we take a close approach as well? Enjoy the waltz of the Galileans and all the fine details on the planet's disk!

Thursday, July 10

It's a stargazer's evening! Long before the Sun sets, look for the Moon to appear in the still-blue sky. As it darkens, watch for brilliant blue–white Spica to appear about two fingerwidths north of the Moon. Now look at the western skyline. Both Mars and Saturn are about a half a degree apart! Just for fun, add bright Regulus nearby and you have a triple treat that any observer will enjoy. Wishing you clear skies!

Have you ever wondered if there were any place on the lunar surface that hasn't seen the sunlight? Then we'll go exploring for one tonight...

Our first order of business will be to identify crater Curtius (Figure 7.25). Directly in the center of the Moon is a dark-floored area known as the Sinus Medii. South of it will be two conspicuously large craters—Hipparchus to the north and ancient Albategnius to the south. Trace along the terminator toward the south until you have almost reached its point (cusp) and you will see a black oval. This normal looking crater with the brilliant west wall is equally ancient crater Curtius. Because of its high southern latitude, we shall never see the interior of this crater—and neither has the Sun! It is believed the inner walls are quite steep, and so crater Curtius' interior has never been illuminated since its formation billions of years ago. Because it has remained dark, we can speculate there may be "lunar ice" (water ice possibly mixed with regolith) pocketed inside its many cracks and rilles which date back to the Moon's formation!

Because our Moon has no atmosphere, the entire surface is exposed to the vacuum of space. When sunlit, the surface reaches up to 385 Kelvin, so any exposed lunar ice would vaporize and be lost because the Moon's gravity could not hold it. The only way for ice to exist would be in a permanently shadowed area. Near Curtius is the Moon's south pole, and imaging from the Clementine spacecraft showed around 15,000 square kilometers of area where such conditions could exist. So where did this ice come from? The lunar surface never ceases to be pelted by meteorites—most of which contain water-related ice. As we know, many craters were formed by just such impacts. Once hidden from the sunlight, this ice could continue to exist for millions of years!

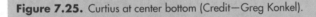

Figure 7.25. Curtius at center bottom (Credit—Greg Konkel).

Friday, July 11

Tonight let's continue our look at the lunar poles by returning to previous study crater Plato. North of Plato you will see a long horizontal area of gray floor—Mare Frigoris—the "Cold Sea." North of it you will note a "double crater." This elongated diamond-shape is Goldschmidt, and the crater which cuts across its western border is Anaxagoras. The lunar north pole isn't far from Goldschmidt, and since Anaxagoras is just about one degree outside of the Moon's theoretical "arctic" area, the lunar sunrise will never go high enough to clear the south-ernmost rim. As proposed with yesterday's study, this "permanent darkness" must mean there is ice! For that very reason, NASA's Lunar Prospector probe was sent to explore here. Did it find what it was looking for? Answer—Yes!

The probe discovered vast quantities of cometary ice which has hidden inside the crater's depths untouched for millions of years. If this sounds rather boring to you, then realize this type of resource may aid our plans to eventually establish a manned base on the lunar surface!

On March 5, 1998 NASA announced that Lunar Prospector's neutron spectrometer data showed water-based ice was discovered at both lunar poles. The first results showed the "ice" mixed in with lunar regolith (soil, rocks, and dust), but long-term data confirmed nearly pure pockets hidden beneath about 40 centimeters of surface material—with the results being strongest in the northern polar region (Figure 7.26). It is estimated there may be as much as 6 trillion

Figure 7.26. Lunar North Pole (Credit—NASA).

kilograms (6.6 billion tons) of this valuable resource! If this still doesn't get your motor running, then realize that without it we could never establish a manned lunar base because of the tremendous expense involved in transporting our most basic human need—water.

The presence of lunar water could also mean a source of oxygen, another vital material we need to survive. In order to return home or voyage onward, these same deposits could provide hydrogen which could be used as rocket fuel. So as you view Anaxagoras tonight, realize you may be viewing one of mankind's future "homes" on a distant world!

Saturday, July 12

Tonight let's take an entirely different view of the Moon as we do a little "mountain climbing!" The most outstanding feature on the visible surface will be the emerging Copernicus, but since we've delved into the deepest areas of the lunar surface, why not climb to some of its peaks?

Using Copernicus as our guide, to the north and northwest of this ancient crater lie the Carpathian Mountains ringing the southern edge of Mare Imbrium. As you can see, they begin well east of the terminator, but look into the shadow! Extending some 40 kilometers beyond the line of daylight, you will continue to see bright peaks—some of which reach a height of 2072 meters. When the area is

Figure 7.27. Lunar mountain peaks (Credit—Greg Konkel, annotation by Tammy Plotner).

fully revealed tomorrow, you will see the Carpathian Mountains disappear into the lava flow that once formed them.

Continuing onward to Plato, which sits on the northern shore of Imbrium, we will look for the disjointed line of (1) Montes Recta—the "Straight Range." Further east you will find the scattered peaks of (2) the Teneriffe Mountains. It is possible these are the remnants of much taller summits of a once stronger range, but only around 1890 meters of them still survive above the surface (Figure 7.27).

To the southeast (3) Mons Pico stands like a monument 2400 meters above the gray sands—a height which places its level with Kindersley Summit at Kootenay Park in British Columbia. Further southeast is the peak of (6) Mons Piton—also standing alone in the barren landscape of Imbrium. Perhaps once a member of the (5) Montes Alpes to the east, Piton still towers 2450 meters above the surface with a base 25 kilometers in diameter still remaining in the lava flow. Yet look closely at the lunar Alpes for (4) Mons Blanc is 3600 meters high!

Just north of shallow Archimedes stand (7) the Montes Spitzbergen whose remaining expanse trails away for 60 kilometers on the southern edge of a rille which begins at the small punctuation of crater Kirch to the north. While they only extend 1500 meters above the surface, that's still comparable with the outer Himalayans!

Sunday, July 13

Today is Friday the 13th. If you're not superstitious, but only having bad luck in finding lunar features, then how about if we take a look at one that's incredibly easy to find? We'll continue our lunar mountain climbing expedition and look at the "big picture" on the lunar surface.

Tonight all of Mare Imbrium is bathed in sunlight and we can truly see its shape (Figure 7.28). Appearing as a featureless ellipse bordered by mountain ranges, let's identify them again. Starting at Plato and moving east to south to west you will find the Alps, the Caucasus, and the Apennines (where Apollo 15 landed) at the western edge of Palus Putredinus. Next come the Carpathian Mountains just north of Copernicus. Look at the form closely: doesn't it appear that perhaps once upon a time an enormous impact created the entire area? This was the Imbrium impact: compare it to the younger Sinus Iridum. Ringed by the Juras Mountains, it may have also been formed by a much later and very similar impact.

And you thought they were just mountains...

Now let's have a look with our eyes first at Delta Ophiuchi (Figure 7.29). Known as Yed Prior ("the hand"), look for its optical double Epsilon to the southeast, symmetrically named Yed Posterior. Try using binoculars or a telescope at absolute minimum power for another undiscovered gem...

Delta Ophiuchi is 170 light-years from us, while Epsilon is 108—but look at the magnificent field they share. Stars of every spectral type are in an area of sky which could easily be covered by a small coin held at arm's length. Enjoy this fantastic field—from the hot, blue youngsters to the old red giants!

Figure 7.28. Gibbous Moon (Credit—Roger Warner).

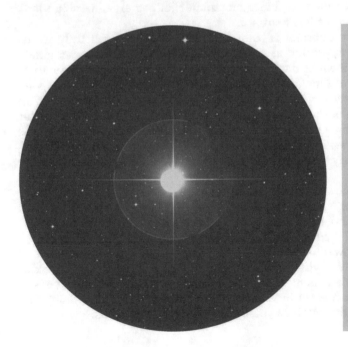

Figure 7.29. Delta Ophiuchi (Credit—Palomar Observatory, courtesy of Caltech).

Monday, July 14

Today in 1965, Mariner 4 became the first spacecraft to perform a flyby of the red planet Mars. Tonight our Moon is going to perform a flyby of its own on the one star known as "Not Mars"—Antares! For almost all observers, the red giant will be less than half a degree away from the lunar limb, but for some lucky few this could be a grand occultation (Figure 7.30)! Be sure to check sources like IOTA for precise times and locations.

Tonight on the lunar surface, let's head north of Sinus Iridum, across Mare Frigoris and northeast of the punctuation of Harpalus for a grand old crater—J. Herschel (Figure 7.31). Although it looks small because it is seen on the curve, this wonderful old-walled plain named for John Herschel contains some very tiny details. Its southeastern rim forms the edge of Mare Frigoris and the small (24 kilometers) Horrebow dots its southwest edge. The crater walls are so eroded with time that not much remains of the original structure. Look for many very small impact craters which dot J. Herschel's uneven basin and exterior edges. Power up! If you can spot the small central crater C, you are resolving a feature only 12 kilometers wide from some 385,000 kilometers away!

While we're out, let's have a look at another astounding system called 36 Ophiuchi located about a thumb's width southeast of Theta (RA 17 15 20 Dec –26 36 10). Situated in space less than 20 light-years from Earth, even small telescopes can split this pair of 5th magnitude K-type giants very similar to our own Sun, and larger telescopes can also pick up the C component as well.

Figure 7.30. Moon occulting Antares on March 3, 2005 (Credit—Tammy Plotner).

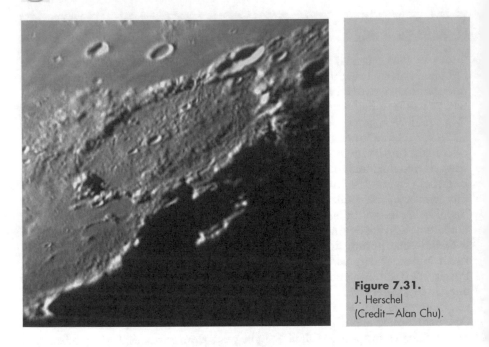

Figure 7.31.
J. Herschel
(Credit—Alan Chu).

Figure 7.32. 36 Ophiuchi (Credit—Palomar Observatory, courtesy of Caltech).

Be sure to mark your lists with both of your observations tonight, because J. Herschel is a Lunar Club challenge and 36 Ophiuchi is on many doubles challenge lists (Figure 7.32).

Tuesday, July 15

Tonight let's start our lunar observations with a subject that's a bit less obvious—crater contrasts. The Oceanus Procellarum is a vast, gray "sea" that encompasses most of the northwestern portion of the Moon. On the terminator to its southwest edge (and almost due west geographically), you will see two craters of near identical size and depth, but not identical lighting (Figure 7.33).

The southernmost is the 45 kilometer wide Billy—one of the darkest-floored areas on the Moon. Named for French mathematician and astronomer Jacques de Billy, it will appear to have a bright ring (the crater rim) around it, but the interior is as featureless as a mare! To the north is 45 kilometer wide Hansteen—note how much brighter and more detailed it is. This far more featured area was named for Dutch geophysicist Christopher Hansteen, and if you power up you'll see the 30 kilometer wide base of Mons Hansteen between the two craters, as well as a 25 kilometer long rima to the west. It's easy to discern that Billy was once filled with smooth lava, while its counterpart Hansteen evolved very differently! Be sure to mark your notes on this lunar challenge.

Figure 7.33. Hansteen and Billy (Credit—Alan Chu).

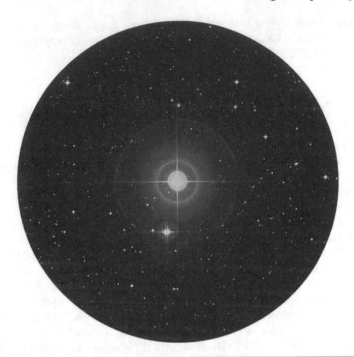

Figure 7.34. Zeta Scorpii 1 (Credit—Palomar Observatory, courtesy of Caltech).

Now that we've looked at contrasting craters, let's have a look at a beautifully contrasting pair of stars—Zeta 1 and 2 Scorpii (Figure 7.34). You'll find them a little less than a handspan south-southeast of Antares and at the western corner of the J of the constellation's shape (RA 16 04 22 Dec –11 22 24).

Although the two Zetas aren't a true physical pair, they are nonetheless interesting. The easternmost, orange sub-giant Zeta 2, appears far brighter for a reason... It's much closer to us at 155 light-years away. But focus your attention on western Zeta 1. It's a blue supergiant that's around 5700 light-years away and shines with the light of 100,000 suns, exceeding even Rigel in sheer power! The colorful pair is easily visible as two separate stars to the unaided eye, but a real delight in binoculars or low-power telescope field. Check them out tonight!

Wednesday, July 16

Today in 1850 at Harvard University, the first photograph of a star (other than the Sun) was made. The honor went to Vega! In 1994, an impact event was about to happen as nearly two dozen fragments of Comet Shoemaker-Levy 9 were speeding their way to the surface of Jupiter (Figure 7.35). The result

Figure 7.35. Shoemaker-Levy 9 impact on Jupiter (Credit—NASA).

was spectacular and the visible features left behind on the planet's atmosphere were the finest ever recorded. Why not take the time to look at Jupiter again tonight while it still holds good sky position? No matter where you observe from, this constantly changing planet offers a wealth of things to look at—be it the appearance of the "Great Red Spot" or just the ever-changing waltz of the Galilean moons.

While we're thinking moons, try looking Selene's way, see if you can spot crater Grimaldi on the western edge without any aid...then grab binoculars or telescope and let's have a look (Figure 7.36)!

Named for Italian physicist and astronomer Francesco Grimaldi, this great old crater is one of the few which actually resembles a mare. It spans 222 kilometers from east to west and 430 kilometers from north to south. Along its southeastern flank, look for a 230 kilometer long rima which extends its way to the double ring of Sirsalis. Grimaldi is a Class V mountain-walled plain whose floor is one of the darkest areas of the Moon, reflecting only 6% of the light. Look carefully at its walls... You'll find the northern area very eroded, while both foothills and mountains edge it to the east and west. Be sure to mark your lunar challenge list as having spotted the great Grimaldi!

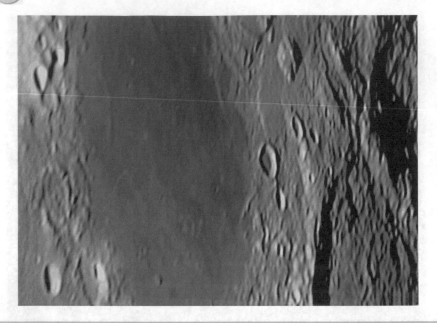

Figure 7.36. Grimaldi (Credit—Alan Chu).

Thursday, July 17

Tonight our Moon stands poised at the edge of becoming full in a matter of hours. If we could see it from space, we would know that it is readying itself to pass either just north or just south of the cone of shadow projected by Earth. Take the time to study the limbs of the Moon for the effects of libration. Follow Tycho's bright ray toward the southwest and see if you can spot the Doerfel Mountains as tiny bumps on the limb's edge. While they might not appear to be much, they are three times higher than Mt. Everest!

Now head about a palm's width east of last night's study star Zeta Scorpii for lovely Theta (RA 17 37 19 Dec –42 59 52). Named Sargas, this 1.8 magnitude star resides around 650 light-years distant in very impressive field of stars for binoculars or a small telescope (Figure 7.37). While all of these are only optical companions, the field itself is worth a look... Remember its position for the future.

About three fingerwidths north (RA 17 33 36 Dec –37 06 13) is a true double— Lambda Scorpii—also known as Shaula ("The Sting"). As the brightest known star in its class, 1.6 magnitude Lambda is a spectroscopic binary which is also a variable of the Beta Canis Majoris type, changing ever so slightly in a matter of a little more than 5 hours (Figure 7.38). Although we can't see the companion star, nearby is yet another that will make learning this starhop "marker" worth your time!

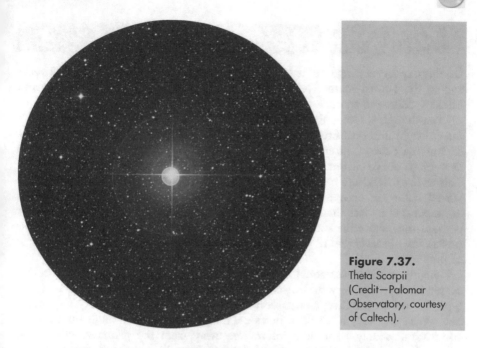

Figure 7.37. Theta Scorpii (Credit—Palomar Observatory, courtesy of Caltech).

Figure 7.38. Lambda Scorpii (Credit—Palomar Observatory, courtesy of Caltech).

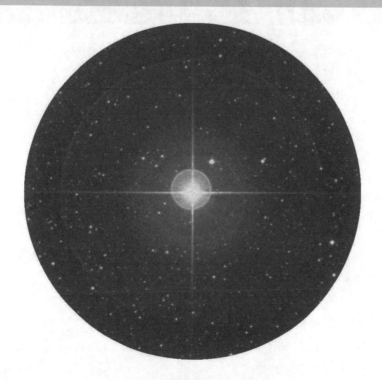

Friday, July 18

On this day 28 years ago, India launched its first satellite (Rohini 1), and 32 years ago in the United States Gemini 10 was launched, carrying John Young and Michael Collins to space.

Even though we have the Full Buck Moon to contend with tonight, there are many bright and beautiful collections to be explored! Using last night's marker of Lambda Scorpii, its time to head due west four fingerwidths (RA 16 52 00 Dec –38 02 00) to encounter Mu Mu (Figure 7.39). That's right: Two Mu's in one! This wide unaided-eye pair holds a place in South Seas legend as "the Inseparable Ones." Believe it or not, these two members of the Scorpius-Centaurus group are separated by less than a light-year, and are considered to be a physical pair because they share the same proper motion. In binoculars or telescopes at low power, the observer will note the westernmost is slightly brighter—at least most of the time.

Hanging out in space some 520 light-years away, western Mu 1 is a spectroscopic binary—the very first discovered to have double lines. This Beta Lyrae–type star has an orbiting companion which eclipses it around every day and a half, yet causes only a 0.3 drop in visual magnitude—even though the orbiting companion is only 10 million kilometers away from the primary. While that sounds like plenty of distance, when the two pass, their surfaces would nearly

Figure 7.39. Mu 1 and Mu 2 Scorpii (Credit—Palomar Observatory, courtesy of Caltech).

Figure 7.40. Iota 1 and Iota 2 Scorpii (Credit—Palomar Observatory, courtesy of Caltech).

touch each other! Watch Mu 1 over a period of time and see if you can detect when the pair are almost matched in brightness.

More? Then head back to Lambda and drop southeast about two fingerwidths (RA 17 47 35 Dec –40 07 37) for Iota Iota. The westernmost of this even wider pair is Iota 1, an F-type supergiant that's around 3400 light-years distant. It also has a 13th magnitude companion, which may or may not be physically related. To the east is Iota 2, an A-type star which also has a questionable companion (Figure 7.40). Who among us doesn't have a least one?!

Saturday, July 19

Today in 1846, Edward Pickering was born (Figure 7.41). Although his name is not well known, he became a pioneer in the field of spectroscopy. Pickering was the Harvard College Observatory Director from 1876 to 1919, and it was during his time there that photography and astronomy began to merge. The archived photographs known as the Harvard Plate Collection remain a valuable source of data to this day.

Now let's have a look at something which would make Edward Pickering proud. He encouraged amateur astronomers and founded the American Association of Variable Star Observers—so set your sights on RR Scorpii (Figure 7.42) about

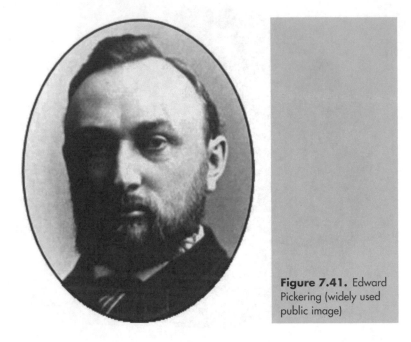

Figure 7.41. Edward Pickering (widely used public image)

Figure 7.42. RR Scorpii in center of field (Credit—Palomar Observatory, courtesy of Caltech).

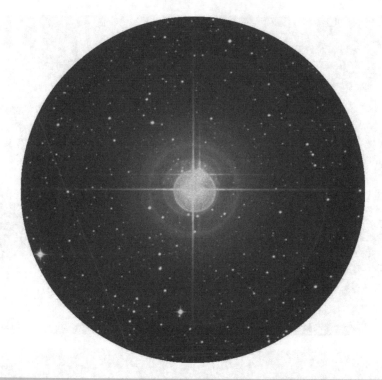

Figure 7.43. Nu Scorpii (Credit—Palomar Observatory, courtesy of Caltech).

two fingerwidths northeast of Eta and less than a fingerwidth southwest of 62 (RA 16 56 37.84 Dec –30 34 48.2). This extremely red Mira-type can reach as high as magnitude 5 and drop as low as 12 in about 280 days!

While you're out, let's sneak a peek at multiple star system Nu Scorpii (Figure 7.43). Located about a fingerwidth east and slightly north of bright Beta (RA 16 11 59 Dec –19 27 38), we find a handsome duo of stars in a field of nebulosity that will challenge telescopic observers much the way that Epsilon Lyrae does. With any small telescope, the observer will easily see the widely separated A and C stars. Add just a little power and take your time... The C star has a D companion to the southwest! For larger telescopes, take a very close look at the primary star. Can you separate the B companion to the south?

Sunday, July 20

Today was a busy day in astronomy history! In 1969, the world held its breath as the Apollo 11 lander touched down and Neil Armstrong and Edwin Aldrin became the first humans to touch the lunar surface (Figure 7.44). We celebrate our very humanity because even Armstrong was so moved that he may have

Figure 7.44. Armstrong's "Small Step" (Credit—NASA).

messed up his lines! The famous words were meant to be "A small step for a man. A giant leap for mankind." That's nothing more than one small error for a man, and mankind's success continued on July 20, 1976 when Viking 1 landed on Mars—sending back the first images ever taken from that planet's surface.

Even though it will be awhile before the Moon rises, the Apollo 11 landing area will be visible tonight as well. All you have to do is look west of the terminator and use what you learned a few days ago!

While we're waiting, let's hop to Xi Scorpii about four fingerwidths north of Beta (RA 16 04 22 Dec –11 22 22).

Discovered by Sir William Herschel in 1782, this 80 light-year distant system poses a nice challenge for mid-sized scopes (Figure 7.45). The yellow-hued A and B pair share a very eccentric orbit about the same distance as Uranus is from our Sun. When observing them in 2008 they should be fairly well spaced, and the slightly fainter secondary should appear to the north. Look a good distance away for the 7th magnitude orange C component and south for yet another closely matched double of 7th and 8th magnitude—the D and E stars.

For the larger scope, this multiple star system does display a little bit of color. Most will see the A and B components as yellow–white, the C star as slightly orange, and the D-E pair as slightly tinged with blue. Be sure to mark your observations because this is one of the finest!

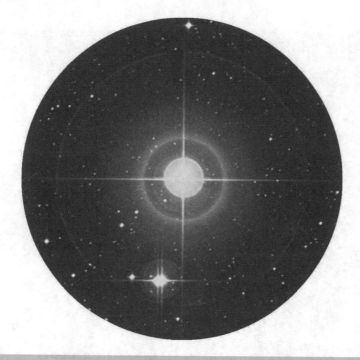

Figure 7.45. Xi Scorpii (Credit—Palomar Observatory, courtesy of Caltech).

Monday, July 21

Today in 1961, Mercury 4 was launched sending Gus Grissom into suborbital space on the USA's second manned flight, and returned him safely in the Liberty Bell 7 (Figure 7.46).

Tonight before moonrise, we will hustle off to explore a single small globular— M80 (RA 16 17 02 Dec –22 58 30). Located about four degrees northwest of Antares (about two fingerwidths), this little globular cluster is a powerpunch (Figure 7.47). Located in a region heavily obscured by dark dust, M80 will shine like an unresolvable star to small binoculars and reveal itself to be one of the most heavily concentrated globulars to the telescope.

Discovered in 1781 within days of each other by Messier and Méchain, respectively, this intense Class I globular cluster is around 36,000 light-years distant. In 1860, M80 became the first globular cluster found to contain a nova... A centrally located star brightened to magnitude 7 over a period of days and became known as T Scorpii. The event then dimmed more rapidly than expected, making observers wonder exactly what they had seen. Because a nova had never been observed before inside a globular cluster, astronomers were stunned to see such activity in a place where no novae were believed to exist.

Figure 7.46. Grissom entering Liberty Bell 7 (Credit—NASA).

Figure 7.47. M80 (Credit—Palomar Observatory, courtesy of Caltech).

Since most globular clusters contain stars all of relatively the same age, the hypothesis was eventually put forward that perhaps the event was an actual collision of stellar members. Given that the cluster contains more than a million stars, there is a reasonable probability that about 2700 collisions of this type may have occurred during M80's lifetime.

Tuesday, July 22

Tonight we will note the work of Friedrich Bessel, who was born on this day in 1784 (Figure 7.48). Bessel was a German astronomer and mathematician whose functions still carry his name, and are used in many areas of mathematical physics. But you may put away your calculator, because Bessel was also the very first person to measure a star's parallax. In 1837, he chose 61 Cygni and the measurement came to no more than a third of an arcsecond, which yielded a distance very close to the currently accepted 11.4 light-years. His work ended the debate about the distance to the stars that had stretched back two millennia to Aristotle's time.

Although you may need to use your finderscope if you observe under bright skies, you'll easily locate 61 between Deneb (Alpha) and Zeta on the eastern side. Look for a small trio of stars and choose the westernmost. Not only is it famous because of Bessel's work, but it is one of the most noteworthy of double stars for a small telescope. Of the stars visible to the unaided eye, 61 Cygni is the fourth closest to Earth (Figure 7.49), with only Alpha Centauri, Sirius, and Epsilon Eridani closer.

Visually, the two components have a slightly orange tint, are less than a magnitude apart in brightness, and have a nice separation of around 30" to the south-southeast. Back in 1792, Piazzi first noticed 61's abnormally large proper motion and dubbed it "The Flying Star." At that time, it was only separated by around 10" and the B star was to the northeast. It takes nearly seven centuries

Figure 7.48.
Friedrich Bessel (widely used public image)

Figure 7.49. 61 Cygni (Credit—Palomar Observatory, courtesy of Caltech).

for the pair to orbit each other. 61 Cygni has had a curious history… At one time the A component was believed to have harbored a planet, but scientists are now doubtful about its existence.

Wednesday, July 23

Tonight for unaided observers, let's begin by identifying Zeta Ophiuchi, the centermost in a line of stars marking the edge of the constellation of Ophiuchus, about a handspan north of Antares. As a magnificent 3rd magnitude blue-white class O, this hydrogen-fusing dwarf is eight times larger than our own Sun. Hanging out some 460 light-years away, it is dulled by the interstellar dust of the Milky Way and would shine two full magnitudes brighter if it were unobstructed. Zeta is a "runaway star"—a product of a one-time supernova event in a double star system. Now roughly halfway through its 8 million year life span, the same fate awaits this star. Now point binoculars or small scopes about three fingerwidths south to have a look at Phi Ophiuchi (Figure 7.50). This is a spectroscopic double star, but it has several delightful visual companions!

Almost in between these two bright stars is our telescopic target for tonight—M107 (RA 16 32 31 Dec –13 03 13). Discovered by Pierre Méchain in 1782, but only added to the catalog in 1947, it's probably one of the last of the Messier

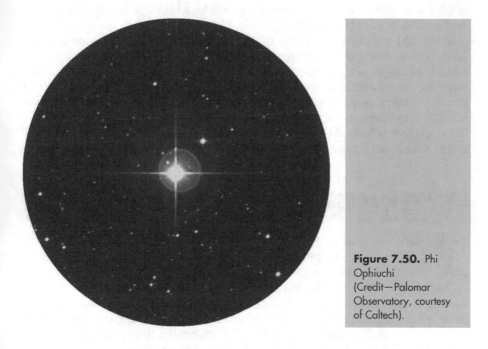

Figure 7.50. Phi Ophiuchi (Credit—Palomar Observatory, courtesy of Caltech).

Figure 7.51. M107 (Credit—Palomar Observatory, courtesy of Caltech).

objects to be discovered, and it wasn't resolved into individual stars until studied by Herschel in 1793 (Figure 7.51).

M107 isn't the most impressive of globulars, but this Class X is notable as a faint, diffuse area with a core region in binoculars, and is surprisingly bright in a small telescope. It's a curious cluster, for some believe it contains dark, dust-obscured areas. Located around 21,000 light-years away, this little beauty contains about 25 known variable stars. Visually, the cluster begins to resolve around the edges to mid-aperture and the structure is rather loose. If sky conditions permit, the resolution of individual chains at the globular's edges make this globular well worth a visit, and you can log it as Herschel IV.40!

Thursday, July 24

Tonight let's continue on our journey through the galactic halo to pick up the Class VIII globular cluster M9 (RA 17 19 12 Dec –18 30 59). You'll find it located around two fingerwidths east of Eta Ophiuchi.

Discovered by Messier in 1764, this particular globular cluster is one of the nearest to our galactic center and is around 2600 light-years away from our solar system (Figure 7.52). Now let's study differences—check out the contrast between this small globular's appearance as compared to last night's M107. In this one we're seeing not only a strong central concentration, but a slightly oval

Figure 7.52. M9 (Credit—Palomar Observatory, courtesy of Caltech).

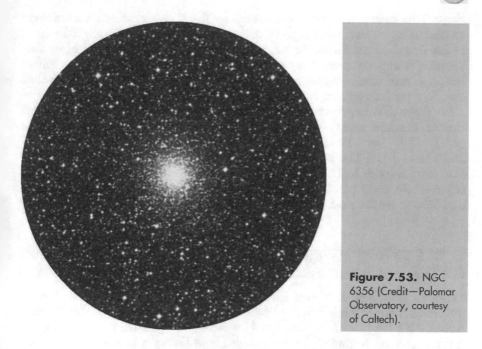

Figure 7.53. NGC 6356 (Credit—Palomar Observatory, courtesy of Caltech).

Figure 7.54. NGC 6342 (Credit—Palomar Observatory, courtesy of Caltech).

shape. This change in structure is caused by the strong absorption of starlight by dust along its northwest edge. Of its huge stellar population, only a dozen or so variable stars are known in M9, which is rather few for a cluster of its size. Visually, it appears more compact than M107, and it is slightly oblate. Rather than chains of stars resolving at the edges, M9 appears to have larger, individual stars in a random pattern—while M107 appears to be a solid core!

For those with larger scopes, you also have the opportunity to study two more clusters that are nearby—Class II NGC 6356 about a degree to the northeast (RA 17 23 34 Dec –17 48 46) and Class IV NGC 6342 to the southeast (RA 17 21 10 Dec –10 35 14). You will find NGC 6356 to be rather small—but bright and concentrated. NGC 6342 appears to be even smaller and far less distinct (Figures 7.53 and 7.54). Compare them both to the structure of M9 and you will find 6356 to be the most concentrated of the three...a "class" act!

Friday, July 25

Today in 1971, Apollo 15 was launched on its way to the Moon, and tonight we'll launch our way north to the Mighty Hercules for a look at another globular study—M92 (Figure 7.55). Although in a relatively open field for starhoppers, it's not too hard to find if you can imagine it as the apex of a triangle with the northern keystone stars—Eta and Pi—as the base (RA 17 17 07 Dec +43 08 11).

Figure 7.55. M92 (Credit—Palomar Observatory, courtesy of Caltech).

Figure 7.56. NGC 6426 (Credit—Palomar Observatory, courtesy of Caltech).

At near magnitude 6, Class IV M92 was discovered by Johann Bode in 1777 and cataloged as Bode 76. Independently recovered by Messier in 1781 and resolved by Herschel in 1783, this bright, compact globular is around 26,700 light-years away and is about 12–14 billion years old. It contains 14 RR Lyrae variables among its 330,000 stars and also a very rare eclipsing binary.

Viewable unaided under the right conditions and very impressive in even small binoculars, M92 is a true delight to even the smallest of telescopes. It has a very bright and unresolvable core with many outlying stars that are easily revealed. Larger scopes will appreciate its fiery appearance!

Now let's hop south to Beta Ophiuchi to have a look at NGC 6426 about a fingerwidth south (RA 17 44 54 Dec +-3 10 12) (Figure 7.56). There's a very good reason why you'll want to at least try with Herschel II.587...

Discovered by Sir William in 1786, this 11th magnitude Class IX globular looks destroyed in comparison to M92. At 67,500 light-years away, it is more than twice the distance from us as M92! Residing 47,600 light-years from the galactic center, NGC 6426 contains 15 RR Lyrae variables (three of which are newly discovered), and is the most metal-poor globular known. So what's the relation to M92? It's even a little bit older!

Forget about finding this one in binoculars and very small telescopes. For the mid-sized scope you'll find it conveniently located about halfway between Beta and Gamma Ophiuchi—but it's not easy. Faint and diffuse, a large telescope is required to begin resolution.

Saturday, July 26

For hardcore observers, tonight's globular cluster study will require at least a mid-aperture telescope, because we're staying up a bit later to go for a pair that can be seen in the same low-power field—NGC 6522 and NGC 6528 (Figures 7.57 and 7.58). You will find them easily at low power just a breath northwest of Gamma Sagittarii (Al Nasl), or the tip of the "teapot's" spout. Once located, switch to higher power to keep the light of Gamma out of the field and let's do some studying.

The brighter, and slightly larger, of the pair to the northeast is Class VI NGC 6522 (RA 18 03 34 Dec –30 02 02). Note its level of concentration compared to Class V NGC 6528 (RA 18 04 49 Dec –30 03 20). Both are located around 2000 light-years from the galactic center, and are seen through a very special area of the sky known as "Baade's Window"—one of the few areas toward our galaxy's core region not obscured by dark dust. While they are similar in concentration, distance, etc., NGC 6522 has a slight amount of resolution toward its edges while NGC 6528 appears more random.

Figure 7.57. NGC 6522 (Credit—Palomar Observatory, courtesy of Caltech).

Figure 7.58. NGC 6528 (Credit—Palomar Observatory, courtesy of Caltech).

Both NGC 6522 and NGC 6528 were discovered by Sir William Herschel on the same night in 1784 and both are the same distance from the galaxy's nucleus. But there the similarities end: NGC 6522 has an intermediate metallicity. At its core, the red giants have been depleted—stripped tidally by evolving blue stragglers. It is possible that core collapse has already occurred. NGC 6528, however, contains one of the highest metal contents of any known globular cluster collected in its bulging core!

Sunday, July 27

Today in 1892, a very special astronomer was born—Sir George Biddell Airy (Figure 7.59). Does that name sound familiar? Anyone who uses a refractor understands the properties of the "Airy disc" as first outlined in his paper "On the Diffraction of an Object-Glass with Circular Aperture." But Sir George achieved a bit more: As Astronomer Royal from 1835 to 1881, his tireless devotion to planetary study led to the discovery by P. A. Hansen of two new irregularities in the moon's motion. Not enough? Airy's calculations also determined the mean density of the Earth. More? Then thank Sir George for giving us Greenwich Mean Time!

Figure 7.59. Sir George Airy (widely used public image)

Figure 7.60. M4 (Credit—Palomar Observatory, courtesy of Caltech).

Are you still having no luck in finding a deep-space object? Then how about one that's simple to locate for all optics. All you have to know is Antares and go west...

Just slightly more than a degree away you'll find a major globular cluster perfectly suited for telescope and binoculars of varying sizes—M4 (RA 16 23 35 Dec –26 31 31). This 5th magnitude Class IX cluster can even be spotted unaided from a dark location! In 1746 Philippe Loys de Chéseaux happened upon this 7200 light-year distant beauty—one of the nearest to us (Figure 7.60). It was also included in Lacaille's catalog as object I.9 and noted by Messier in 1764. Much to Charles' credit, he was the first to resolve it!

As one of loosest globular clusters, M4 would be tremendous if we were not looking at it through a heavy cloud of interstellar dust. To binoculars, it is easy to pick out a very round, diffuse patch—yet it will begin resolution with even a small telescope. Large telescopes will also easily see a central "bar" of stellar concentration across M4's core region, which was first noted by Herschel.

As an object of scientific study, the first millisecond pulsar was discovered within M4 in 1987—one which spins 10 times faster than the Crab Nebula pulsar. Photographed by the Hubble Space Telescope in 1995, M4 was found to contain white dwarf stars—the oldest in our galaxy—with a planet orbiting one of them! A little more than twice the size of Jupiter, this planet is believed to be as old as the cluster itself. At 13 billion years, it would be three times the age of the Sol system!

Figure 7.61. M12 (Credit—Palomar Observatory, courtesy of Caltech).

Figure 7.62. M10 (Credit—Palomar Observatory, courtesy of Caltech).

Monday, July 28

Tonight let's head on out toward two more giant globulars that appear different from the rest (and each other)—the same-field binocular pair M10 and M12 (Figures 7.61 and 7.62). Located about half a fistwidth west of Beta Ophiuchi (RA 16 47 14 Dec –01 56 52), each is easily seen as a hazy round spots in binoculars. Now let's go to the telescope to find out what makes them tick… Northernmost M12 is much more loosely concentrated, so smaller scopes will begin to resolve individual stars from this 24,000 light-year distant Class IX cluster. Note there is a slight concentration toward the core region, but for the most part the cluster appears fairly even. Large instruments will resolve out individual chains and knots of stars.

Now head about 3.5 degrees southeast and check out Class VII M10 (RA 16 57 08 Dec –04 05 57). What a difference in structure! Although they appear to be close together and close in size, the pair is actually separated by some 2000 light-years. M10 is a much more concentrated globular showing a brighter core region to even the most modest of instruments. This compression of stars is what classifies one type of globular cluster from another, and M10 appears brighter not because of this compression, but because it is about 2000 light-years closer.

Now grab a comfortable seat because the Delta Aquarid meteor shower reaches its peak tonight. It is not considered a prolific shower, and the average fall rate is about 25 per hour—but who wouldn't want to take a chance on observing a meteor about every 4 to 5 minutes? These travelers are considered to be quite slow, with speeds of about 24 kilometers per second, and they are known to leave yellow trails. One of the most endearing qualities of this annual shower is its broad stream of around 20 days before and 20 days after peak. This will allow it to continue for at least another week and overlap the beginning stages of the famous Perseids.

The Delta Aquarid stream is a complicated one, and a mystery not quite yet solved. Its possible gravity split the stream from a single comet into two parts, and each may very well be a separate stream. One thing we know for certain is they will seem to emanate from the area around Capricornus and Aquarius, so you will have best luck facing southeast and getting away from city lights. Just relax and enjoy a warm summer night... It's time to catch a "shooting star!"

Tuesday, July 29

Tonight we're going to move back toward Ophiuchus and a globular cluster unlike any that we've seen so far—M19 (RA 17 02 37 Dec –26 16 04). Locate Antares and about a fistwidth to the east you will see Theta Ophiuchi with fainter star 44 to its northwest and our previously studied multiple system 36 to the southeast. Move around two degrees to the west of 36 and let's check it out.

With a visual magnitude of 6.8, this class VIII globular cluster can be seen with small binoculars, but requires a telescope to begin to take on form. Discovered by Charles Messier in 1764, M19 is the most oblate globular known (Figure 7.63). Harlow Shapely, who studied globular clusters and cataloged their elliptical natures, estimated that there were about twice as many stars along the major axis as along the minor. This stretching of the cluster from its expected round shape may very well have to do with its proximity to the Galactic Center—a distance of only about 5200 light-years. This makes it only a tiny bit more remote from us than the very center of the Milky Way!

Very rich and dense, even smaller telescopes can discern that this globular cluster has a faint blue tinge to it. It is definitely one of the more interesting, due to its shape, but for the adventurous? There are two more (Figures 7.64 and 7.65). The Class VI NGC 6293 is about a degree and a half to the east-southeast (RA 17 10 10 Dec –26 34 54) and is far brighter than you might expect. Note how much rounder and more concentrated at the core this cluster is. Now move about a degree and a half to the north-northeast of M19 to find dimmer Class IX NGC 6284 (RA 17 04 28 Dec –24 45 53). Although it is the same size as NGC 6293, look at how much more "loosely" this one is constructed!

Wednesday, July 30

Today's history celebrates the 2001 flyby of the Moon by the Wilkinson Microwave Anisotropy Probe (WMAP) on its way to LaGrange Point 2 to study the cosmic microwave background radiation (Figure 7.66).

Now it's time to kick back and watch the peak of the Capricornid meteor shower! Although it is hard for the casual observer to distinguish these meteors from the Delta Aquarids, no one minds. Again, face the general direction of southeast and enjoy! The fall rate for this shower is around 10–35 per hour, but unlike the Aquarids, this stream produces those great "fireballs" known as bolides.

So where has Sir William Herschel been lately? Looking much further to the east than we were while watching the Capricornids! So be glad we took along our binoculars tonight, and we can pick up one of the members of the Herschel "400" list he discovered on this night in 1784.

NGC 6755 (Figure 7.67) is located about mid-way between Theta Serpentis and Delta Aquilae (RA 19 07 48 Dec +04 14 00). At magnitude 7.5, it is easily visible in binoculars as a small hazy patch of stars with little to no resolution. There's good reason why. It's obscured by the dust from the spiral arm of our own Milky Way. Although it might not seem terribly spectacular, knowing we are looking at the light of a cluster which left some 2300 years before Herschel cataloged it as H VII.19 is fun!

Figure 7.63. M19 (Credit—Palomar Observatory, courtesy of Caltech).

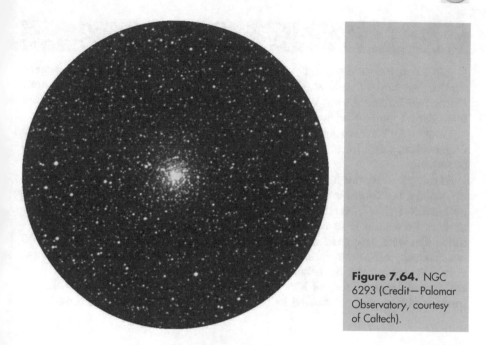

Figure 7.64. NGC 6293 (Credit—Palomar Observatory, courtesy of Caltech).

Figure 7.65. NGC 6284 (Credit—Palomar Observatory, courtesy of Caltech).

Thursday, July 31

If you're up before dawn this morning, be sure to have a look at the eastern skyline just as the Sun rises. Can you spot the ultra-slim crescent of the Moon? How about Pollux nearby? Within hours, the Moon will move in front of the Sun from Earth's point of view...

Tonight let's return again to the oblate and beautiful M19 and drop two fingerwidths south for another misshapen globular—M62 (RA 17 01 12 Dec –30 06 44).

At magnitude 6, this 22,500 light-year distant Class IV cluster can be spotted in binoculars, but comes to wonderful life in the telescope (Figure 7.68). Discovered by Messier in 1771, Herschel was the first to resolve it and report on its deformation. Because it is so near the galactic center, tidal forces have crushed it—much like what happened with M19. When studying it in the telescope, you will note its very off-center core. Unlike M19, M62 has at least 89 known variable stars—85 more than its neighbor—and the dense core may have undergone collapse. A large number of X-ray binaries have also been discovered within its inner structure, perhaps caused by the close proximity of stellar members.

Figure 7.66. WMAP (Credit—NASA).

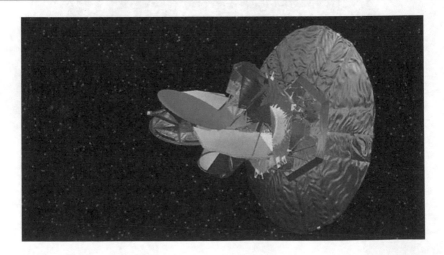

Figure 7.67. NGC 6755 (Credit—Palomar Observatory, courtesy of Caltech).

Figure 7.68. M62 (Credit—Palomar Observatory, courtesy of Caltech).

CHAPTER EIGHT

August, 2008

Friday, August 1

Mark your calendar! A total solar eclipse occurs today in northern Canada, the Arctic, and Asia (Figure 8.1). Totality will begin at 09:21:07 UT in Canada, with the path crossing Greenland, the Arctic Ocean, Russia, and Mongolia—ending in China at 11:21:28 UT. Maximum occurs at 10:21:08 UT. For those not in the path, a partial eclipse will be visible over northeastern Canada, most of Asia and Europe, and the Middle East, between 08:04:07 UT and 12:38:28 UT. Be sure to consult with online sources such as Mr. Eclipse for accurate locations of the path of totality. And please...**NEVER** look at the Sun without taking proper precautions. Wishing you clear skies for this event!

Today is also the birthdate of Maria Mitchell (Figure 8.2). Born in 1818, Mitchell became the first woman to be elected as an astronomer to the American Academy of Arts and Sciences. She later rocketed to worldwide fame when she discovered a bright comet in 1847.

Tonight we'll continue our exploration of globular clusters. These gravitationally bound concentrations of stars contain anywhere from 10,000 to 1 million members, and attain sizes of up to 200 light-years in diameter. At one time, these fantastic members of our galactic halo were believed to be round nebulae; perhaps the very first to be discovered was M22 in Sagittarius by Abraham Ihle in 1665 (Figure 8.3). This particular globular is easily seen in even small binoculars and can be easily located just slightly more than two degrees northeast of Sagittarius' "teapot lid" star—Kaus Borealis (RA 18 36 24 Dec −23 54 12).

Figure 8.1. Total eclipse (Credit—NASA).

Ranking third amidst the 151 known globular clusters in total light, M22 is probably the nearest of these incredible systems to our Earth, with an approximate distance of 9600 light-years. It is also one of the nearest globulars to the galactic plane. Since it resides less than a degree from the ecliptic, it often shares

Figure 8.2. Maria Mitchell (widely used public image).

Figure 8.3. M22 (Credit—Palomar Observatory, courtesy of Caltech).

Figure 8.4. NGC 6642 (Credit—Palomar Observatory, courtesy of Caltech).

the same eyepiece field with a planet. At magnitude 6, the class VII M22 will begin to show individual stars to even modest instruments and will burst into stunning resolution for larger aperture. About a degree west-northwest, mid-sized telescopes and larger binoculars will capture the smaller 8th magnitude NGC 6642 (RA 18 31 54 Dec –23 28 34) (Figure 8.4). At class V, this particular globular will show more concentration toward the core region than M22. Enjoy them both!

Saturday, August 2

If you're out tonight at sunset, be sure to watch the horizon in hopes of catching a glimpse of the very beginning of the Moon's return. Both Regulus and Venus are nearby!

Tonight let's return again to look at two globular giants so that we might compare roughly equal sizes, but not equal classes. To judge them fairly, you must use the same eyepiece. Start first by re-locating previous study M4. This is a class IX globular cluster. Notice the powder-like qualities. It might be heavily populated, but it is not dense. Now return to another previous study, M13, which is of class V. Most telescopes will achieve at least some resolution and show a distinct core region. It is the level of condensation that creates the

Figure 8.5. M55 (Credit—Palomar Observatory, courtesy of Caltech).

Figure 8.6. M71 (Credit—Palomar Observatory, courtesy of Caltech).

different classes. Judging a globular's concentration is no different from judging magnitudes, and simply takes practice.

Now try your hand at M55 (RA 19 39 59 Dec –30 57 43) along the bottom of the Sagittarius teapot—it's a class XI (Figure 8.5). Although it is a full magnitude brighter than the class I cluster M75 we looked at earlier in the week, can you tell the difference in concentration? For those with GoTo systems, take a quick hop through Ophiuchus and look at the difference between NGC 6356 (class II) and NGC 6426 (class IX). If you want to try one that science can't even classify? Look no further than M71 in Sagitta (RA 19 53 46 Dec +18 46 42) (Figure 8.6). It's all a wonderful game and the most fun comes from learning!

Sunday, August 3

For skywatchers tonight, be sure to catch the tender crescent Moon pairing with lovely Saturn just after sunset! Now let's return to earlier evening skies as we continue our studies with one of the globulars nearest to the galactic center—M14 (Figure 8.7). Located about 16 degrees (less than a handspan) south of Alpha Ophiuchi (RA 17 37 36 Dec –03 14 45), this 9th magnitude, class VIII cluster can be spotted with larger binoculars, but will only be fully appreciated with the telescope.

Figure 8.7. M14 (Credit—Palomar Observatory, courtesy of Caltech).

When studied spectroscopically, globular clusters are found to be much lower in heavy element abundance than stars such as our own Sun. These earlier generation stars (Population II) began their formation during the birth of our galaxy, making globular clusters the oldest formations an amateur can study. Globulars are distributed in a spherical halo around the galaxy center. In contrast, stars in the disk are mostly much younger, their populations having gone through cycles of starbirth and supernovae, which in turn have enriched the heavy element concentration in nearby star-forming clouds. Of course, as you may have guessed, M14 breaks the rules! It contains an unusually high number of variable stars—in excess of 70—with many of them known to be the W Virginis type. In 1938, a nova appeared in M14, but it went undiscovered until 1964 when Amelia Wehlau of the University of Ontario was surveying the photographic plates taken by Helen Sawyer Hogg. The nova was revealed on eight of these plates taken on consecutive nights and showed itself as a 16th magnitude star—and at its peak, it was believed to be about five times brighter than other cluster members. So unlike 80 years earlier with T Scorpii in M80, actual photographic evidence of this event existed. In 1991, the eyes of the Hubble were turned its way, but neither the suspect star nor traces of a nebulous remnant were discovered. But six years later, a *rare* carbon star was discovered in M14.

To a small telescope, M14 will offer little to no resolution and will appear almost like an elliptical galaxy, lacking any central condensation. Larger scopes will show hints of resolution, with a gradual fading toward the cluster's slightly oblate edges. A true beauty!

Monday, August 4

Tonight as the skies darken, be sure to look for the close pairing of Mars and the "Earthshine" Moon!

As we explore globular clusters, we usually assume they are all part of the Milky Way galaxy, but this might not always be the case. We know they are mostly concentrated in orbit around the galactic center, but there may be four of them which actually belong to other galaxies! Tonight we'll look at one such cluster, which is being drawn into the Milky Way's halo. Set your sights just about one and a half degree west-southwest of Zeta Sagittarii for M54 (RA 18 55 03 Dec –30 28 42).

At around magnitude 7.6, M54 is definitely bright enough to be spotted in binoculars, but its rich class III concentration is more notable in a telescope (Figure 8.8). Despite its brightness and deeply concentrated core, M54 isn't exactly easy to resolve. At one time we thought it to be around 65,000 light-years distant and to have a large number of variables, with a known number (82) of RR Lyrae types. We knew it was receding from us, but when the Sagittarius Dwarf Elliptical Galaxy (SagDEG) was discovered in 1994, it was noted M54 was receding at almost precisely the same speed! When more accurate distances were measured,

Figure 8.8. M54 (Credit—Palomar Observatory, courtesy of Caltech).

M54 was found to nearly coincide with the SagDEG distance of 80,000–90,000 light-years, and M54's distance is now calculated at 87,400 light-years. No wonder it's hard to resolve!

Tuesday, August 5

Today we celebrate the 78th birthday of Neil Armstrong, the first human to walk on the Moon (Figure 8.9). Congratulations! Also on this date in 1864, Giovanni Donati made the very first spectroscopic observations of a comet (Tempel, 1864 II). His observations of three absorption lines led to what we now know as the Swan bands, originating from a form of molecular carbon (C_2).

Tonight it's going to be very hard to ignore the Moon, so why don't we start by returning to a previously studied Lunar Club challenge? Your mission is to locate crater Petavius (Figure 8.10) along the southeast shore of Mare Fecunditatis and have a look at the Petavius Wall...

While you're admiring Petavius and its rima, keep in mind this 80 kilometer long crack is a buckle in the lava flow across the crater floor. Now look along the terminator for the long, dark runnel which is often called the Petavius Wall but is actually the fascinating crater Palitzsch. This 41 kilometer wide crater is confluent with a 110 kilometer long valley that is outstanding at this phase!

Our study continues tonight as we move away from the galactic center in search of a remote globular cluster that can be viewed by most telescopes. As we have learned, radial velocity measurements show us that the majority of globulars are involved in highly eccentric elliptical orbits—one which takes them far outside the plane of the Milky Way. These orbits form a sort of spherical "halo," which is more concentrated toward our galactic center. Reaching out several thousands of light-years, this halo is actually larger than the disk of our own galaxy. Since globular clusters don't take part in our galaxy's disk rotation, they can possess very high relative velocities. Tonight let's head toward the constellation of Aquila and look at one such globular—NGC 7006 (RA 21 02 29 Dec +16 11 14).

Located about half a fistwidth east of Gamma Aquilae, NGC 7006 is speeding toward us at a velocity of around 350 kilometers per second (Figure 8.11).

Figure 8.9. Neil Armstrong (widely used public image).

Figure 8.10. Petavius (Credit—Alan Chu).

Figure 8.11. NGC 7006 (Credit—Palomar Observatory, courtesy of Caltech).

At 150,000 light-years from the center of our galaxy, this particular globular could very well be an extra-galactic object, like our earlier study M54. At magnitude 11.5, it's not for the faint of heart, but can be spotted in scopes as small as 6", and requires larger aperture to look like anything more than a suggestion.

Given its tremendous distance from the galactic center, it's not hard to realize this is a class I (the most concentrated), even though it is quite faint. Even the largest of amateur scopes will find it unresolvable!

Wednesday, August 6

Today in 2001 the Galileo spacecraft made its flyby of Jupiter's moon Io, sending back incredible images of the surface (Figure 8.12). For Southern Hemisphere observers, be on watch as the Iota Aquarid meteor shower peaks on this universal date.

Tonight as the Sun sets and stars begin to appear, look for Spica no more than a fingerwidth north of the Moon. As the skies darken, we'll venture to the lunar surface near previous study Posidonius to have a look at the incredible Serpentine Ridge (Figure 8.13). Known more properly as Dorsa Smirnov, it meanders for 130 kilometers north to south across Mare Serenitatis. Can you spot the very tiny crater Very in its center? Very good!

Figure 8.12. Galileo approaching Jupiter (computer-generated image) (Credit—NASA).

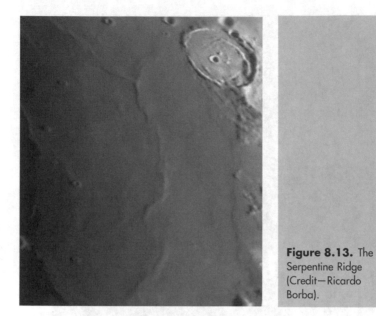

Figure 8.13. The Serpentine Ridge (Credit—Ricardo Borba).

Let's continue with our look at globular clusters as we delve more deeply into their structure. As a rule, globular clusters contain a large number of variable stars, and most usually they are of the RR Lyrae type, such as we saw with M54. At one time such variables were known as "cluster variables"—with the total amount varying from one cluster to another. Many globulars contain a large number of white dwarfs, while some have neutron stars which we can detect as pulsars. But out of all 151 of the galaxy's globular cluster population, only four contain a very unusual kind of object—a planetary nebula...

Look toward the emerging constellation of Pegasus to find the magnitude 6.5, class IV M15 (Figure 8.14). Easily located with even small binoculars about four degrees northwest of Enif, this magnificent globular cluster is a true delight in a telescope. Amongst the globulars, M15 ranks third in variable star population with 112 identified. As one of the densest of clusters, it is surprising that it is considered to be only class III. Its deeply concentrated core is easily apparent, and has begun the process of core collapse. The central core itself is very small compared to the cluster"s true size, and almost half of M15's mass is contained within it. Although it has been studied by the Hubble, we still do not know if this density is caused by the mutual gravity of the core's stars, or if it might disguise a supermassive object similar to those found in galactic nuclei.

As you might suspect from the above, M15 was the first globular cluster in which a planetary nebula, known as Pease 1, was identified. Larger aperture scopes can easily see it at high power. Surprisingly enough, M15 also is home to nine known pulsars—neutron stars left behind from previous supernovae occurring during the cluster's evolution—one of which is a double neutron star. While total resolution is impossible, a handful of bright stars can be picked out against the magnificent core region, and wonderful chains and streams of members await your investigation tonight!

Figure 8.14. M15 (Credit—NOAO/AURA/NSF).

Thursday, August 7

On this date in 1959, Explorer 6 became the first satellite to transmit photographs of the Earth from its orbit (Figure 8.15).

So... Are you ready to do a lunar walk for a challenge crater we haven't listed yet? Then look to the northwest shore of Mare Serenitatis for the pair of Aristoteles and Eudoxus. What's that? You see more? Then mark your notes for Eudoxus and let's have a look at many other studies you may not have observed before (Figure 8.16)!

Now let's return to our studies and wait for the Moon to set... It's time to take one last look at this year's globular cluster studies. As we know, the major distribution of globular clusters centers on our galaxy's core region in the Ophiuchus/Sagittarius area. Tonight let's explore what creates a globular cluster's form—we'll start with the "head of the class," M75 (Figure 8.17).

Orbiting the galactic center for billions of years, globular clusters have endured a wide range of disturbances. Their component stars escape when accelerated by mutual encounters, and the tidal force of our own Milky Way pulls them apart when they are at periapsis, or nearest to the galactic center. Even close encounters with other masses, such as other clusters and nebulae, can act upon them! At

Figure 8.15. Explorer 6 (Credit—NASA).

Figure 8.16. Eudoxus area (Credit—Greg Konkel, annotation by Tammy Plotner). (1) Burg, (2) Barrow, (3) Grove, (4) Daniel, (5) Posidonius, (6) Apollo 17 Landing Area, (7) Plinius, (8) Bessel, (9) Menelaus, (10) Manilius, (11) Apennine Mountains, (12) Conon, (13) Palus Putredinus, (14) Mons Hadley, (15) Archimedes, (16) Autolycus, (17) Aristillus, (18) Mons Piton, (19) Cassini, (20) Caucasus Mountains, (21) Calippus, (22) Alexander, (23) Eudoxus, (24) Mare Serenitatis, (25) Linné, (26) Haemus Mountains.

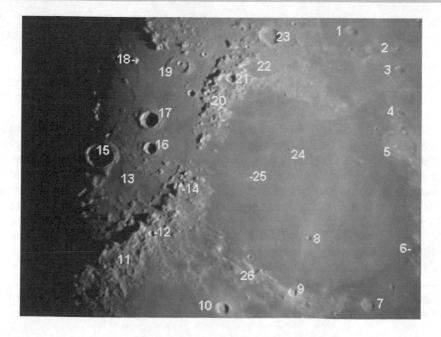

Figure 8.17. M75 (Credit—Palomar Observatory, courtesy of Caltech).

the same time, their stellar members are also evolving, and the resulting loss of gas can contribute to mass loss and the deflation of these magnificent objects. Although this happens far less quickly than in open clusters, our observable globular friends may be the last survivors of a once larger population, whose stars have been spread throughout the halo. This process of destruction is never-ending, and it is believed that globular clusters will entirely cease to exist in about 10 billion years.

Although it will be later in the evening when M75 appears on the Sagittarius/Capricornus border, you will find the journey of about eight degrees southwest of Beta Capricorni (RA 20 06 04 Dec –21 55 16) worth the wait. At magnitude 8, it can be glimpsed as a small round patch in binoculars, but a telescope is needed to see its true glory. Residing around 67,500 light-years from our solar system, M75 is one of the more remote of Messier's globular clusters. Since it is so far from the galactic center—possibly 100,000 light-years distant—M75 has survived billions of years to remain one of the few Class I globulars. Although resolution is possible in very large scopes, note that this cluster is one of the most concentrated in the sky, with only the outlying stars resolvable to most instruments.

In the meantime, don't forget all those other wonderful globular clusters such as 47 Tucanae, Omega Centauri, M56, M92, M28, and a host of others!

Friday, August 8

Our first order of business for the evening will be to pick up a Lunar Club challenge we haven't noted yet—Hipparchus (Figure 8.18). Located just slightly south of the central point of the Moon and very near the terminator, this is not truly a crater—but a hexagonal mountain-walled plain. Spanning about 150 kilometers in diameter with walls around 3320 meters high, it is bordered just inside its northern wall by crater Horrocks. This deep appearing "well" is 30 kilometers in diameter, and its rugged interior drops down an additional 2980 meters below the floor. To the south and just outside the edge of the plain is crater Halley. Slightly larger at 36 kilometers in diameter, this crater named for Sir Edmund is a little shallower at 2510 meters—but it has a very smooth floor. To the east you'll see a series of three small craters—the largest of which is Hind.

On this date in 2001, the Genesis Solar Particle Sample Return mission was launched on its way toward the Sun (Figure 8.19). On September 8, 2004, it returned with its sample of solar wind particles—unfortunately a parachute failed to deploy, causing the sample capsule to plunge unchecked into the Utah soil. Although some of the specimens were contaminated, many did survive the mishap. So what is "star stuff?" Mostly highly charged particles generated from a star's upper atmosphere flowing out in a state of matter known as plasma.

Figure 8.18. Hipparchus (Credit—Alan Chu).

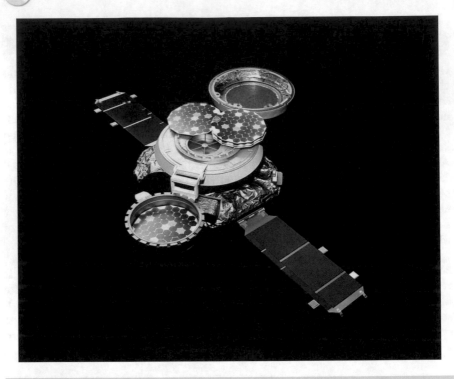

Figure 8.19. Genesis spacecraft (Credit—NASA).

Figure 8.20. M8: The Lagoon Nebula (Credit—Peter Ward).

Despite tonight's Moon, let's study one of the grandest of all solar winds as we seek out an area about three fingerwidths above the Sagittarius teapot's spout as we have a look at the magnificent M8 (Figure 8.20).

Visible to the unaided eye as a hazy spot in the Milky Way, fantastic in binoculars, and an area truly worth study in scopes of any size, this 5200 light-year diameter area of emission, reflection, and dark nebulae has a rich history. Its involved star cluster—NGC 6530—was discovered by Flamsteed around 1680, and the nebula by Le Gentil in 1747. Cataloged by Lacaille as III.14 about 12 years before Messier listed it as number 8, its brightest region was recorded by John Herschel, and dark nebulae were discovered within it by Barnard.

Tremendous areas of starbirth are taking place in this region, while young, hot stars excite the gas in a region known as the "Hourglass" around the stars Herschel 36 and 9 Sagittarii. Look closely around cluster NGC 6530 for Barnard Dark Nebulae B 89 and B 296 at the nebula's southern edge...and try again on a darker night. No matter how long you choose to swim in the "Lagoon" you will surely find more and more things to delight both the mind and the eye!

Saturday, August 9

Today in 1976, the Luna 24 mission was launched on a return mission of its own (Figure 8.21)—not to retrieve solar wind samples, but lunar soil! Remember this mission as we take a look at it in the weeks ahead.

When we begin our observations tonight, we'll start by having a look at another great study crater—Archimedes (Figure 8.22). You'll find it located in the Imbrium plain north of the Apennine Mountains and west of Autolycus.

Under this lighting, the bright ring of this class V walled plain extends 83 kilometers in diameter. Even though it looks to be quite shallow, it still has impressive 2150 meter high walls. To its south is a feature not often recognized— the Montes Archimedes. Though this relatively short range is heavily eroded, it still shows across 140 kilometers of lunar topography. Look for a shallow rima that extends southeast across Palus Putredinus toward the Apennines. Mark your challenge notes!

Now let's go have a look at a star buried in one of the spiral arms of our own galaxy—W Sagittarii...

Located less than a fingerwidth north of the tip of the teapot spout (Gamma), W Sagittarii (RA 18 05 01 Dec −29 34 48) is a Cepheid variable that's worth keeping an eye on (Figure 8.23). While its brightness varies only by less than a magnitude, it does so in less than 8 days! Normally holding close to magnitude 4, nearby field stars will help you correctly assess when minimum and maximum occur. While it's difficult for a beginner to see such changes, watch it over a period of time. At maximum, it will be only slightly fainter than Gamma to the south. At minimum, it will be only slightly brighter than the stars to its northeast and southwest.

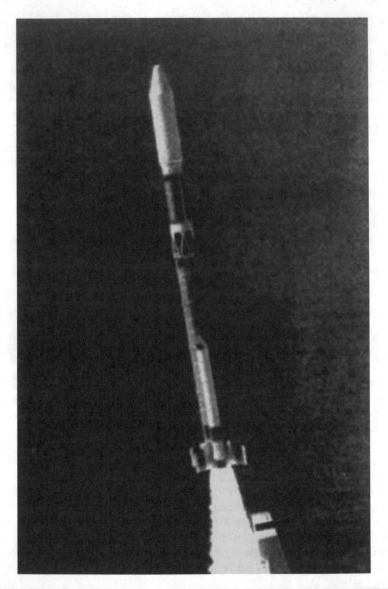

Figure 8.21. Luna 24 launch (press release photo).

While you watch W go through its changes—think on this. Not only is W a Cepheid variable (a standard for the cosmic distance scale), but it is also one that periodically changes its shape. Not enough? Then think twice... Because W is also a Cepheid binary. Still not enough? Then you might like to know that recent research points toward the W Sagittarii system having a third member as well!

Figure 8.22. Archimedes (Credit—Wes Higgins).

Figure 8.23. W Sagittarii (Credit—Palomar Observatory, courtesy of Caltech).

Sunday, August 10

Today in 1966 Lunar Orbiter 1 was successfully launched on its mission to survey the Moon. In the days ahead, we'll take a look at what this mission sent back! Tonight keep a very close watch on Selene as Antares is less than a degree away. Check for an occultation event!

Our lunar mission for tonight is to move south, past the crater rings of Ptolemaeus, Alphonsus, Arzachel, and Purbach, until we end up at the spectacular crater Walter (Figure 8.24). Named for Dutch astronomer Bernhard Walter, this 132 by 140 kilometer wide lunar feature offers up amazing details at high power. It is perhaps most fascinating to take the time to study the differing levels, which drop to a maximum of 4130 meters below the surface. Multiple interior strikes

Figure 8.24.
Albategnius to Walter
(Credit—Alan Chu).

Figure 8.25. Eta Sagittarii (Credit—Palomar Observatory, courtesy of Caltech).

abound, but the most fascinating of all is the wall crater Nonius. Spanning 70 kilometers, Nonius would also appear to have a double strike of its own—one that's 2990 meters deep!

Although it will be tough to locate with the unaided eye, thanks to the Moon, let's take a closer look at one of the most unsung stars in this region of sky—Eta Sagittarii (RA 18 17 37 Dec –36 45 42) (Figure 8.25). This M-class giant star will display a wonderful color contrast in binoculars or scopes, showing up as slightly more orange than the stars in the surrounding field. Located 149 light-years away, this irregular variable is a source of infrared radiation and is a little larger than our own Sun—yet 585 times brighter. At around three billion years old, Eta has either expended its helium core or just began to use it to fuse carbon and oxygen—creating an unstable star capable of changing its luminosity by about 4%. But have a closer look...for Eta is also a binary system with an 8th magnitude companion!

Monday, August 11

On this date in 1877, Asaph Hall of the US Naval Observatory was very busy. This night would be the first time he would see Mars' outer satellite Deimos! Six nights later, he observed Phobos, giving Mars its grand total of just two moons.

Tonight let's do a little Moon discovering of our own as we head to the western shore of Mare Cognitum and look along the terminator for the Montes Riphaeus—the "Mountains in the Middle of Nowhere" (Figure 8.26). But are they really mountains? Let's take a look...

At its widest, this unusual range spans about 38 kilometers and runs for a distance of around 177 kilometers. Less impressive than most lunar mountain ranges, some peaks reach up to 1250 meters high, making these summits about the same height as our earthly volcanoes Mounts St. Augustine and Kilauea. While we are considering volcanic activity, consider that these peaks are the only things left of Mare Cognitum's walls after the lava filled it in. At one time this area may have been among the tallest of lunar features!

Tonight after midnight is the peak of the Perseid meteor shower, and this year there's Moon. Obstruct it and let's sit back and talk about the Perseids while we watch...

The Perseids are undoubtedly the most famous of all meteor showers, and never fail to provide an impressive display (Figure 8.27). They appear in Chinese records dating all the way back to 36 ad. In 1839, Eduard Heis was the first observer to give an hourly count, and discovered their maximum rate was about 160 per hour. He and other observers continued their studies in subsequent years and found this number varied considerably.

Giovanni Schiaparelli was the first to relate the orbit of the Perseids to periodic comet Swift-Tuttle (1862 III). Recently the Perseid stream has been studied more deeply, and many complex variations in it have been discovered. There

Figure 8.26. Montes Riphaeus (Credit—Greg Konkel).

Figure 8.27. Perseids (Credit—Sirko Molau, IMO, Archenhold-Sternwarte/NASA).

are actually four individual streams derived from the comet's 120 year orbit. Although these streams peak on different nights, your best bet for viewing the great Perseids is between tonight and tomorrow morning at dawn.

Meteors from this shower enter Earth's atmosphere at a speed of 60 kilometers per second (134,000 miles per hour), from the general direction of the border between the constellations Perseus and Cassiopeia. While they can be seen anywhere in the sky, if you extend their paths backward, all the true members of the stream will point back to this region of the sky. For best success, position yourself so you are generally facing northeast and get comfortable. The radiant will continue to climb higher in the sky as dawn approaches and the Moon begins to set. If you are clouded out, don't worry. The Perseids will be around for a few more days yet, so continue to watch!

Tuesday, August 12

Did you mark your calendar to be up before dawn to view the Perseid meteor shower? Good!

Tonight let's start our lunar observations with features that can be seen with both binoculars and telescopes. Just slightly north of center along the terminator, look for the bright point of Kepler. Watch as this feature develops a bright ray system in the coming days. To the north you will see equally bright Aristarchus—quite probably one of the youngest of the prominent features at about 50 million years of age. This crater will also develop a ray system.

Now grab your telescope and look west of Aristarchus for the less-prominent crater Herodotus. Just to the north you will see a fine white thread known as Vallis Schroteri or Schroter's Valley (Figure 8.28). Winding its way across the Aristarchus plain, this feature is about 160 kilometers long, from 3 to 8 kilometers wide, and about 1 kilometer deep—but what is it?

Schroter's Valley is a prime example of a collapsed lava tube—created when molten rock flowed over the surface. This may have been from early volcanic activity, or possibly a major meteor strike, like the one which formed the Aristarchus crater itself. What is left is a long, narrow cave on the surface which shows well only when the lighting is correct. Like many sinuous rilles on the surface, Schroter's Valley has collapsed. If any intact lava tubes can be found on the Moon, they could conceivably provide shelter for future settlers!

Now let's have a look at the brightest star in the "Archer," Epsilon Sagittarii (RA 18 24 10 Dec −34 23 04). Known as Kaus Australis, or the "Southern Bow," Epsilon holds a respectable magnitude 1.8 and is located around 120 light-years from Earth (Figure 8.29) . This sparkling blue–white star is 250 times brighter than our own Sun. While a major challenge would be to spot Epsilon's 14th magnitude companion located about 32" away, even the smallest of telescopes and most binoculars can try for the 7th magnitude visual companion, which is widely spaced to the north-northwest.

Figure 8.28. Schroter's Valley (Credit—Wes Higgins).

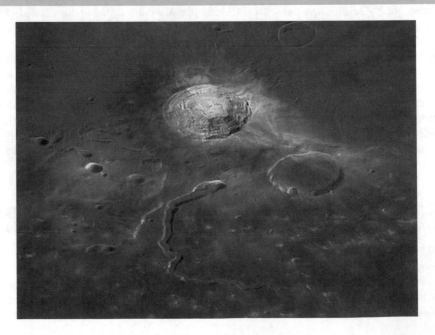

Figure 8.29. Epsilon Sagittarii: Kaus Australis (Credit—Palomar Observatory, courtesy of Caltech).

Wednesday, August 13

If you see sunlight today, be sure to wish a happy birthday to Anders Ångström, born in 1814 (Figure 8.30). He investigated magnetism, heat conduction, and practiced spectroscopy—specializing in the Sun. Ångström was the first to announce the existence of hydrogen in the Sun's atmosphere in 1862, and in 1868 he published his authoritative chart of the solar spectrum. He was also the first to study the spectra of other stars and of the Aurora Borealis. Have you ever heard of an Ångström unit?

Today also celebrates the birth (in 1861) of Herbert Hall Turner (Figure 8.31). As a professor of astronomy, he was a leader in the effort to produce an astrographic chart of the sky—developing better ways of obtaining positions and magnitudes of stars from survey plates. He's often been credited with coining the term "parsec," and was the man who (upon a suggestion by an 11 year old girl) proposed the name Pluto for our questionable ninth planet!

Look...on the western horizon... Is it a bird? Is it a plane? No. It's Venus and it's back on the scene less than half a degree away from Saturn! No special equipment is needed to see it, but you could fit both planets in a low-power field

Figure 8.30. Anders Ångström (widely used public image).

Figure 8.31. Herbert Hall Turner (Credit—Royal Astronomical Society (historical image).

of view. While you're at it, the dance of the ecliptic continues as Jupiter and the Moon are waltzing together on the other horizon!

Tonight we'll start our observations in the lunar southwest as we look along the terminator to identify the challenge crater Schickard (Figure 8.32). Look for an elongated gray oval that's more than just another cool crater...

Named for Dutch mathematician and astronomer Wilhelm Schickard, this 227 kilometer diameter feature is a ringed plain and is very old. At high power you'll see a variegated floor and dark areas near the walls—yet the center is creased by a lighter coloration. It is believed Schickard was formed by an early impact before Mare Nectaris formed. Its floor may have contained vents which allowed it to fill with lava during the Imbrium period. As it cooled and matured, another impact formed the Orientale basin and splashed material Schickard's way. But Schickard wasn't done evolving yet... Lava continued to flow and left even more dark evidence for us to observe. How do we know this is so? If you're able to resolve Schickard's tiny interior impacts you'll see that far fewer occur over the newer material. Older formations bear the scars of time and impact, while younger features are fresh and unmarked!

For observers of the southern skies, it's your turn to take a look at a specific star! Your mission? Gamma Arae (RA 17 25 23 Dec −56 33 39)...

Visually, this 680 light-year distant 3rd magnitude B-type has a nearby star (Figure 8.33). But have you looked telescopically? A well-separated 10th

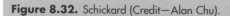

Figure 8.32. Schickard (Credit—Alan Chu).

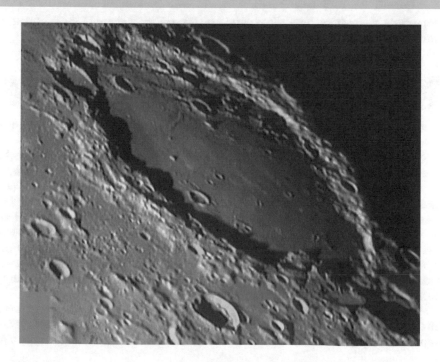

Figure 8.33. Gamma Arae (Credit—Palomar Observatory, courtesy of Caltech).

magnitude companion was discovered by John Herschel in 1835. Discover it yourself tonight!

Thursday, August 14

Tonight we'll continue our study of lunar evolution as we have a look at another walled plain just south of Grimaldi.

Named for English naturalist Charles Darwin, this old feature also bears the scars of the impact that created the Orientale Basin (Figure 8.34). Look carefully at the slopes in the northeast, for this may very well be material that was thrown there and left to slide back down to the crater floor. Spanning about 130 kilometers in diameter, Darwin's actual size is only diminished by the fact that we view it on a curve. Its northern and southern shores have almost completely eroded, yet some evidence remains of its eastern margin, broken in places by the Rima Darwin which stretches for 280 kilometers. Was there lava here as well? Yes. Evidence of this still exists in the form of a dome along Darwin's battered western edge.

Mirror, mirror on the wall... Who's the brightest Cepheid of them all? Let's find out as we head way down south again for a look at another unusual star—Kappa Pavonis (RA 18 56 57 Dec –67 14 00). This normal-appearing G-type star

Figure 8.34. Darwin
(Credit—Alan Chu).

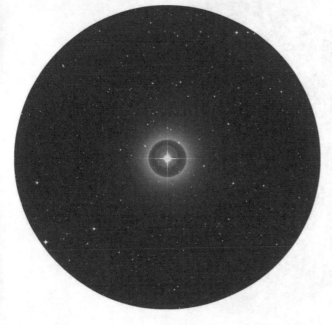

Figure 8.35.
Kappa Pavonis
(Credit—Palomar
Observatory, courtesy
of Caltech).

is actually a Cepheid variable in disguise (Figure 8.35). Watch for it to change by almost a full magnitude in a period of just about nine days!

Friday, August 15

It's going to be very hard to ignore the presence of the Moon tonight—or the conjunction that's about to happen in the west! Just in case you get clouded out tomorrow, be sure to have a look at Mercury, Venus, and Saturn getting closer by the minute. But, oh my... It's going to get even better yet because the Moon is also creeping closer and closer to Earth's shadow!

Think having all this Moon around is the pits? Then let's venture to Zeta Sagittarii (RA 19 02 36 Dec –29 52 48) and have a look at Ascella—the "Armpit of the Centaur" (Figure 8.36). While you'll find Zeta easily as the southern star in the handle of the teapot formation, what you won't find is an easy double. With almost identical magnitudes, Ascella is one of the most difficult of all binaries. Discovered by W. C. Winlock in 1867, the components of this pair orbit each other very quickly: in little more than 21 years. While they are about 140 light-years away, this gravitationally bound pair waltz no further apart than do our own Sun and Uranus!

Figure 8.36. Zeta Sagittarii: Ascella (Credit—Palomar Observatory, courtesy of Caltech).

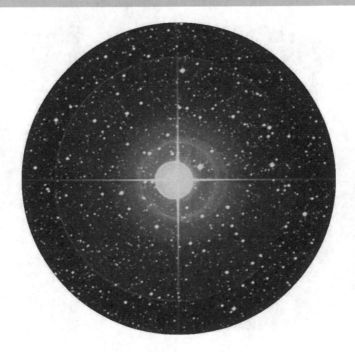

Figure 8.37. Nu Sagittarii: The Eye of the Archer (Credit—Palomar Observatory, courtesy of Caltech).

Too difficult? Then have a look at Nu Sagittarii (Ain al Rami), or "The Eye of the Archer" (RA 18 54 10 Dec −22 44 41) (Figure 8.37). It's one of the earliest known double stars and was recorded by Ptolemy. While Nu 1 and Nu 2 are not physically related to one another, they are an easy split in binoculars. Eastern Nu 2 is a K-type giant which is about 270 light-years from our solar system, but take a very close look at the western Nu 1. While it appears almost as bright, this one is 1850 light-years away! As a bonus, power up the telescope, because this is one very tight triple star system.

Saturday, August 16

Today is the birthday of none other than Pierre Méchain (1744), Charles Messier's well-known assistant (Figure 8.38). As a cartographer and astute mathematician, Méchain was a comet hunter as well and, much to his credit, was able to calculate the orbits of his discoveries. This quiet man contributed nearly a third of the objects found in what we now refer to as the "Messier Catalog," and was quite probably one of the first to realize just how many galaxies reside in the Virgo region. Although war and disease would bring an early end to this distinguished astronomer's life, Méchain became the director of the Paris Observatory and traveled to England where he met Sir William Herschel.

Figure 8.38. Pierre Méchain (widely used public image).

Mark your stargazer calendar for tonight, because it's going to be one awesome show! Starting off just after sunset on the western horizon, look for Mercury, Saturn, and Venus to gather together in a tight triangle to watch as the Moon heads quietly for the Earth's shadow (Figure 8.39). A lunar eclipse is about to occur!

Figure 8.39. Saturn, Venus, and Mercury conjunction (Credit—R. Jay GaBany).

Figure 8.40. Total eclipse (Credit—Alan Chu).

Although it will only be partial, the event will be visible over most of Asia, Australia, Europe, Africa, and South America. But, don't sell it short. This is a significant event since the Moon will pass deep inside Earth's umbral shadow at its maximum—an umbral magnitude of 0.8! The eclipse will begin at 19:35:45 UT and will end about 3 hours later, with 21:10:08 as the moment of greatest eclipse.

Begin watching at 18:23:07 UT as the Moon begins to enter the shadow; it will not fully exit the shadow until 23:57:06 UT. One of the most breathtaking adventures you can undertake is to watch the Moon through a telescope during an eclipse—both in ingress and in egress (Figure 8.40) . Craters take on new dimensions and subtle details light up as the shadow seems to race across the surface. And if you are lucky enough to see it at maximum, be sure to look at the stars near Moon. What a wonder it is to behold what is normally hidden by the light!

And what else is about to be hidden? Neptune! Less than a degree to the north of the lunar limb, the "King of the Sea" is about to be occulted. Check out IOTA for times and locations... Or just have a look for yourself. Enjoy your eclipse experience and remember to try your hand at photography!

Sunday, August 17

Today in 1966 Pioneer 7 was launched. It was the second in a series of satellites sent to monitor the solar wind, and it also studied cosmic rays and the interplanetary magnetic field.

Although the Moon will be along soon, return to previous study star Lambda Scorpii and hop three fingerwidths northeast to NGC 6406 (RA 17 40 18 Dec –32 12 00)... We're hunting the "Butterfly!"

Easily seen in binoculars and tremendous in the telescope, this brilliant 4th magnitude open cluster was discovered by Hodierna before 1654 and

Figure 8.41. M6: The Butterfly Cluster (Credit—Palomar Observatory, courtesy of Caltech).

independently found by de Chéseaux as his object 1, before being cataloged by Messier as M6 (Figure 8.41). Containing about 80 stars, the light you see tonight left its home in space around the year 473 AD. M6 is believed to be around 95 million years old and contains a single yellow supergiant—the variable BM Scorpii. While most of M6's stars are hot, blue, and belong to the main sequence, the unique shape of this cluster gives it not only visual appeal, but wonderful color contrast as well!

So where has the master observer, Sir William Herschel, been lately? Believe it or not, almost all of his discoveries during the month of August over the years occurred on the seventeenth. While we're out, it would be a great time to have a look at one such and we'll start by traveling to the far eastern edge of Sagittarius for NGC 6818 (RA 19 43 57 Dec –14 09 11)...

Best known as the "Little Gem" nebula for its distinctive green color, this 9th magnitude object can be seen as nearly stellar in binoculars, and as annular with a telescope using high power (Figure 8.42). Small, but bright, it's located about 6000 light-years away and was a prime target of the Hubble Space Telescope. Research showed it to have two distinct regions—the round shell-like outer layer and an elongated interior "bubble." Scientists believe the central star to be responsible for this unusual characteristic. Fast winds from this fueling core may have elongated the inner region and ruptured the outer shell!

Figure 8.42. NGC 6818: The Little Gem (Credit—Palomar Observatory, courtesy of Caltech).

Remember the Little Gem's position, for we'll be back under darker skies to take a closer look...

Monday, August 18

On this day in 1868, Norman Lockyer was very busy, becoming the first person to see helium absorption lines in the Sun's spectrum. Tonight we'll take a walk from helium-rich Lambda Scorpii about three fingerwidths east-northeast to an even more prominent area of stars, dating back to Ptolemy in 130 AD (Figure 8.43).

Astronomers throughout the ages have spent time with this cluster: Hodierna knew it as Ha II.2, Halley (in 1678) as number 29, Denham in 1733 as number 16, De Chéseaux as Number 10, Lacaille as II.14, Bode as 41, John Herschel as h 3710, and Dreyer as NGC 6475... But we know it best as Messier Object 7 (RA 17 53 48 Dec −34 47 00). (William Herschel noted it, but didn't give it a number, as was his practice with Messier objects.)

Set against the backdrop of the Milky Way, even the smallest of binoculars will enjoy this bright open cluster, while telescopes can resolve all of its 80 members. Roughly 800 light-years away, it contains many different spectral types in various stages of evolution, giving the cluster an approximate age of 260 million years.

Figure 8.43. M7: Ptolemy's Cluster (Credit—Palomar Observatory, courtesy of Caltech).

Full of binaries and close doubles, an extreme test of tonight's soon-to-be moonlit conditions would be to see if you can spot 11th magnitude globular cluster NGC 6543 to the northwest!

Tuesday, August 19

Born today in 1646, let's reflect on the achievements of John Flamsteed, an English astronomer with a passion for what he did. Despite a rather difficult childhood and no formal education, he went on to become First Observer at the Royal Observatory, and his catalog of 3000 stars was perhaps the most accurate published up to his time. Flamsteed star numbers are still in use.

Also born on this day was Orville Wright, in 1871 and Milton Humason, in 1891, a colleague of Edwin Hubble at Mt. Wilson and Mt. Palomar. The latter was instrumental in measuring the faint spectra of galaxies, which in turn provided evidence for the expansion of the universe.

Tonight let's return to the "Little Gem" and head less than a degree south to have a look at one of the summer's most curious galaxies—NGC 6822 (RA 19 44 57 Dec –14 48 11) (Figure 8.44). This study is a telescopic challenge even for skilled observers, but not an impossible one. Why bother? Because this

Figure 8.44. NGC 6822: Barnard's Galaxy (Credit—Palomar Observatory, courtesy of Caltech).

distant dwarf galaxy is bound to our own Milky Way by invisible gravitational attraction...

Named after its discoverer, E. E. Barnard—1884, "Barnard's Galaxy" is a not-so-nearby member of our local galaxy group. Discovered with a 6" refractor, this 1.7 million light-year distant galaxy is not easily found, but can be seen with very dark sky conditions and at the lowest possible power. Due to large apparent size, and overall faintness (magnitude 9), low power is essential in larger telescopes to give a better sense of the galaxy's frontier. Observers using large scopes will see faint regions of glowing gas (HII regions) and unresolved concentrations of bright stars. To distinguish them, try a nebula filter to enhance the HII and downplay the star fields. Barnard's Galaxy appears like a very faint open cluster overlain with a sheen of nebulosity, but by using the above technique the practiced eye will clearly see that the "shine" behind the stars is extragalactic in nature!

Wednesday, August 20

Tonight go south of Sagittarius to look for the delicate crescent of stars called the Southern Crown—Corona Australis. Its hidden jewel is the 7.3 magnitude, 28,000 light-year distant globular cluster NGC 6723 (RA 18 59 33 Dec −36 37 53). Discovered on June 3, 1826 by James Dunlop of New South

Wales, Australia, NGC 6723 can be best found by heading less than seven degrees due south of Zeta Sagittarii. This mid-sized cluster gives a surprising view, but if you're in a more northerly location, best catch it at its highest (Figure 8.45).

But if you're in the south? Then drop just a bit further southeast and have a look at the reflection-emission nebulae NGC 6726/27/29 (RA 19 01 39 Dec −36 53 28) (Figure 8.46). The three nebulae were discovered by Johann Schmidt, during his observations at Athens Observatory in 1861, and were independently recovered by Marth three years later in Malta. Even with a small scope, the brighter portions are visible as a faint haze and come to life with aperture. Be sure to note the easily split double BSO 14 in the same field!

Now relax... Tonight is the peak of the Kappa Cygnid meteor shower. Although the Moon will interfere late in the evening, start early by watching the area near Deneb. Discovered in the late 1800s, the Kappa Cygnids are often overlooked because the grander, more prolific Perseids get more attention. Although the stream has been verified, peak dates and fall rates vary from year to year. The average rate is usually no more than five per hour, but it is not uncommon to see 12 or more per hour, with many fireballs. The stream's duration is about 15 days. Clear skies!

Figure 8.45. NGC 6723 (Credit—Palomar Observatory, courtesy of Caltech).

Figure 8.46. NGC 6726 (Credit—Stephen Boyd).

Thursday, August 21

Tonight begin with just your eyes as you gaze about four fingerwidths above the top of the Sagittarius teapot dome for an open window on the stars and the magnificent M24...

This huge, hazy patch of stars is in reality an area of space known as "Baade's Window"—an area free of obscuring gas and dust. Cataloged by Messier in 1764 as object 24, even small binoculars will reveal the incredible vista of the Sagittarius Star Cloud. Although it's not truly a cluster, but rather a clean view of an area of our own galaxy's spiral arm, that will not lessen its impact when viewed through a telescope. Spanning a degree and a half of sky, it is one of the few areas where even a novice can easily perceive areas of dark dust.

For larger telescopes, look for the dim open cluster NGC 6603 (RA 18 18 26 Dec –18 24 21), located in the northeastern corner of M24 (Figure 8.47). There are two very notable dark nebulae, B 92 and B 93 (Figure 8.48), also located in the northern segment. Near teardrop-shaped B 92 and its single central star you should spot open cluster Collinder 469; and Markarian 38 is located south of B 93 (RA 18 16 54 Dec –18 04 00). You'll find B 86 near Y Sagittarii. At the

Figure 8.47. NGC 6603 region (Credit—Palomar Observatory, courtesy of Caltech).

Figure 8.48. Barnard 93 (Credit—Palomar Observatory, courtesy of Caltech).

Figure 8.49. NGC 6589/90/95 (Credit—Palomar Observatory, courtesy of Caltech).

southern edge of the star cloud, look for emission nebula IC 1283-1284, along with reflection nebulae NGC 6589 (Figure 8.49) and NGC 6590 and the open cluster NGC 6595 (RA 18 17 00 Dec −19 53 54). Still up for more? Then head to the southwestern edge to see if you can find the 12th magnitude planetary nebula NGC 6567.

Even if you don't accept these challenges, you can still enjoy looking at a 560 light-year swatch of stars from one of the Milky Way's loving arms! (If you're out late, look for Mira... It was discovered by Fabricius on this date in 1596.)

Friday, August 22

Tonight let's have a look at NGC 6302, a very curious planetary nebula located around three fingerwidths west of Lambda Scorpii: it is better known as the "Bug" nebula (RA 17 13 44 Dec −37 06 16).

With a rough visual magnitude of 9.5, the Bug belongs to the telescope—but it's history as a very extreme planetary nebula belongs to all of us (Figure 8.50). At its center is a 10th magnitude star, one of the hottest known. Appearing in the telescope as a small bowtie, or figure 8 shape, huge amounts of dust lie within it—very special dust. Early studies showed it to be composed of hydrocarbons, carbonates, and iron. At one time, carbonates were believed to be associated

Figure 8.50. NGC 6302: The Bug Nebula (Credit—Palomar Observatory, courtesy of Caltech).

with liquid water, and NGC 6302 is one of only two regions known to contain carbonates—perhaps in a crystalline form.

Ejected at a high speed in a bi-polar outflow, further research on the dust has shown the presence of calcite and dolomite, making scientists reconsider the kind of places where carbonates might form. The processes that formed the Bug may have begun 10,000 years ago—meaning it may now have stopped losing material. Hanging out about 4000 light-years from our own solar system, we'll never see NGC 6302 as well as the Hubble Telescope presents its beauty, but that won't stop you from enjoying one of the most fascinating of planetary nebulae!

Saturday, August 23

Do you remember a few days ago in history when Lunar Orbiter 1 was launched? Well, on this day in history it made headlines as it sent back the very first photo of Earth as seen from space (Figure 8.51)!

Tonight let's venture about three fingerwidths northeast of Lambda Sagittarii to visit a well-known but little-visited galactic cluster—M25 (RA 18 31 42 Dec –19 07 00) (Figure 8.52).

Discovered by de Chéseaux and then cataloged by Messier, it was also observed and recorded by William Herschel, Elert Bode, Admiral Smythe, and T. W. Webb... but was never added to the catalog of John Herschel. Thanks to

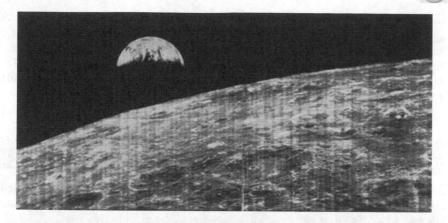

Figure 8.51. Lunar Orbiter's first photo (Credit—NASA).

J.L.E. Dreyer, it did make the second Index Catalog as IC 4725. Seen with even the slightest optical aid, this 5th magnitude cluster contains two G-type giants and well as a Cepheid variable with the designation of U. This star varies by about one magnitude in a period of less than a week. M25 is a very old cluster, perhaps

Figure 8.52. M25 (Credit—Palomar Observatory, courtesy of Caltech).

90 million years old, and the light you see tonight left the cluster over 2000 years ago. While binoculars will see a double handful of bright stars overlaying fainter members, telescopes will reveal more and more as aperture increases. At one time it was believed to have only around 30 members, later thought to have 86... But recent studies by Archinal and Hynes indicate it may have as many as 601 member stars!

Sunday, August 24

Today in 1966 from an Earth-orbiting platform, the Luna 11 mission was launched for a three day trip (Figure 8.53). After successfully achieving orbit, the mission went on to study lunar composition and nearby meteoroid streams.

On this date in 2006, 424 members of the International Astronomical Union shocked the world as they officially declared Pluto "to no longer be a planet." Discovered in 1930, Pluto enjoyed its planetary status for 76 years before being

Figure 8.53. Luna 11 (Credit—NASA).

retired. While textbooks will have to be re-written and the amateur science community will continue to recognize it as a solar system body, it is now considered to be a "dwarf planet."

So far in our southern expedition we've mined for gems, had our head in the clouds, and squashed a bug. What's left? Let's head over to the dark side as we take a look at the "Snake"...

Barnard Dark Nebula 72 is located about a fingerwidth north of Theta Ophiuchi (RA 17 23 02 Dec −23 33 48) (Figure 8.54). While sometimes dark nebulae are hard to visualize because they are simply an absence of stars, patient observers will soon learn to "see in the dark." The trained eye often realizes the presence of unresolved stars as a type of background "noise" that most of us simply take for granted—but not E. E. Barnard. He was sharp enough to realize there were at least 182 areas of the sky where these particular areas of nothingness existed, and he correctly assumed they were nebulae which were obscuring the stars behind them.

Unlike bright emission and reflection nebulae, these dark clouds are interstellar masses of dust and gas which remain unilluminated. We would probably not even know they were there except for the fact they eradicate star fields we know to be present! It is possible one day they may form stars of their own, but until that time we can enjoy these objects as splendid mysteries—and one of the most fascinating of all is the "Snake."

Put in a widefield eyepiece and relax... It will come to you. Barnard 72 is only a few light-years in expanse and a relatively short 650 light-years away. If at first

Figure 8.54. B 72: The Snake Nebula (Credit—Palomar Observatory, courtesy of Caltech).

you don't see it, don't worry. Like many kinds of object, spotting dark nebulae takes some practice.

Monday, August 25

Tonight we'll hop about two fingerwidths north of Nu Sagittarii to have a look at an open cluster that's a bit off the beaten path—NGC 6716 (RA 18 54 30 Dec –19 54 00). Comprised of around 75 genuine cluster members, this 100 million year old cluster will appear almost like a loose globular, with brighter stars superimposed over the field to the mid-sized telescope in a distinctive horseshoe pattern (Figure 8.55). At magnitude 7.5, it's not only within range of larger binoculars, but part of challenge lists as well. Be sure not to confuse it with the far more open Collinder 394 about a half degree southwest. Like all Collinder clusters, this is a large, sparse open that contains only a handful of stars in a V-shape.

If you're still feeling adventurous with a larger scope, drop back and take a much closer look at the Nu Sagittarii system. On the southern edge of eastern Nu-2's influence, you just might catch globular cluster NGC 6717 (RA 18 55 06 –22 42 02). If not, keep trying because you need a Palomar globular for your studies (Figure 8.56)! At very near magnitude 10, this loose, class VIII globular was discovered by Sir William Herschel in this month in 1784 and listed as H

Figure 8.55. NGC 6716 (Credit—Palomar Observatory, courtesy of Caltech).

Figure 8.56. NGC 6717 (Credit—Palomar Observatory, courtesy of Caltech).

III.143. Although it will appear as nothing more than a faint, round unresolved area, it truly is a globular cluster. At one time, this small cluster of stars was designated as IC 4802 with surrounding nebulosity—but tonight we'll log it as Palomar 9!

Tonight is also the peak of the Northern Iota Aquarid meteor shower. With no Moon to interfere, you might catch a bright streak!

Tuesday, August 26

On this date in 1981, Voyager 2 made a flyby of Saturn. Eight years later in 1989, Voyager 2 flew by Neptune on this date (Figure 8.57). Why don't we make a "date" tonight to have a look at this distant blue world? You'll find it on the ecliptic plane at sunset, hanging low in the east-southeast, at an altitude of only 24°. Neptune will be the lowest point of a nearly equilateral triangle with Altair above, and Jupiter to the west. While large binoculars can pick up Neptune's very tiny blue orb, you'll need a telescope to see it as a distant planet. Tonight let's head back to Scorpius to have a look at three pristine-open clusters. Begin your starhop at the colorful southern Zeta pair and head north less than 1° for NGC 6231 (RA 16 54 08 Dec −41 49 36).

Wonderfully bright in binoculars and well resolved to the telescope, this tight open cluster was discovered by Hodierna before 1654 (Figure 8.58). De Chéseaux

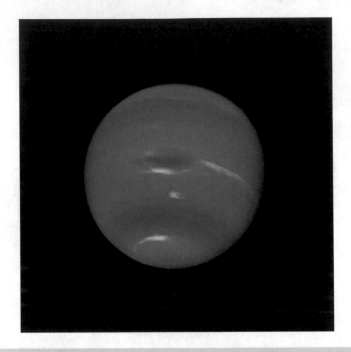

Figure 8.57. Voyager 2 image of Neptune (Credit—NASA).

Figure 8.58. NGC 6231 (Credit—Palomar Observatory, courtesy of Caltech).

Figure 8.59. Collinder 316 (Credit—Palomar Observatory, courtesy of Caltech).

cataloged it as object 9, Lacaille as II.13, Dunlop as 499, Melotte as 153, and Collinder as 315. No matter what catalog number you choose to put in your notes, you'll note this 3.2 million year young cluster as the "Northern Jewelbox!" For high power fans, look for the brightest star in this group—it's van den Bos 1833, a splendid binary.

About another degree north is the loose open cluster Collinder 316 (RA 16 55 30 Dec −40 50 00) with its stars scattered widely across the sky (Figure 8.59). Caught on its eastern edge is another cluster known as Trumpler 24—a place where new variables might lurk. This entire region is encased in a faint emission nebula called IC 4628—making this low-power journey through southern Scorpius a red hot summer treat!

Wednesday, August 27

If you did not get a chance to look at the Northern Jewelbox region in Scorpius, return again and sweep the area tonight. For those with larger telescopes, we're going to hop about a degree and a half south of the twin Nu Sagittarii for NGC 6242 (RA 16 55 36 Dec −39 28 00).

Discovered by Lacaille and cataloged as I.4, it is also known as Dunlop 520, Melotte 155, and Collinder 317 (Figure 8.60). At roughly magnitude 6, this open

Figure 8.60. NGC 6242 (Credit—Palomar Observatory, courtesy of Caltech).

Figure 8.61. NGC 6268 (Credit—Palomar Observatory, courtesy of Caltech).

cluster is within binocular range, but truly needs a telescope to appreciate its fainter stars. While NGC 6242 might seem like nothing more than a pretty little cluster with a bright double star, it also contains an X-ray binary which is a "runaway" black hole. It is surmised that it formed near the galactic center and was vaulted into an eccentric orbit when the progenitor star exploded. Its kinetic energy is much like that of a neutron star or a millisecond pulsar, and it is the first black hole to be confirmed to be in motion across the sky.

Now head a little more than a degree east-southeast for NGC 6268 (RA 17 02 40 Dec –39 44 18). At a rough magnitude of 9, this small open cluster can be easily observed in smaller scopes and resolved in larger ones (Figure 8.61). The cluster itself is somewhat lopsided, with more of its members concentrated on the western side. While it, too, might not seem particularly interesting, this young cluster is highly evolved and contains magnetic, chemically peculiar stars, and also Be class, or metal-weak, members.

Thursday, August 28

In 1789 on this day, Sir William Herschel discovered Saturn's moon Enceladus. But if Sir William were around tonight, he'd be looking at a star-forming region, as we head about a palm's width north of the lid star (Lambda) in the Sagittarius teapot as we seek out "Omega"...

Figure 8.62. M17: The Omega Nebula (Credit—Palomar Observatory, courtesy of Caltech).

Easily viewed in binoculars of any size and outstanding in every telescope, the 5000 light-year distant Omega Nebula was discovered by Philippe Loys de Chéseaux in 1745–46 and later (1764) cataloged by Messier as object 17 (RA 18 20 26 Dec –16 10 36). This beautiful emission nebula is the product of hot gases excited by the radiation of newly born stars. As part of a vast region of interstellar matter, many of its embedded stars don't show up in photographs, but reveal themselves beautifully to the eye at the telescope. As you look at its unique shape, you realize many of these areas are obscured by dark dust, and this same dust is often illuminated by the stars themselves.

Often known as "The Swan," M17 will appear as a huge, glowing check mark or ghostly "2" in the sky—but power up if you use a larger telescope and look for a long, bright streak across its northern edge with extensions to both the east and the north (Figure 8.62). While the illuminating stars are truly hidden, you will see many glittering points in the structure itself and at least 35 of them are true members of this region, which spans up to 40 light-years and could form up to 800 solar masses. It is awesome...

Friday, August 29

Tonight we'll return to the nebula hunt as we head about a fingerwidth north and just slightly west of M8 for the "Trifid"...

Figure 8.63. M20: The Trifid Nebula (Credit—Palomar Observatory, courtesy of Caltech).

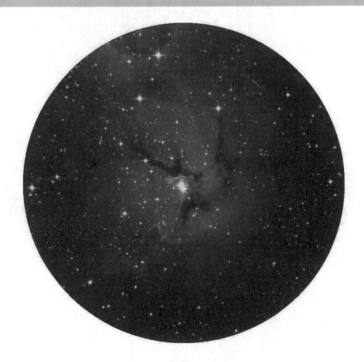

M20 (RA 18 02 23 Dec –23 01 48) was discovered by Messier on June 5, 1764, and much to his credit he described it as a cluster of stars encased in nebulosity (Figure 8.63). This is truly a wonderful observation since the Trifid could not have been easy given his equipment. Some 20 years later William Herschel (although he tried to avoid repeating Messier objects) found M20 of enough interest to assign separate designations to parts of this nebula—IV.41, V.10, V.11, and V.12. The word "Trifid" was first used by John Herschel to describe its beauty.

While M20 is a very tough call in binoculars, it is not impossible with good conditions to see light from an area which left its home nearly a millennium ago. Even smaller scopes will pick up the round, hazy patch of both emission and reflection, but you will need aversion to see the dark nebula which divides it; this was cataloged by Barnard as B 85. Larger telescopes will find the Trifid as one of the very few objects that actually appears much in the eyepiece as it does in photographs—with each lobe containing beautiful details, rifts, and folds best seen at lower powers. Look for its cruciform star cluster and its fueling multiple star system while you enjoy this triple treat tonight!

Saturday, August 30

Today (in 1991) celebrates Yohkoh (Figure 8.64). The Yohkoh Mission was a joint effort of both Japan and the United States to launch a satellite to monitor the Sun's corona and study solar flares. While the mission was quite successful, on December 14, 2001, the spacecraft's signal was lost during a total eclipse. Controllers were unable to point the satellite back toward the Sun, so its batteries discharged and Yohkoh became inoperable.

Tonight is New Moon and while the darkest skies are on our side, we'll fly with the "Eagle" as we hop another fingerwidth north of M17 to M16 (RA 18 18 48 Dec –13 49 00) and head for one of the most famous areas of starbirth, IC 4703...

While the open cluster NGC 6611 was discovered by Chéseaux in 1745–46, it was Charles Messier who cataloged the object as M16. And he was the first to note the nearby nebula IC 4703, now commonly known as the Eagle (Figure 8.65). At 7000 light-years distant, this roughly 7th magnitude cluster and nebula can be spotted in binoculars, but at best it is only a hint. As part of the same giant cloud of gas and dust as neighboring M17, the Eagle is also a place of starbirth illuminated by these hot, high-energy stellar youngsters which are only about 5.5 million years old.

In small to mid-sized telescopes, the cluster of around 20 brighter stars comes alive with a faint nebulosity that tends to be brighter in three areas. For larger telescopes, low power is essential. With good conditions, it is very possible to see areas of dark obscuration and the wonderful notch where the "Pillars of Creation" are located. Immortalized by the Hubble Space Telescope, they won't be nearly as grand or as colorful as the HST saw them, but what a thrill to know they are there!

Figure 8.64. Yohkoh (Credit—NASA).

Figure 8.65. M16 and the Eagle Nebula (Credit—R. Jay GaBany).

Sunday, August 31

Tonight we will begin entering the stream of the Andromedid meteor shower, which peaks off and on for the next couple of months. For those of you in the Northern Hemisphere, look for the lazy "W" of Cassiopeia to the northeast. This is the radiant—or relative point of origin—for this meteor stream. At times, this shower has been known to be spectacular, but let's stick with an accepted fall rate of around 20 per hour. These are the offspring of Biela's Comet, one that split apart in 1846 leaving radically different streams—much like 73/P Schwassman-Wachmann in 2006. These meteors have a reputation for red fireballs with spectacular trains, so watch for them in the weeks ahead. While there's still no Moon to interfere with the dark, let's take another look at the "dark" as we head toward open cluster NGC 6520...

Located just slightly more than a fingerwidth above Gamma Sagittarii and 5500 light-years away, NGC 6520 (RA 18 03 24 Dec –27 53 00) is a galactic star cluster which formed millions of years ago (Figure 8.64). Its blue stars are far younger than our own Sun, and may very well have formed from what you don't see nearby—a dark, molecular cloud. Filled with dust, Barnard 86 literally blocks the starlight coming from our galaxy's own halo area in the direction of the core. To get a good idea of just how much light is blocked by B 86, take a look at the

Figure 8.66. NGC 6520 and B 86 (Credit—Palomar Observatory, courtesy of Caltech).

star SAO 180161 on the edge. Behind this obscuration lies the densest part of our Milky Way! This one is so dark that it's often referred to as the "Ink Spot."

While both NGC 6520 and B 86 are about the same distance away, they don't reside in the hub of our galaxy, but in the Sagittarius Spiral Arm. Seen in binoculars as a small area of compression, and delightfully resolved in a telescope, you'll find this cluster is on the Herschel "400" list and many others as well. Enjoy this rare pair tonight!

CHAPTER NINE

September, 2008

Monday, September 1

On this day in 1859, solar physicist Richard Carrington (who first assigned sunspot rotation numbers) observed the first recorded solar flare. Naturally enough, an intense aurora followed the next day. In 1979, 120 years later, Pioneer 11 made history as it flew by Saturn. And where is Saturn now? Hanging out with the Sun!

As the nights begin to cool and darken earlier for those in the north, it's time for us to fly with the "Swan," as the graceful arch of the Milky Way turns overhead. Tonight we'll start by taking a look at a bright star cluster that's equally great in either binoculars or telescope—M39 (RA 21 32 58 Dec +48 27 00).

Located about a fistwidth northeast of Deneb (Alpha Cygni), you will easily see a couple of dozen stars in a triangular pattern. M39 is particularly beautiful because it will seem almost three dimensional against its backdrop of fainter stars (Figure 9.1). Younger than the Coma Berenices cluster, and older than the Pleiades, it is estimated to be 230 million years old. This loose, bright, galactic cluster is around 800 light-years away. Its members are all main sequence stars, with the brightest of them beginning to evolve into giants.

For more of a challenge, try dropping about a degree south-southwest for NGC 7082 (RA 21 29 00 Dec +47 08 00)—also known as H VII.52 (Figure 9.2). While it is a less rich, less bright, and far less studied open cluster, at magnitude 7.5 it's within range of binoculars and is on many open cluster observing lists. With

Figure 9.1. M39 (Credit—Palomar Observatory, courtesy of Caltech).

Figure 9.2. NGC 7082 (Credit—Palomar Observatory, courtesy of Caltech).

only a handful of bright stars to its credit, larger telescopes are needed to resolve many of the fainter members. Be sure to mark your notes for both objects!

Tuesday, September 2

Tonight we'll hunt with the "Fox" as we head to Vulpecula to try two more open cluster studies. The first can be seen easily with large binoculars or a low-power scope. It's a rich beauty that resides officially in the constellation of Vulpecula, but is more easily found by moving around three degrees southeast of Beta Cygni.

Known as Stock 1 (RA 19 35 48 Dec +25 13 00), you will often return to this stellar swarm, which contains 50 or so members of varying magnitudes (Figure 9.3). Combining and averaging the light of all its stars, Stock 1 has a visual magnitude near 5. Loose associations of stars—like Stock clusters—are currently being studied: recent information indicates that the members of Stock 1 are truly associated with one another.

A little more than a degree to the northeast is NGC 6815 (RA 19 40 44 Dec +26 45 32). While this slightly more compressed open cluster has no particular status among deep sky objects, it is another one to add to your collections of things to do and see (Figure 9.4)!

Figure 9.3. Field of Stock 1 (Credit—Palomar Observatory, courtesy of Caltech).

Figure 9.4. NGC 6815 (Credit—Palomar Observatory, courtesy of Caltech).

Wednesday, September 3

Today in 2006, SMART 1 went into the history books as it impacted the lunar surface in Lacus Excellentiae. Launched on September 27, 2003 by ESA, it entered lunar orbit over a year later on November 13, 2004 (Figure 9.5). After operating 5000 hours through 483 starts and stops, the xenon-ion engines ran out of fuel in September 2005, but not before the mission successfully completed its mapping studies. Although the craft caused a brief flash on the surface as it ended its life, it left an indelible mark on history—one almost as important as left in 1959 by the first craft to impact the Moon, Luna 2.

For skywatchers, look no further than the crescent Moon tonight to help you identify bright Spica nearby. Tonight your telescopic lunar mission is to journey to the edge of the east limb and slightly south of central to identify crater Humboldt (Figure 9.6). Seen on the curve, this roughly 200 kilometer wide crater holds a wealth of geographical details. Its flat, cracked floor has central peaks and a small mountain range, as well as radial rille structure. If libration and steadiness of skies are in your favor some time, power up and look for dark pyroclastic areas and a concentric inner crater.

Tonight we'll start with an asterism known as the "Coat Hanger," but which is also known as Brocchi's Cluster, or Collinder 399 (RA 19 25 24 Dec +20 11 00) (Figure 9.7). Let the colorful double star Beta Cygni (Albireo) be your guide as

Figure 9.5. SMART 1 (Credit—NASA).

you move about four degrees to its south-southwest. You will know this cluster when you see it, because it really does look like a coat hanger! Enjoy its red stars.

Discovered by Al Sufi in 964 AD, this 3.5 magnitude collection of stars was again recorded by Hodierna. Thanks to its extensive size of more than 60 arc minutes, it escaped the catalogues of both Messier and Herschel. Only a half dozen or so stars share the same proper motion, which may make it a cluster much like the Pleiades, but studies suggest it is merely an asterism...but one with two binary stars at its heart.

For larger scopes? Fade east to the last prominent star in the Coat Hanger and power up. NGC 6802 (RA 19 30 36 Dec +20 16 00) awaits you! At near magnitude 9, Herschel VI.14 is a well-compressed open cluster composed of faint stars (Figure 9.8). It continues to be useful in studies of stellar evolution and the properties of the galaxy's disc population. And this 100,000 year old cluster is on many observing challenge lists!

Thursday, September 4

Twenty-four hours ago in 1976, the Viking 2 lander successfully touched down on Mars. Tonight we'll land our eyes (through our telescopes!) on the lunar surface. If you were unable to identify Humboldt last night, try again tonight with Petavius as your guide. Although we have studied Petavius before, now is your chance once again to mark your studies and the Petavius Wall (Figure 9.9). Look for unusual features, such as 57 kilometer diameter Wrottesley on Petavius'

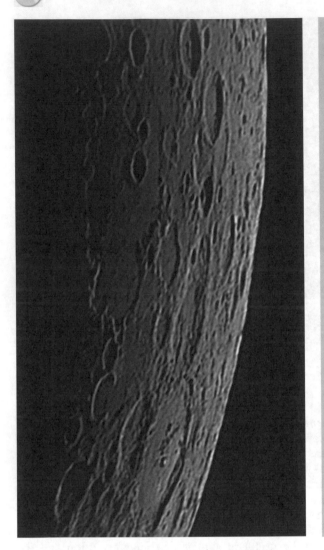

Figure 9.6. Crater Humboldt (Credit—Ricardo Borba).

northwest wall, or 83 kilometer wide Hase to the south with its deep interior impact...or how about long, shallow Legendre...and Phillips on Humboldt's west wall? If libration is good, you might even spot the edge of Barnard on Humboldt's southeast edge!

While the Moon will dominate tonight's sky, we can still take a very unusual and beautiful journey to a bright and very colorful pair of stars known as Omicron 1 Cygni (Figure 9.10). Easily located about halfway between Alpha (Deneb) and Delta on the western side (RA 20 13 28 Dec +46 46 40), this is a pure delight in binoculars or in telescopes of any size. The striking gold color of 3.7 magnitude Omi 1 A is easily highlighted against the blue of its same field companion, 5th magnitude Omi 1 B. Although this wide pairing is only an optical one, the K-type giant (Omi 1 A) is indeed a double star—an eclipsing variable about 150 times larger than or own Sun—and is surrounded by a gaseous corona more than

Figure 9.7. Collinder 399 (Credit—Palomar Observatory, courtesy of Caltech).

Figure 9.8. NGC 6802 (Credit—Palomar Observatory, courtesy of Caltech).

Figure 9.9. Petavius (Credit—Ricardo Borba).

Figure 9.10. Omicron 1 Cygni (Credit—Palomar Observatory, courtesy of Caltech).

double the size as the star itself. If you are using a scope, you can easily spot its blue-tinted, 7th magnitude companion star about one-third the distance between the two giants. Although our true pair are some two billion kilometers apart, they are oriented nearly edge-on from our point of view—allowing the smaller star to be totally eclipsed during each revolution. This total eclipse lasts for 63 days and happens about every 10.4 years, but don't stay up too late... We've still got more than five years to wait!

Friday, September 5

Tonight we'll discover beauty on our own Moon as we have a look at one of the last lunar challenges which occurs during the first few days of the Moon's appearance—Piccolomini (Figure 9.11). You'll find it to the southwest of the

Figure 9.11.
Piccolomini
(Credit—Alan Chu).

shallow ring of Fracastorius on Mare Nectaris' southern shore. Piccolomini is a standout lunar feature—mainly because it is a fairly fresh impact crater. Its walls have not yet been destroyed by later impacts, and the interior is nicely terraced. Power up and look carefully at the northern interior wall where a rock slide may have rumbled toward the crater floor. While the floor itself is fairly featureless, the central peak is awesome. Rising a minimum of two kilometers above the floor, it is even higher than the White Mountains in New Hampshire!

When you've caught up on your studies, let's have a look at Beta Lyrae (Figure 9.12) and Gamma Lyrae, the lower two stars in the "Harp." Beta is actually a quickly changing variable which drops to less than half the brightness of Gamma in around 12 days. For a few days the pair will seem of almost equal brightness; then you will notice the star closest to Vega begins to fade away. Beta is one of the most unusual spectroscopic stars in the sky, and it is possible that its eclipsing binary companion may be a prototypical "collapsar" (Yep—a black hole!) rather than an actual luminous body.

Now use the telescope for a pair of stars which are very close—Epsilon Lyrae (RA 18 44 20 Dec +39 40 12) (Figure 9.13). Known to most of us as the "Double Double," look about a fingerwidth northeast of Vega. Even the slightest optical aid will reveal this tiny star as a pair, but the real treat is with a telescope— because each component is a double star! Both sets of stars appear as primarily white, and each pair is very close in magnitude. What is the lowest power that you can use to split them?

Figure 9.12. Beta Lyrae (Credit—Palomar Observatory, courtesy of Caltech).

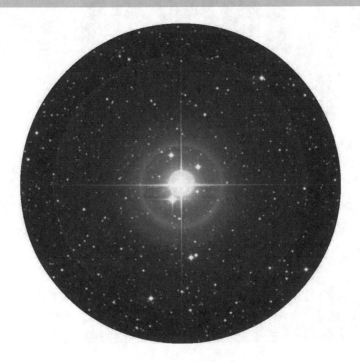

Figure 9.13. Epsilon 1 & 2 Lyrae (Credit—Palomar Observatory, courtesy of Caltech).

Saturday, September 6

Today celebrates the founding of the Astronomical and Astrophysical Society of America. Started in 1899, it is now known as the American Astronomical Society. Also on this date, in 2006, the milestone 1500th episode of Jack Horkheimer's Star Gazer series aired. The long-running short program on public television has led thousands of people, young and old, to "keep on looking up!" For a lifetime of achievement in public outreach, we salute you Mr. Horkheimer (Figure 9.14)!

Tonight when you have had a look at the Serpentine Ridge, drop south along the terminator and see if you can identify the very old crater Abulfeda (Figure 9.15), west of Theophilus.

This charming crater was named for Prince Ismail Abu'l Feda, who was a Syrian geographer and astronomer born in the late 13th century. Spanning 62 kilometers, its rocky walls show what once was a great depth, but the crater is now filled-in by lava, and drops to a mere 3110 meters below the surface. While it doesn't appear very large to the telescope, that's quite big enough to entirely hide Mt. Siple—one of the highest peaks in Antarctica! If conditions are steady, power up and take a look at Abulfeda's smooth-appearing floor. Can you see many smaller strikes? If the lighting is correct, you might even spot one far younger than the others!

Figure 9.14. Jack Horkheimer (widely used public image).

Figure 9.15. Abulfeda (Credit—Wes Higgins).

Sunday, September 7

For binoculars and telescopes, tonight's Moon will provide a piece of scenic history as we take an in-depth look at crater Albategnius. This huge, hexagonal, mountain-walled plain will appear near the terminator about one-third the way up from the south limb. This 136 kilometer wide crater is approximately 4390 meters deep, and its west wall will cast a black shadow on the dark floor. Albategnius is a very ancient formation, which was partially filled with lava at one point during its development. It is home to several wall craters like Klein (which will appear telescopically on its southwest wall). Albategnius holds more than just the distinction of being a prominent crater—it holds a place in history. On May 9, 1962 Louis Smullin and Giorgio Fiocco of the Massachusetts Institute of technology aimed a red laser toward the lunar surface and Albategnius became the first lunar object to be illuminated by a laser and then detected from Earth (Figure 9.16)!

On March 24, 1965 Ranger 9 took this "snapshot" of Albategnius (in the lower right) from an altitude of approximately 2500 kilometers. Companion craters in the image are Ptolemaeus and Alphonsus, which will be revealed for us tomorrow night. Ranger 9 was designed by NASA for one purpose—to achieve a lunar impact trajectory and send back high-resolution photographs and high-quality

Figure 9.16. Ranger 9 Image of Lunar Surface (Credit—NASA).

Figure 9.17. Ranger 9's Vidicon Cameras (Credit—NASA).

video images of the lunar surface. It carried no other scientific experiments, and its only destiny was to take pictures right up to the moment of final impact (Figure 9.17). It is interesting to note that Ranger 9 slammed into Alphonsus approximately 18.5 minutes after this photo was taken. They called that...a "hard landing."

Monday, September 8

Today in 1966, a legend was born as the television program *Star Trek* premiered. Created by Gene Roddenberry, its enduring mission inspired several generations' interest in space, astronomy, and technology. The long-running series still airs, along with many movie and series sequels. May it continue to "live long and prosper."

Let's walk upon the Moon this evening as we take a look at sunrise over one of the most often studied and mysterious of all craters—Plato (Figure 9.18). Located on the northern edge of Mare Imbrium and spanning 95 kilometers in diameter, Class IV Plato is simply a feature that all lunar observers check because of the many reports of unusual happenings there. Over the years, mists, flashes of light,

Figure 9.18. Sunrise over Plato (Credit—Wes Higgins).

areas of brightness and darkness, and the sudden appearance of small craters have become a part of Plato's lore.

On October 9, 1945 an observer sketched and reported "a minute, but brilliant flash of light" inside the western rim. Lunar Orbiter 4 photos later showed where a new impact may have occurred. While Plato's interior craterlets average about 1–2 kilometers in diameter, many times they can in fact be observed by amateurs—and not always under similar lighting conditions. No matter how many times you observe this crater, it is ever changing—and very worthy of your attention!

Now let's journey to a very pretty starfield as we head toward the western wingtip in Cygnus to have a look at Theta—also known as 13 Cygni (Figure 9.19). It is a beautiful main sequence star, also considered by modern catalogs to be a double. For large telescopes, look for a faint (13th magnitude) companion to the west... But it's also a wonderful optical triple!

Figure 9.19. Theta Cygni (Credit—Palomar Observatory, courtesy of Caltech).

Also in the field with Theta to the southeast is the Mira-type variable R Cygni which ranges in magnitude from 7 to 14 in slightly less than 430 days. This pulsating red star has a really quite interesting history that can be found at the AAVSO website, and is circumpolar for far northern observers. Check it out!

If you see a shooting star while you're out, it may belong to the Piscid meteor stream, which will reach its peak tonight with an expected maximum of around five meteors per hour. This particular shower favors the Southern Hemisphere. While this branch of the Piscids is largely unstudied, it is an unusual and diffuse stream that is active all month.

Tuesday, September 9

Tonight's featured lunar crater is located on the south shore of Mare Imbrium, right where the Apennine mountain range meets the terminator. Eratosthenes is a 58 kilometer diameter, and 375 meter deep, crater which is unmistakable. Named for the ancient mathematician, geographer, and astronomer Eratosthenes, this splendid Class I crater will display a bright west wall and a deep interior which includes its massive crater-capped central mountain, which reaches up to 3570 meters high (Figure 9.20)! Extending like a tail, an 80 kilometer long mountain ridge angles away to the southwest. As beautiful as Eratosthenes appears tonight,

Figure 9.20. Eratosthenes (Credit—Alan Chu).

it will fade away to total obscurity as the Moon becomes more nearly full. See if you can spot it in 5 days!

On this day in 1839, John Herschel made the very first glass plate photograph— and we're very glad he did! The photo was of the famous 40-foot telescope of John's father, William Herschel (Figure 9.21). The scope had not been used in decades and was disassembled shortly after its photograph was taken.

So, have you ever wondered if Sir William had a bad night or ever made an error? Considering how much the technology of astronomy has changed in the over 200 years since he did his work, you'd be surprised at the uncanny knack Herschel had for making correct calls.

Tonight take the journey to Vulpecula in an area roughly between M27 and NGC 6940. We're looking for a small grouping of brighter stars, so first focus your attention (and telescope) on star 20. This is the region for a magnitude 6 open cluster-within-a-cluster known by the names NGC 6882 (RA 20 11 48 Dec +26 49 00) and NGC 6885 (RA 20 12 00 Dec +26 29 00). It's significant because they were discovered on September 9–10, 1784 by Sir William and logged as H

Figure 9.21. The Herschel Scope Photographic Plate (archival image).

Figure 9.22. NGC 6882/NGC 6885 (Credit—Palomar Observatory, courtesy of Caltech).

VIII.20 and H VIII.22 (Figure 9.22). His notes indicate the clusters were north and south of each other, yet according to early research, his descriptions don't precisely match the starfield.

For many years, this was widely considered an error—assuming the only true cluster (NGC 6885) was the one around star 20... Yet another larger grouping does exist. Scientific studies have proved that two distinct physical clusters of stars are paired together in this region. While errors and disagreements followed later because of cataloging both objects under different names, one fact remains. On this night 224 years ago? Herschel was right.

Wednesday, September 10

Today is the birthday of James E. Keeler. Born in 1857, the American was a pioneer in the fields of spectroscopy and astrophysics. In 1895, Keeler proved that different parts of Saturn's rings rotate at different velocities. This clearly showed the Saturn's rings were not solid, but were instead a collection of smaller particles in independent orbits.

Tonight on the Moon, let's take an in-depth look at one of the most impressive of the southern lunar features—Clavius (Figure 9.23). Although you cannot help but be drawn visually to this crater, let's start at the southern limb near the terminator and work our way up. Your first sightings will be the large and shallow dual rings of Casatus, with its central crater, and Klaproth adjoining it. Further north is Blancanus with its series of very small interior craters—but wait until

Figure 9.23. Clavius and its craterlets (Credit—Wes Higgins, annotation by Tammy Plotner).

you see Clavius. Caught on the southeast wall is Rutherford, with its central peak, and crater Porter on the northeast wall. Look between them for the deep depression labeled D. West of D you will also see three outstanding impacts C, N, and J, while CB resides between D and Porter. The southern and southwest walls are also home to many impacts, and look carefully at the floor for many, many more! It has often been used as a test for a telescope's resolving power to see just how many more craters you can find inside tremendous old Clavius. Power up and enjoy!

Now let's head for the northeast corner of the little parallelogram that is part of Lyra for easy unaided eye and binocular double Delta 1 and 2 Lyrae (RA 18 54 30 Dec +36 53 55) (Figure 9.24). This is not an actual physical double (as was once suspected).

The westernmost Delta 1 is about 1100 light-years away and is a class B dwarf, but take a closer look at brighter Delta 2. This M class giant is only 900 light-years away. Perhaps 75 million years ago, it too was a B class star, but it now has a dead helium core, which nevertheless keeps on growing. While it is slightly variable, it may in the future become a Mira type. A closer look will show that it also has a true binary system nearby—a tightly matched 11th magnitude system, whose components *are* believed to be physically related. Oddly enough, this pair is about the same distance from Delta 2 as is Delta 1 itself.

Figure 9.24. Delta 1 and 2 Lyrae (Credit—Palomar Observatory, courtesy of Caltech).

Thursday, September 11

Today celebrates the birthday of Sir James Jeans (Figure 9.25). Born in 1877, English-born Jeans was an astronomical theoretician. At the beginning of the 20th century, Jeans worked out the fundamentals of gravitational collapse. This was an essential contribution to the understanding of the formation of solar systems, stars, and galaxies.

For skywatchers, Mercury has now reached its greatest elongation for early evening. Although the trio will be quite low and hard to pick out of twilight skies, look for Venus, Mercury, and Mars gathered with bright Spica just after the Sun dips below the horizon.

Tonight exploring the Moon will be in order as one of the most graceful and recognizable lunar features will be prominent—Gassendi (Figure 9.26). As an ancient mountain-walled plain that sits proudly at the northern edge of Mare Humorum, Gassendi sports a bright ring and triple central mountain peaks which are in range of binoculars.

Telescopic viewers will appreciate Gassendi at high power to view how its southern border has been eroded by lava flow and to see the many rilles and ridges inside the crater. Also note the presence of the younger crater Gassendi A on the north wall. While viewing the Mare Humorum region, keep in mind that we are looking at an area about the size of the state of Arkansas. It is believed that a planetoid collision originally formed Mare Humorum. This incredible impact crushed the surface layers of the Moon, resulting in a concentric "anticline" that can be traced to twice the size of the original impact area. The floor of this huge crater then filled in with lava. It was once thought to have a greenish appearance, but in recent years it has more accurately been described as reddish. That's one mighty big crater!

Figure 9.25. Sir James Jeans (widely used public image).

Figure 9.26. Gassendi (Credit—Wes Higgins).

Figure 9.27. Arthur Auwers (historical image).

Friday, September 12

Arthur Auwers was born today in 1838 (Figure 9.27). His life's work included unifying the world's observational catalogs, and he also calculated the orbits of Sirius and Procyon long before their companions were discovered. There's a lunar crater named for him!

Also today, in 1959, the USSR's Luna 2 became the first manmade object to hit the Moon (Figure 9.28). The successful mission landed in the Paulus Putredinus area. Today also celebrates the launch of Gemini 11 in 1966.

Tonight our primary lunar study is crater Kepler (Figure 9.29). Look for it as a bright point, slightly north of lunar center near the terminator. Its home is the Oceanus Procellarum—a sprawling dark mare composed primarily of minerals of low reflectivity (low albedo), such as iron and magnesium. Bright, young Kepler will display a wonderfully developed ray system. The crater rim is very

Figure 9.28. Luna 2 (Credit—NASA).

Figure 9.29. Lunar Orbiter 3 image of Kepler (Credit—NASA).

bright, consisting mostly of a pale rock called anorthosite. The "lines" extending from Kepler are fragments that were splashed out and flung across the lunar surface when the impact occurred. The region is also home to features known as "domes"—seen between the crater and the Carpathian Mountains. So unique are Kepler's geological formations that it became the first crater mapped by US Geological Survey in 1962.

Saturday, September 13

Today in 1922, the highest air temperature ever recorded on the surface of the Earth occurred. The measurement, taken in Libya, burned in at a blistering 136° F (58°C)—but did you know that the temperatures in the sunlight on the Moon double that? If you think the surface of the Moon is a bit too warm for comfort, then know that surface temperatures on the closest planet to the Sun can reach up to 800°F (430°C) at the equator during the day! As odd as it may sound, and even as close to the Sun as Mercury is, it could very well have ice deposits hidden below the surface at its poles.

Get out your telescope, because tonight we're going to have a look at a lunar feature that goes beyond simply incredible—it's downright weird. Start your journey by identifying Kepler, and head due west across Oceanus Procellarum until you encounter the bright ring of crater Reiner (Figure 9.30). Spanning 30 kilometers, this crater isn't anything showy...just shallow-looking walls with a

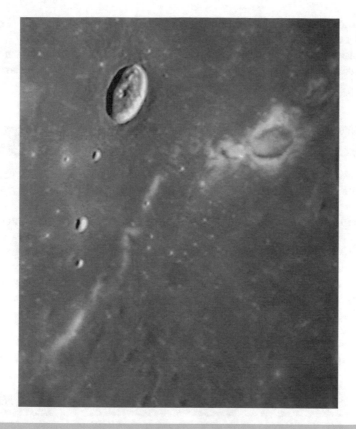

Figure 9.30. Reiner and Reiner Gamma (Credit—Alan Chu).

little hummock in the center. But look further west and a little more north for an anomaly—Reiner Gamma.

Well, it's bright. It's slightly eye shaped. But what exactly is it? Having no appreciable elevation or depth, Reiner Gamma could very well be an extremely young feature caused by a comet. Only three other such features are known to exist—two on the lunar far side and one on Mercury. They are high albedo surface deposits with magnetic properties. Unlike a lunar ray, consisting of material ejected from below the surface, Reiner Gamma can be spotted during the daylight hours—when ray systems disappear. And, unlike other lunar formations, it never casts a shadow.

Reiner Gamma is also a magnetic deviation on a barren world that has no magnetic field, so how did it form? Many ideas have been proposed, such as solar storms, volcanic activity, or even seismic waves. But the best explanation? It is the result of a cometary strike. Evidence exists that a split-nucleus comet, or cometary fragments, once impacted the area, and the swirl of gases from the high-velocity debris may have somehow changed the regolith. On the other hand, ejecta from such an impact could have formed around a magnetic "hot spot," much like a magnet attracts iron filings.

No matter which theory is correct, the simple act of viewing Reiner Gamma and realizing it is different from all other features on the Moon's Earth-facing side makes this journey well worth the time!

Sunday, September 14

With a nearly Full Moon, skies are light-trashed tonight. But would you like to have some fun? Then invite someone along for the ride and let's take a look at how differently people perceive stellar color!

Although it will be a bit on the low side, start first with Alpha Canes Venaticorum (RA 12 56 01 Dec +38 19 06), which is better known as Cor Caroli (Figure 9.31). The "Heart of Charles" is about 130 light-years away and is an easy double for even a small telescope. While many very noteworthy observers fail to see color in this pair, many of us can! Take a close look... Do you think the primary star is tinged a bit more on the copper side, while the secondary is faintly blue?

Now head up to Beta Cygni (RA 19 30 43 Dec +27 57 34)—Albireo (Figure 9.32). Again, this star is an easy and colorful split. Or is it? Well noted for its color contrast, almost every person this author has shared the eyepiece with sees it

Figure 9.31. Alpha CVn: Cor Caroli (Credit—Palomar Observatory, courtesy of Caltech).

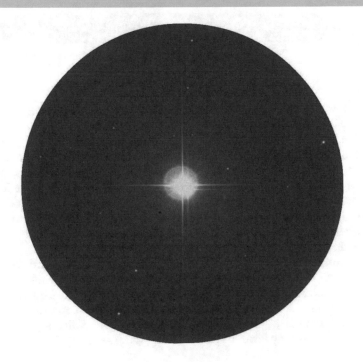

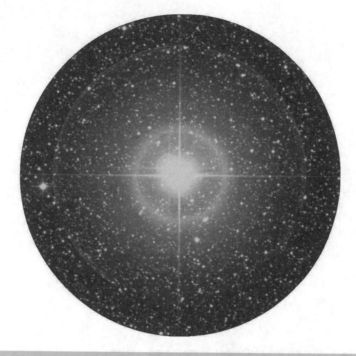

Figure 9.32. Beta Cygni: Albireo (Credit—Palomar Observatory, courtesy of Caltech).

differently. The primary star is often touted as a golden yellow and the secondary as blue...but in whose eyes? While I perceive them as orange and almost purple, many folks have reported seeing no color at all, or radical differences between them!

Now move on to Alpha Herculis (RA 17 14 38 Dec +14 23 25)—Ras Algethi (Figure 9.33). While it's a lot tougher to split, the suggestion that the M-type primary should be red to the sight isn't always correct. Also usually noted as a colorful pair, the companion star is supposed to be quite green—a color sensed well by the dark-adapted human eye. Perhaps some of my observing companions haven't been quite "human," because most see it as a very pale blue. Me? I see red and green. It would seem the answers aren't quite black and white!

So, what do all of these stars have in common? None of them are "normal." The A component of Cor Caroli is a magnetic and spectroscopic variable which has periodic changes in its metallic absorption lines. It is the most blue at minimum. Both the A and B stars are enveloped in an intense magnetic field. Albireo's primary star has a composite spectrum and is actually a binary—a K-type star with a spectroscopic B-type companion. The B component of Albireo is also odd—it shows strong hydrogen absorption lines. And what of Ras Algethi? Believe it or not, the red giant primary is a variable star which is shedding a huge envelope of a gas, engulfing its B companion in the process. A companion star which itself is a binary with a composite spectrum!

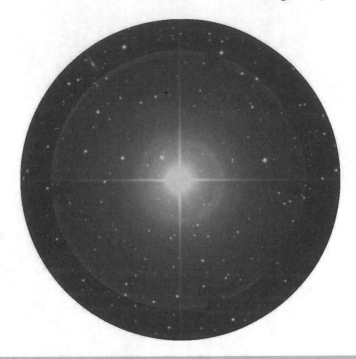

Figure 9.33. Alpha Herculis: Ras Algethi (Credit—Palomar Observatory, courtesy of Caltech).

Take a look at all of these stars in black and white...before I "color" your opinion of what you yourself can see!

Monday, September 15

In 1991 the Upper Atmosphere Research Satellite (UARS) was launched from Space Shuttle Discovery. The successful mission lasted well beyond its expected lifetime—sending back critical information about our ever-changing environment. After completing 14 years and 78,000 orbits, UARS remains a scientific triumph (Figure 9.34).

This is the universal date the Moon will become full and it will be the one closest to the Autumnal Equinox. Because its orbit is more nearly parallel to the eastern horizon, it will rise nearly at dusk for the next several nights in a row. On the average, the Moon rises about 50 minutes later each night, but at this time of the year it's only about 20 minutes later for mid-northern latitudes. Because of this added extra light, the name "Harvest Moon" arose because it allowed farmers more time to work in the fields (Figure 9.35).

Often times we perceive the Harvest Moon as being more orange than at any other time of the year. The reason is not only scientific but true. Coloration is caused by the scattering of the light by particles in our atmosphere. When the

Figure 9.34. UARS (Credit—NASA).

Moon is low, like now, we get more of that scattering effect, and it truly does appear more orange. The very act of harvesting itself produces more dust and often times that coloration will last the whole night through. And we all know that size is only an "illusion"...

So, instead of cursing the Moon for hiding the deep sky gems tonight, enjoy it for what it is...a wonderful natural phenomenon that doesn't even require a telescope!

And if you'd like to visit another object that only requires your eyes, then look no further than Eta Aquilae (RA 19 52 28 37 Dec +01 00 20), about one fistwidth due south of Altair...

Figure 9.35. Full Moon (Credit—John Chumack).

Figure 9.36. Eta Aquilae (Credit—Palomar Observatory, courtesy of Caltech).

Discovered by Pigot in 1784, this Cepheid variable varies by over a magnitude in a period of 7.17644 days (Figure 9.36). During this time it will reach a maximum of magnitude 3.7, and then decline slowly over five days to a minimum of 4.5... Yet it only takes two days to brighten again! This period of expansion and contraction makes Eta unique. To help gauge these changes, compare Eta to Beta on Altair's same southeast side. When Eta is at maximum, it will just about equal Beta in brightness.

Tuesday, September 16

Tonight we'll have a look at the central star of the "Northern Cross"—Gamma Cygni (Figure 9.37). Also known as Sadr, this beautiful main sequence star lies at the northern edge of the "Great Rift." Surrounded by a field of nebulosity known as IC 1310, second magnitude Gamma is very slowly approaching us, but still maintains an average distance of about 750 light-years. It is here in the rich, starry fields that the great dust cloud begins its stretch toward southern Centaurus—dividing the Milky Way into two streams. The dark region extending north of Gamma toward Deneb is often referred to as the "Northern Coalsack," but its true designation is Lynds 906.

Figure 9.37. Cygnus (Credit—John Chumack).

If you take a very close look at Sadr, you will find it has a well-separated 10th magnitude companion star, which is probably not related—yet in 1876 S. W. Burnham found that it is itself a very close double. Just to its north is NGC 6910, a roughly 6th magnitude open cluster which displays a nice concentration in a small telescope. To the west is Collinder 419, another bright gathering that is nicely concentrated. South is Dolidze 43, a widely spaced group with two brighter stars on its southern perimeter. East is Dolidze 10, which is far richer in stars of various magnitudes and contains at least three binary systems.

Whether you use binoculars or telescopes, chances are you won't see much nebulosity in this region—but the sheer population of stars and objects in this area makes your visit with Sadr worth the time!

Wednesday, September 17

Today in 1789, William Herschel discovered Saturn's moon Mimas. Tonight let's head about a fingerwidth south of Gamma Cygni to discover an open cluster well suited for all optics—M29 (RA 20 23 56 Dec +38 31 23).

Found and cataloged in 1764 by Charles Messier, this type D cluster has an overall brightness of about magnitude 7, but isn't exactly rich in stars (Figure 9.38). Hanging out anywhere from 6000 to 7200 light-years away, one would assume this to be a very rich cluster and it may very well be home to hundreds of stars—but their light is blocked by a dust cloud that is a thousand times denser than average.

Approaching us at around 28 kilometers per second, this loose grouping could be as old as 10 million years, and appears much like a miniature of the constellation Ursa Major at low powers. Even though it isn't the most spectacular cluster in star-rich Cygnus, it is another Messier object to add to your list!

Now let's have a look at Delta Cygni (RA 19 44 58 Dec +45 07 50). Located around 270 light-years away, Delta is known to be a difficult binary star (Figure 9.39). Its duplicity was discovered by F. Struve in 1830, and it is a very tough test for smaller optics. Located no more than 220 AU away from the magnitude 3 parent star, the companion orbits in 300 to 540 years, and is often

Figure 9.38. M29 (Credit—Palomar Observatory, courtesy of Caltech).

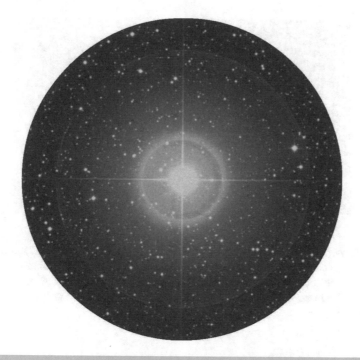

Figure 9.39. Delta Cygni (Credit—Palomar Observatory, courtesy of Caltech).

rated to be as dim as 8th magnitude. Delta is on many challenge lists, so if skies aren't steady enough to split it tonight, try again!

Thursday, September 18

While we are studying some of the summer's finest objects, we'd be remiss if we didn't look at another cosmic curiosity—the "Blinking Planetary" (RA 19 44 48 Dec +50 31 30). Located a couple of degrees east of visible star Theta Cygni, and in the same lower power field as 16 Cygni, it is formally known as NGC 6826 (Figure 9.40).

Viewable in even small telescopes at mid to high power, you'll learn very quickly how it earned its name. When you look directly at it, you can only see the central 9th magnitude star. Now look away. Focus your attention on visual double 16 Cygni. See that? When you avert, the nebula itself becomes visible. This is actually a trick of the eye. The central portion of our eye is more sensitive to detail and will only see the central star. At the edge of our vision, we are more likely to see dim light, and the planetary nebula appears. Located around 2000 light-years from our solar system, it doesn't matter if the "Blinking Planetary" is a trick of the eye or not... Because it's cool!

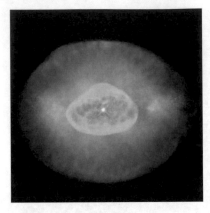

Figure 9.40.
NGC 6826: The
Blinking Planetary
(Credit—Hubble Space
Telescope/NASA).

Also known as Herschel IV.73 and Caldwell 16, this tiny planetary contains an abundance of carbon, and has many dust pockets in its structure. It skyrocketed to fame when viewed by the Hubble Space Telescope, which revealed the mysterious red "fliers," whose bow shocks point toward this planetary nebula—instead of away from it.

Friday, September 19

On this day in 1848, William Boyd was watching Saturn—and discovered its moon Hyperion. Also today in 1988, Israel launched its first satellite.

This evening will feature possibly the best occultation of the Pleiades by the Moon this year, so be sure to check with IOTA for details!

How long has it been since you've watched the ISS pass overhead, or seen an Iridium flare? Both are terrific events that don't require any special equipment to be seen—even in the daytime! Be sure to check with www.heavens-above.com for accurate times in your location—and enjoy. While you're out SkyWatching, be sure to have a look for Spica on the southwestern horizon after sunset. You just might discover a few planets joining the show!

When skies are dark, it's time for us to head directly between the two southernmost stars in the constellation of Lyra and grab the "Ring" (Figure 9.41). What summer would be complete without it?

Discovered by French astronomer Antoine Darquier in 1779, the Ring Nebula was cataloged later that year by Charles Messier as M57 (RA 18 53 35 Dec +33 01 45). In binoculars the Ring will appear as slightly larger than a star, yet it cannot be focused to a sharp point. To a modest telescope at even low power, M57 turns into a glowing donut against a wonderful stellar backdrop. The accepted distance to this unusual structure is about 1400 light-years, and how you see the Ring on any given night is highly dependent on conditions. As aperture and power increase, so do details, and it is not impossible to see braiding in the nebula's structure with scopes as small as 8" on a fine night, or to pick up the star caught on the edge in even smaller apertures.

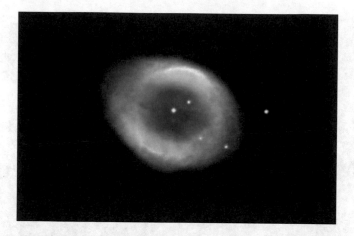

Figure 9.41. M57: The Ring Nebula (Credit—John Chumack).

Like all planetary nebulae, seeing the central star is considered the ultimate achievement in viewing. The central itself is a peculiar bluish dwarf which gives off a continuous spectrum, and might very well be a variable. At times, this shy, near 15th magnitude star can be seen with ease with a 12.5" telescope, yet be elusive to even 31" in aperture weeks later. No matter what details you may see, reach for the "Ring" tonight. You'll be glad you did.

Saturday, September 20

On this night in 1948, the 48" Schmidt telescope at Mt. Palomar was busy taking pictures (Figure 9.42). Its very first photographic plate was being exposed by the same man who ground and polished the corrector plate for this scope—Don Hendricks. His object of choice was reproduced as panel 18 in the Hubble Atlas of Galaxies, and tonight we'll join his vision as we take a look at the fantastic M31, the Andromeda Galaxy.

Seasoned amateur astronomers can literally point to the sky and show you the location of M31, but perhaps you have never tried to find it. Believe it or not, this is an easy galaxy to spot even under the moonlight. Simply identify the large diamond-shaped pattern of stars that is the Great Square of Pegasus. The northernmost star is Alpha, and it is here we will begin our hop. Stay with the northern chain of stars and look four fingerwidths away from Alpha for an easily seen star. The next along the chain is about three more fingerwidths away... And we're almost there. Two more fingerwidths to the north and you will see a dimmer star that looks like it has something smudgy nearby. Point your binoculars there, because that's no cloud—it's the Andromeda Galaxy (Figure 9.43)!

Now aim your binoculars or small telescope its way... Perhaps one of the most outstanding of all galaxies to the novice observer, M31 spans so much sky that

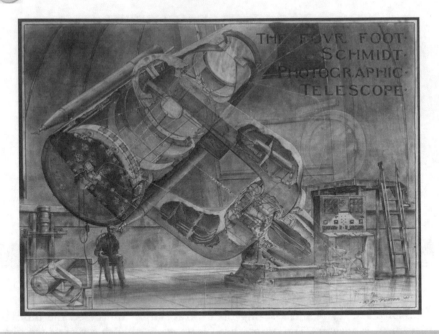

Figure 9.42. The 48 inch Schmidt Scope (Drawing by Russell Porter) (Credit—Palomar Observatory, courtesy of Caltech).

Figure 9.43. The Andromeda Galaxy (Credit—Anonymous).

it takes up several fields of view in a larger telescope, and even contains its own clusters and nebulae with New General Catalog designations. If you have a slightly larger telescope, you may also be able to pick up M31's two companions—M32 and M110. Even with no scope or binoculars, it's pretty amazing that we can see something—anything!—that is over two million light-years away!

Sunday, September 21

On this date in 2003, the Galileo spacecraft bravely entered the atmosphere of Jupiter as it completed its final mission (Figure 9.44). Launched in 1989 and orbiting the giant planet since 1995, the hugely successful Galileo taught us much about Jupiter's lethal radiation belts, magnetic field, atmosphere, and moons—but it had one last command to execute, self-destruction. Although it was still performing flawlessly (despite its lack of fuel and with its instrumentation badly scarred by radiation), scientists feared it might contaminate other possible life-sustaining moons such as Europa, and the decision was made to aim it into Jupiter's certain oblivion. We salute its final moments!

And what was Sir William Herschel doing on this date a couple of centuries ago? You can bet he was out telescoping; and his discoveries on this night were many. How about if we take a look at two logged on September 21 which made the Herschel "400" list?

Figure 9.44. Artist's concept of Galileo (Credit—NASA).

Our first stop is northern Cygnus for NGC 7086 (RA 21 30 30 Dec +51 35 00). Located on the galactic equator about five degrees west of Beta Cephei, our target is an open cluster (Figure 9.45). At magnitude 8.4, this loose collection will be difficult for the smaller scope, and show as not much more than an arrow-like asterism. However, larger scopes will be able to resolve many more stars, arrayed in long loops and chains around the brighter members. Although it's sparse, NGC 7086 has been studied for metal abundance, galactic distance, membership richness, and its luminosity function. Be sure to mark your notes for H VI.32, logged by Herschel in 1788.

Now hop on over to Andromeda for NGC 752 (RA 01 57 41 Dec +37 47 06). You'll find it just a few degrees south of Gamma and in the field north of star 56 (Figure 9.46). Located 1300 light-years away, there's a strong possibility this cluster was noted first by Hodierna before being cataloged by Herschel on this night (1786). At near magnitude 5, this "400" object is both large and bright enough to be seen in binoculars or small telescopes, and people have often wondered why Messier did not discover it. The star-studded field containing about 70 members of various magnitudes belong to H VII.32—a very old cluster which has more recently been studied for its metallicity and the variations in the magnetic fields of its members. Enjoy them both tonight! Sir William did...

Figure 9.45. NGC 7086 (Credit—Palomar Observatory, courtesy of Caltech).

Figure 9.46. NGC 752 (Credit—Palomar Observatory, courtesy of Caltech).

Monday, September 22

Today is the Autumnal Equinox. This marks the first day of the fall season for the Northern Hemisphere and we astronomers welcome back earlier dark skies!

Tonight we are going to take a journey once again toward an area which has intrigued this author since I first laid eyes on it with a telescope. Some think it difficult to find, but there is a very simple trick. Look for the primary stars of Sagitta just to the west of bright Albireo. Make note of the distance between the two brightest and look exactly that distance north of the "tip of the arrow" and you'll find M27 (RA 19 59 36 Dec +22 43 16).

Discovered in 1764 by Messier in a 3.5 foot focal length telescope, I discovered this 48,000 year old planetary nebula for the first time in a 4.5″ telescope (Figure 9.47). I was hooked immediately. Here before my eager eyes was a glowing green "apple core" which had a quality about it that I did not understand. It somehow moved... It pulsated... It appeared "living."

For many years I quested to understand the 850 light-year distant M27, but no one could answer my questions. I researched and learned the green glow was caused by doubly ionized oxygen, but that didn't explain the pulsations I saw. Like all amateurs, I became the victim of "aperture fever," and I continued to study M27 with a 12.5″ telescope, never realizing the answer was right there—I just hadn't powered up enough.

Figure 9.47. M27: The Dumbbell Nebula (Credit—R. Jay GaBany).

Several years later while studying at the Warren Rupp Observatory, I was viewing through a friend's identical 12.5" telescope and, as chance would have it, he was using about twice the magnification that I normally used on the "Dumbbell." Imagine my total astonishment as I realized for the very first time the faint central star had an even fainter companion which made it seem to wink! At smaller apertures or low power, this was not revealed. Still, the eye could "see" movement within the nebula—the central radiating star and its companion.

Do not sell the Dumbbell short. It can be seen as a small, unresolved area in common binoculars, is easily picked out with larger binoculars as an irregular planetary nebula, and turns astounding with even the smallest of telescopes. In the words of Robert Burnham,

The observer who spends a few moments in quiet contemplation of this nebula will be made aware of direct contact with cosmic things; even the radiation reaching us from the celestial depths is of a type unknown on Earth...

Tuesday, September 23

On this day in 1846, Johann Galle of the Berlin Observatory made a visual discovery. While at the telescope, Galle saw and identified the planet Neptune for the first time. Also on this day in 1962, the prime time cartoon *The Jetsons* premiered. Think of all the technology it inspired!

This evening, let's take the opportunity to have a look at one of the prettiest clusters in the night—M11 (RA 18 51 00 Dec –06 16 00) (Figure 9.48).

Figure 9.48. M11: The Wild Duck Cluster (Credit—Palomar Observatory, courtesy of Caltech).

Discovered in 1681 by German astronomer Gottfried Kirch at the Berlin Observatory, M11 was later cataloged by Charles Messier in 1764 and first dubbed the "Wild Ducks Cluster" by Admiral Smyth. To our modern telescopes and binoculars, there is little doubt as to how this rich galactic cluster earned its name—for it has a distinctive wedge-shaped pattern that closely resembles a flight of ducks. This fantastic open cluster of several thousand stars (about 500 of them are magnitude 14 or brighter) is approximately 250 million years old.

M11 is easily located by identifying Altair, the brightest star in Aquila. By counting two stars down the "body" of Aquila and stopping on Lambda you will find your starhop guide. Near Lambda you will see three stars, the centermost is Eta Scuti. Now just aim! Even small binoculars will have no problem finding M11, but a telescope is required to start resolving individual stars. The larger the telescope's aperture the more the stars will be revealed in this most concentrated of all open clusters!

Now grab a blanket and kick back to watch the Alpha Aurigid meteor shower. Relax, face northeast and look for the radiant near Capella. The fall rate is around 12 per hour, and they are fast and leave trails!

Wednesday, September 24

In 1970, the first automated return of lunar material to the Earth occurred on this day when the Soviets' Luna 16 returned with 85 grams of the Moon.

Tonight we'll start with the brightest star in Vulpecula—Alpha (Figure 9.49). Although it is not a true binary star, it is quite attractive in the telescope and an

Figure 9.49. Alpha Vulpeculae (Credit—Palomar Observatory, courtesy of Caltech).

Figure 9.50. NGC 6800 (Credit—Palomar Observatory, courtesy of Caltech).

Figure 9.51. NGC 6793 (Credit—Palomar Observatory, courtesy of Caltech).

easy split for binoculars. Alpha itself is a 4.4 magnitude red giant which makes a nice color contrast with the unrelated yellow field star which is two magnitudes dimmer.

Now head around a half degree northwest for the open cluster NGC 6800 (RA 19 27 06 Dec +25 08 00). Also known as Herschel VIII.21, this cluster is suitable for even smaller scopes but requires aperture to resolve completely (Figure 9.50). Discovered by Sir William in 1784, you'll like this ring-like arrangement of stars!

Now drop 2.7 degrees southwest of Alpha for yet another open cluster—NGC 6793 (RA 19 23 12 Dec +22 08 00). Discovered by Herschel in 1789 and logged as catalog object H VIII.81, you'll find a few more bright stars here (Figure 9.51). The challenge in this cluster is not so much being able to see it in a smaller telescope—but being able to distinguish a cluster from a starfield!

Thursday, September 25

Tonight let's take the time to hunt down an often overlooked globular cluster—M56 (Figure 9.52).

Located roughly midway between Beta Cygni and Gamma Lyrae (RA 19 15 35 Dec +30 11 04), this class X globular was discovered by Charles Messier in 1779 on the same night he discovered a comet, and it was later resolved by Herschel.

Figure 9.52. M56 (Credit—Palomar Observatory, courtesy of Caltech).

At magnitude 8 and small in size, it's a tough call for a beginner's binoculars but a very fine telescopic object. With an approximate distance of 33,000 light-years, this globular resolves well with larger scopes, but doesn't show as much more than a faint, round area with small aperture. However, the beauty of the chains of stars in the field makes it quite worth the visit!

While you're there, look carefully: M56 is one of the very few objects whose variable stars have been studied photometrically primarily by amateurs. While one bright variable was previously known, up to a dozen more have recently been discovered. Of those, six had their variability periods determined by CCD photography and telescopes just like yours!

Friday, September 26

Tonight we'll return again to Vulpecula—but with a different goal in mind. What we're after requires dark skies—but can be seen in both binoculars and a small telescope. Once you've found Alpha begin about two fingerwidths southeast and right on the galactic equator for NGC 6823 (RA 19 43 10 Dec +23 17 54).

The first thing you will note is a fairly large, somewhat concentrated, magnitude 7 open cluster. Resolved in larger telescopes, the viewer may note these stars are the hot, blue–white variety (Figure 9.53). For good reason... NGC 6823 only

Figure 9.53. NGC 6823/6820 (Credit—Palomar Observatory, courtesy of Caltech).

formed about two million years ago. Although it is some 6000 light-years away and occupies about 50 light-years of space, it's sharing the field with something more—a very large emission/reflection nebula called NGC 6820.

In the outer reaches of the star cluster, new stars are being formed in masses of gas and dust as hot radiation is shed from the brightest of the stellar members of this pair. Fueled by emission, NGC 6820 isn't always an easy visual object—it's faint and covers almost four times as much area as the cluster. But trace the edges very carefully, since the borders are much more illuminated than the region of the central cluster. It's like a whisper against your eyes. Take the time to really observe this one! The processes going on are very much like those occurring in the Trapezium area of the Orion nebula.

Be sure to mark your observing notes. NGC 6823 is Herschel VII.18, and NGC 6820 is also known as Marth 401!

Saturday, September 27

Today we celebrate the birth of Daniel Kirkwood in 1814 (Figure 9.54). In 1866, this American astronomer was the first to publish his discovery of gaps in the distances of asteroids from the Sun—"Kirkwood Gaps." Not only did he study the orbits of asteroids, but he was the first to suggest that meteor showers

Figure 9.54. Daniel Kirkwood (widely used public image).

Figure 9.55. Hickson Compact Group 87 (Credit—Palomar Observatory, courtesy of Caltech).

were caused by orbiting debris from comets. Known as "the American Kepler," Kirkwood went on to author 129 publications, including three books.

Tonight it's time to break out the muscle and challenge big telescope users to hone their skills. It's galaxy hunting time and our destination for tonight is Hickson Compact Group 87 (RA 20 48 11 Dec –19 50 24) (Figure 9.55)...

Several billion years ago, on the ecliptic plane about four degrees west-southwest of Theta Capricorni, and around 400 million light-years from our solar system, a galactic association decided to form their own "Local Group." Orbiting around a common center about every 100 million years, their mutual gravity is pulling each of them apart—creating starbursts and feeding their active galactic nuclei. Small wonder they're shredding each other... They're only 170,000 light-years apart! One day they may even form a single elliptical galaxy bright enough for the average telescope to see—because as they are now, this group isn't going to be seen with anything less than 20" in aperture.

So, shall we try something a little more within the realm of reality? Then go ahead and drop about eight degrees south of Theta and try picking up on the NGC 7016/17/18 group (RA 21 07 20 Dec –25 29 15). Are they faint? Of course! It wouldn't be a challenge if they were easy, would it? With an average magnitude of 14, this tight trio known as Leavenworth 1 is around 600 million light-years away (Figure 9.56). They're very small, and not very easy to locate... But for those who like something a bit different?

I dare you...

Figure 9.56. NGC 7016/17/18 (Credit—Palomar Observatory, courtesy of Caltech).

Sunday, September 28

As your starry mission this evening, we'll continue our studies in Vulpecula with a spectacular open cluster—NGC 6940 (RA 20 34 24 Dec +28 17 00). At close to magnitude 6, you'll find this unsung symphony of stars around three fingerwidths southwest of Epsilon Cygni (Figure 9.57).

Discovered by Sir William Herschel in 1784 and logged as H VIII.23, this intermediate-aged galactic cluster will blow your mind in large aperture. Although visible in binoculars, as aperture increases the field explodes into about 100 stars in a highly compressed, rich cloud. Although not visited often, NGC 6940 is on many observing challenge lists. Use low power to get the full effect of this stunning starfield!

Now move on to Aquila and look at the hot central star of an interesting planetary nebula—NGC 6804 (RA 19 31 35 Dec +09 13 32). You'll find it almost four degrees due west of Altair (Figure 9.58). Discovered by Herschel and classed as open cluster H VI.38, it wasn't until Pease took a closer look that its planetary nature was discovered. Interacting with clouds of interstellar dust and gases, NGC 6804 is a planetary in decline, with its outer shell around magnitude 12 and the central star at about magnitude 13. While only larger telescopes will get a glimpse of the central star, it's one of the hottest objects in space—with a temperature of about 30,000 Kelvin!

Figure 9.57. NGC 6940 (Credit—Palomar Observatory, courtesy of Caltech).

Figure 9.58. NGC 6804 (Credit—Palomar Observatory, courtesy of Caltech).

Monday, September 29

Tonight is New Moon and a great opportunity to have another look at some of the things we've studied during the last week or so. However I would encourage those of you with larger binoculars and telescopes to head for a dark-sky location, because tonight we are going on a quest...the quest for the holy "Veil."

By no means is the Veil Nebula Complex an easy one. The brightest portion, NGC 6992 (RA 20 56 20 Dec +31 41 48), can be spotted in large binoculars and you can find it just slightly south of the central point between Epsilon and Zeta Cygni (Figure 9.59). NGC 6992 is much better in a 6–8 inch scope, however, and low power is essential to see the long, ghostly filaments which span more than a degree of the sky.

About 2.5 degrees west-southwest, and incorporating star 52 (RA 20 45 39 Dec +30 43 11), is another long narrow ribbon of what may be classified as a supernova remnant (Figure 9.60). When aperture reaches the 12 inch range, the true breadth of this fascinating complex is revealed. It is possible to trace these filaments across several fields of view. They sometimes dim and at other times widen, but like a surreal solar flare, you will not be able to tear your eyes away from this area. Another undesignated area lies between the two NGCs, and the whole 1500 light-year distant complex spans over 2.5 degrees. Sometimes known as the Cygnus Loop, it's definitely one of the summer's finest objects.

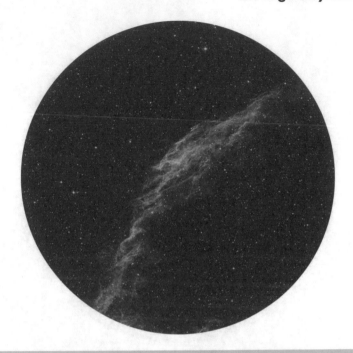

Figure 9.59. NGC 6992 (Credit—Palomar Observatory, courtesy of Caltech).

Figure 9.60. 52 Cygni region (Credit—Palomar Observatory, courtesy of Caltech).

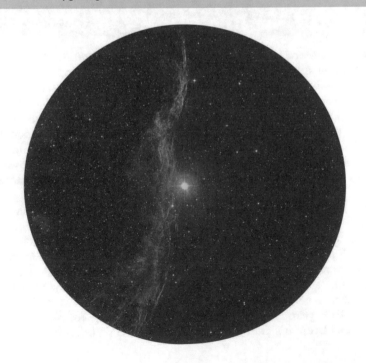

Figure 9.61. IC 5070: The Pelican Nebula (Credit—Palomar Observatory, courtesy of Caltech).

Tuesday, September 30

Today in 1880, Henry Draper must have been up very early indeed when he took the first photo of the Great Orion Nebula (M42). Although you might not wish to set up equipment before dawn, you can still use a pair of binoculars to view this awesome nebula! You'll find Orion high in the southeast for viewers in the Northern Hemisphere, and M42 in the center of the "sword" that hangs below its bright "belt" of three stars.

Tonight before we leave Cygnus for the year, try your luck with IC 5070 (RA 20 51 00 Dec +44 22 00), also known as the "Pelican Nebula" (Figure 9.61). You'll find it just about a degree southeast of Deneb and the surrounding binary star 56 Cygni.

Located around 2000 light-years away, the Pelican is an extension of the elusive North American Nebula, NGC 7000. Given its great expanse and faintness, catching the Pelican does require clean skies, but it can be spotted best with large binoculars. As part of this huge star-forming region, look for the obscuring dark dust cloud Lynds 935 to help you distinguish the nebula's edges. Although it is every bit as close as the Orion Nebula, this star hatchery isn't quite as easy!

CHAPTER TEN

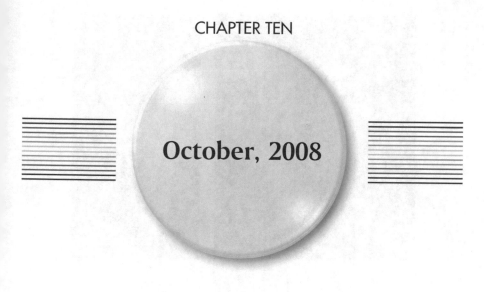

October, 2008

Wednesday, October 1

In 1897, the world's largest refractor (40") was debuted at the dedication of the University of Chicago's Yerkes Observatory (Figure 10.1). The immense telescope was 64 feet (19.5 meters) long and weighed six tons. Also today, in 1958, NASA was established by an act of Congress. More? In 1962, the 300 foot radio telescope of the National Radio Astronomy Observatory (NRAO) went live at Green Bank, West Virginia (Figure 10.2). It held its place as the world's second largest radio scope until it collapsed in 1988. (It was rebuilt as a 100 meter dish in 2000.)

Although first light for the 40" was Jupiter, E. E. Barnard later discovered the third companion star to Vega using the Yerkes refractor. First "light" studies at Green Bank included a radio source galaxy and pulsar for NRAO. Tonight we're going to turn our attention toward Pegasus. Our destination is not an easy one, but if you have a 6" or larger scope, you'll fall in love at first sight! Let's head for Eta Pegasi and slightly more than four degrees north-northeast for NGC 7331 (RA 22 37 04 Dec +34 24 58).

This beautiful, 10th magnitude, tilted spiral galaxy is very much how our own Milky Way would appear if we could travel 50 million light-years away and look back (Figure 10.3). Very similar in structure to both our galaxy and the Great Andromeda Galaxy, this particular galaxy gains more and more interest as scope size increases—yet it can also be spotted with larger binoculars. At around

Figure 10.1. The 40 inch refractor at Yerkes (widely used public image).

8" in aperture, a bright core appears at the beginnings of wispy arms. In the 10–12 inch range, spiral patterns begin to emerge and, with good seeing conditions, you can see patchiness in structure as nebulous areas are revealed, and the western half is deeply outlined with a dark dustlane. But hang on...

Because the best is yet to come!

Figure 10.2. 300 foot Radio Telescope at Green Bank (Credit—NRAO/AUI/NSF).

Figure 10.3. NGC 7331 (Credit—R. Jay GaBany).

Thursday, October 2

As the skies darken tonight, have a look at the Moon! If the lunar terminator has not advanced too far at your viewing time, scan the southeast shoreline of Mare Crisium for Agarum Promontorium. Look how boldly it progresses northward across the dark plain before it disappears beneath the once-molten lava. There were times in the past when great lunar observers noted a mist-like appearance in this area—another transient lunar phenomenon.

While time and the stars appear to stand still, and astronomical twilight begins earlier each night, return to NGC 7331 with all the aperture you have. What we are about to look at is truly a challenge and requires dark skies, optimal position, and excellent conditions. Now breathe the scope about a half degree south-southwest (RA 22 36 03 Dec +33 56 52) and behold one of the most famous galaxy clusters in the night…

In 1877, French astronomer Edouard Stephan was using the first telescope designed with a coated mirror when he discovered something a bit more with NGC 7331. He found a whole group of nearby galaxies! This faint gathering of five is now known as "Stephan's Quintet" and its members are no further apart than the diameter of our own Milky Way galaxy (Figure 10.4).

Visually in the average large scope, these members are all rather faint, but their proximity is what makes them such a curiosity. The Quintet is made up of five galaxies numbered NGC 7317, 7318, 7318A and B, 7319 and 7320—which is the largest. Even with a 12.5" telescope, this book's author has never seen them as much more than tiny, barely-there objects that look like ghosts of rice grains on a dinner plate. So why bother? Because I've seen them with really large aperture…

What our backyard equipment can never reveal is what else exists within this area—more than 100 star clusters and several dwarf galaxies. Some 100 million

Figure 10.4. Stephan's Quintet (Credit—Palomar Observatory, courtesy of Caltech).

years ago, the galaxies collided and left long streamers of material, which created star-forming regions of their own, and this tidal pull keeps them connected. The stars within the galaxies themselves are nearly a billion years old, but in the region between them are much younger stars. Although we cannot see them, you can make out the soft sheen of the galactic nuclei of our interacting group.

Enjoy their faint mystery!

Friday, October 3

Tonight we'll begin on the lunar surface and go out on a limb—the southeastern limb—to have a look at an unusual crater. Named for the French agrochemist and botanist Jean-Baptiste Boussingault, this elliptical-appearing crater actually spans a handsome 71 kilometers (Figure 10.5). What makes Boussingault so unusual is that it is home to its own large interior crater—A. This double-ring formation gives it a unique stepped, concentric look that's worth your time!

Now wait for the Moon to wester a bit and we'll return to Pegasus and the incredible M15. Although skies are a bit bright, you can still have a very satisfactory look at M15 through binoculars or telescope of any size (Figure 10.6).

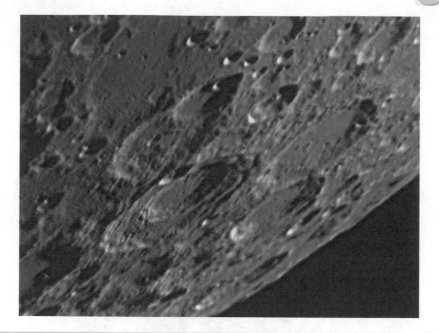

Figure 10.5. Boussingault (Credit—Ricardo Borba).

Figure 10.6. M15 (Credit—Palomar Observatory, courtesy of Caltech).

Figure 10.7. Gamma Aquilae (Credit—Palomar Observatory, courtesy of Caltech).

You can find it easily just about two fingerwidths northwest of red Epsilon Pegasi—Enif (RA 21 29 58 Dec +12 10 00). Shining brightly at magnitude 6.4, low-power users will find it a delightfully tight ball of stars, but scope users will find it unique. As resolution begins, sharp-eyed observers will note the presence of a planetary nebula—Pease 1. This famous X-ray source you have just seen with your eyes may have supernova remnants buried deep inside...

When we're done? Let's go have a look at Gamma Aquilae just for the heck of it (Figure 10.7). Just northwest of bright Altair, Gamma (RA 19 46 15 Dec +10 36 47) has the very cool name of Tarazed and is believed to be over 300 light-years away. This K3-type giant will show just a slight yellow coloration—but what really makes this one special is the low-power field!

Saturday, October 4

Today in 1957, the USSR's Sputnik 1 made space history as it became the first manmade object to orbit the Earth (Figure 10.8). The Earth's first artificial satellite was tiny, roughly the size of a basketball, and weighed no more than the average man. Every 98 minutes it swung around Earth in its elliptical orbit...and changed everything. It was the beginning of the "Space Race." Many of us old enough to remember Sputnik's grand passes will also recall just how inspiring it was. Take

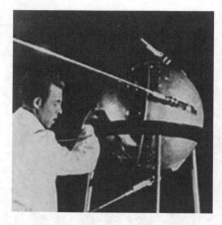

Figure 10.8. Sputnik 1 (Credit—NASA).

the time with your children or grandchildren to check www.heavens-above.com for visible passes of the ISS, and think about how much our world has changed in just over half a century!

Do you remember the Professor Burg who discovered Antares' companion during an occultation? Well, tonight we're going to become a whole lot more familiar with the good professor, because it's about to happen again! For almost

Figure 10.9. 2005 Antares Occultation (Credit—John Chumack).

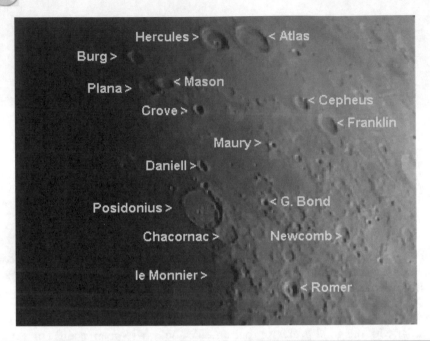

Figure 10.10. Detail view of Posidonius (Credit—Greg Konkel, annotation by Tammy Plotner).

all observers, brilliant red Antares is less than half a degree to the north of tonight's crescent Moon. For many of you, this will be a spectacular occultation, so be sure to check IOTA for precise times and locations (Figure 10.9).

While we're waiting for the event, let's have a look at the crater named for Burg as we begin by using past study crater Posidonius as our guide. How many more of these craters can you identify?

If you walk along the terminator to the northwest, you'll see the punctuation of 40 kilometer wide Burg just emerging from the shadows. While it doesn't appear to be a grand crater like Posidonius (Figure 10.10), it has a redeeming feature— it's deep, real deep. If Burg were filled with water here on Earth, it would require a deep submergence vehicle like ALVIN to reach its 3680 meter floor! This class II crater stands nearly alone on an expanse of lunarscape known as Lacus Mortis. If the terminator has advanced enough at your time of viewing, you may be able to see this walled plain's western boundary peeking out of the shadows.

Sunday, October 5

Today marks the birthdate of Robert Goddard (Figure 10.11). Born in 1882, Goddard is known as the father of modern rocketry—and with good reason. In 1907, Goddard came into the public eye when a cloud of smoke erupted from the basement of the physics building in Worcester Polytechnic Institute,

Figure 10.11.
Robert Goddard
(Credit—NASA).

from which he had just fired a powder rocket. By 1914, he had patented the use of liquid rocket fuel, and the design of two- or three-stage solid fueled rockets. His work continued as he sought methods of lofting equipment ever higher, and by 1920 he had envisioned his rockets reaching the Moon. Among his many achievements, he proved that a rocket would work in a vacuum. By 1926 the first scientific equipment went along for the ride; by 1932 Goddard was guiding those flights; and by 1937 his motors were pivoting on gimbals and being controlled gyroscopically. His lifetime of work went pretty much unnoticed until the dawn of the Space Age, but in 1959 (14 years after his death) he received acclaim at last as NASA's Goddard Space Flight Center was established in his memory.

Tonight let's rocket to the Moon to explore a binocular curiosity located on the northeast shore of Mare Serenitatis. Look for the bright ring of Posidonius, which contains several equally bright points both around and within it. Now look at Mare Crisium and get a feel for its size. A little more than one Crisium's length west of Posidonius you'll meet Aristotle and Eudoxus. Drop a similar length south and you will be at the tiny, bright crater Linne on the expanse of Mare Serenitatis (Figure 10.12). So what's so cool about this little white dot? With only binoculars you are resolving a crater that is 1 mile wide, in a 7 mile wide patch of bright ejecta—from close to 400,000 kilometers away!

Tonight in 1923, Edwin Hubble was also busy as he discovered the first Cepheid variable in the Andromeda Galaxy. Hubble's discovery was crucial in proving that the objects once classed as "spiral nebulae" were actually independent and external stellar systems like our own Milky Way.

While we're out, let's have a look at a Mira type variable, as we look about halfway between Beta and Gamma Cygni for Chi (RA 19 50 33 Dec +32 54 51).

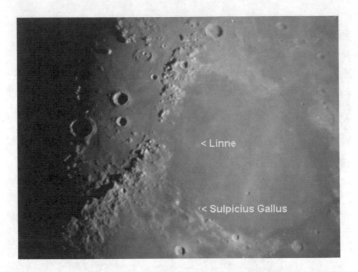

Figure 10.12. Crater Linne (Credit—Greg Konkel, annotation by Tammy Plotner).

Figure 10.13. Chi Cygni at minimum (center of field) (Credit—Palomar Observatory, courtesy of Caltech).

Noted for being the second long-term variable discovered (by Gottfried Kirch in 1868), Chi is visible to the unaided eye when at maximum—but demands a telescope at minimum (Figure 10.13). Fluxing between magnitude 4 and 12, you'll know if you've caught it at its lowest point when you can't distinguish it from background stars! Of course, this is another wonderful cosmic joke on Bayer—for it was several years until the star he classed as visual returned to view. Maximum or minimum? Enjoy your own perceptions of this lovely red star!

Monday, October 6

Tonight let's walk on the Moon and check off a few more features on your lunar list! Look for the prominent pair of Aristillus and Autolycus caught just east of the Apennine Mountain range (Figure 10.14). If you haven't logged the shallow Archimedes, tonight is your chance. Take the time to closely inspect the differing lava flow patterns on the floor of Palus Putredinus—you can't miss the bright ring of Manilius.

Now point those binoculars toward the northwestern corner of Capricornus and have a look at the spectacular Alpha (RA 20 18 00 Dec –12 32 00)!

Although the Alpha 1 and 2 pairing is strictly visual, that won't stop you from enjoying their slightly yellow and orange colors (Figure 10.15). Collectively they are named Al Giedi, and the brighter of the pair is Alpha 2 about 100

Figure 10.14. Moon walk (Credit—Roger Warner).

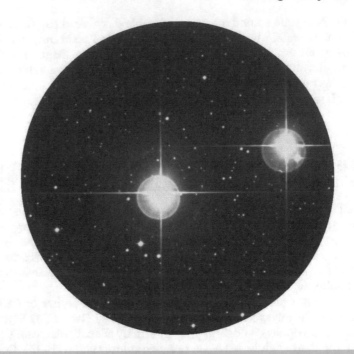

Figure 10.15. Alpha Capricorni (Credit—Palomar Observatory, courtesy of Caltech).

Figure 10.16. Beta Capricorni (Credit—Palomar Observatory, courtesy of Caltech).

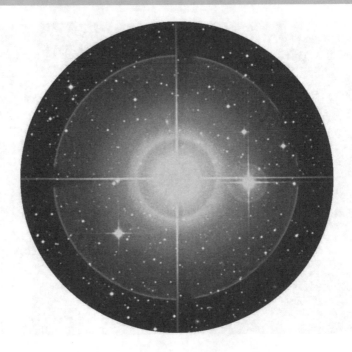

light-years distant, while Alpha 1 is about five times further away. Now power up with a telescope and you'll find that both stars are also visual doubles! While the companion stars to both are nearly the same magnitude, you'll find that the Alpha 2 pair is separated by three times as much distance. Be sure to mark your observation lists and enjoy!

Now let's travel just south of Alpha to beautiful Beta (RA 20 21 00 Dec –14 46 52)...

Named Dabih, this lovely white 3rd magnitude star has a very easily split 6th magnitude companion which will appear slightly blue (Figure 10.16). Over 100 times brighter than our own Sun, the primary star is also a spectroscopic triple—one whose unseen companions orbit in a little over 8 days and 1374 days, respectively. Oddly enough the B star is also a very tight binary as well—yet the two major stars of this system are separated by about a trillion miles! If you have a large aperture telescope—power up. According to T. W. Webb, a 13th magnitude unrelated double is also found in between the two brighter stars. No matter if you choose binoculars or a telescope, I'm sure you'll find the 150 light-year trip worth your time to add to your doubles list!

Tuesday, October 7

Today celebrates the birthday of Niels Bohr (Figure 10.17). Born in 1885, Bohr was a pioneer Danish atomic physicist. If Niels were alive today, he'd be out early this evening looking at the beautiful sight of Jupiter and the Moon gracing the sky.

Figure 10.17. Niels Bohr (widely used public image).

Figure 10.18.
Purbach and
Regiomontanus
(Credit—Wes Higgins).

Figure 10.19. Delta Capricorni (Credit—Palomar Observatory, courtesy of Caltech).

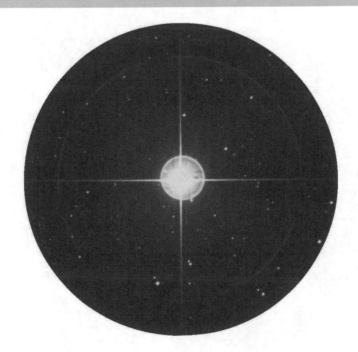

Our lunar mission for tonight is to move south past the crater rings of Ptolemaeus, Alphonsus, and Arzachel until we reach the remains of crater Purbach. Named for 15th century astronomer Georg von Peuerbach, this highly eroded ring spanning 118 kilometers is nearly obliterated on its western side—yet the eastern walls still stand a respectable 2980 meters high. Power up a take a close look at the W crater in Purbach's center (Figure 10.18). It, too, has eroded with time!

Sharing a common wall, you will find the very unique Regiomontanus to the south. Unlike the ring of Purbach, this one is oblate. With dimensions of 120 by 110 kilometers, Regiomontanus' beauty is in the central peak. Again, use high magnification. On a steady night you'll see the A crater sitting atop a nearly six kilometer wide base mountain which reaches 1200 meters above the crater floor. Careful examination reveals what appears to be a volcano that's blown its top! Can you trace the lava flow?

Now let's go have a look at the northeastern corner of Capricornus as we learn about Delta (RA 21 47 02 Dec −16 07 38)...

Its proper name is Deneb Algedi and this nearly 3rd magnitude star is a stunning blue—white (Figure 10.19). Curiously enough, it's a rather close star—only about 50 light-years from Earth. Hovering so close to it that we cannot even correctly assess its spectral type is a binary companion whose eclipsing orbit causes Delta to be a very slight variable—with a period of just about one day. In its own way, Delta is rather historic...because it was only four degrees north of this star that Uranus was first sighted by Galle in 1846!

Wednesday, October 8

It's bold. It's beautiful. You've looked at it hundreds of times...and tonight? It's Copernicus...

While Copernicus is not the oldest, deepest, largest, or brightest crater on the Moon, it certainly is one of the most detailed. Visible in binoculars toward Plato and near the terminator, this youthful crater gives a highly etched appearance. Its location in a fairly smooth plain near the center of the Moon's disc, and its prominent "splash" ray system, all combine to make Copernicus visually stunning in a small telescope.

Spanning 100 kilometers, with 23 kilometer thick walls, the "Mighty One" is most definitely an impact crater that left its impression down to 3840 meters below the surface. Geologist Gene Shoemaker cited many features of Copernicus which mirror our own terrestrial impact features. Many of these Copernican features could have been caused by a large meteoritic body—a body about the size of Comet Halley's nucleus. No matter what optical aid you use, mid-placed Copernicus simply rocks (Figure 10.20)!

Today marks the birthday of Ejnar Hertzsprung (Figure 10.21). Born in 1873, Hertzsprung was a Danish astronomer who first proved the existence of giant and dwarf stars in the early 1900s. His discoveries included the relationship between

Figure 10.20. Copernicus (Credit—Greg Konkel).

color and luminosity, which wasn't truly recognized until it was recovered by Henry Russell. Now it is a familiar part of all our studies as the Hertzsprung–Russell diagram. His use of absolute magnitudes will come into play tonight as we have a look at the age-old mystery of M73.

Figure 10.21. Ejnar Hertzsprung (widely used public image).

Located about three fingerwidths north-northwest of Theta Capricorni (RA 20 59 00 Dec –12 38 00), this 9th magnitude open cluster consisting of four stars was discovered by Charles Messier on October 4, 1780. He described it as a "Cluster of three or four small stars, which resembles a nebula at first glance..." It has been hotly debated whether the grouping is a genuine cluster or simply an asterism. M73 was also included in John Herschel's catalog (GC 4617) and was given the NGC 6994 designation by Dreyer (Figure 10.22). In 1931 Collinder cataloged it as Cr 426, with an estimated distance of 12,000 light-years. Still, the debate about its authenticity as a physically related group continues.

At least two stars show the same proper motion, leading some scientists to believe M73 may be the remnant of a much older and now dispersed cluster. On the other hand, these two stars may be the only ones actually related. But another study of 140 stars in the region concluded that 24 may be real members—including those from Messier's original observations. Thanks to the work of Hertzsprung and Russell, these candidates fall within the color–magnitude diagram of a 2–3 billion year old cluster, with Messier's suspect four being evolved giants.

Even though more recent data once again indicate M73 may simply be an asterism—sharing no common proper motion—you can still enjoy this unusual Messier in even a small telescope!

Figure 10.22. M73 (Credit—Palomar Observatory, courtesy of Caltech).

Thursday, October 9

Tonight is the peak of the Draconid meteor shower, whose radiant is near the westering constellation of Hercules. This particular shower can be quite impressive when comet Giacobini-Zinner passes near Earth. When this happens, the fall rate jumps to 200 per hour and has even been known to reach 1000. So what am I going to tell you about this year? Comet Giacobini-Zinner reached perihelion on July 2, 2005, passing within 8 million kilometers of Earth, but has now greatly distanced itself from our solar system. Chances are the Draconids will only produce about 3–5 per hour, but no one knows for sure.

While we're cursing the Moon for blinding us, let's take a look through the telescope as we first identify Palus Epidemiarum to the south. Not much more than a slightly darker extension of Mare Nubium to the northeast, this innocent looking area holds a wealth of rimae. Power up and look at the south shore for the peninsula-like feature of crater Capuanus (Figure 10.23). Named for the Italian theologian and astronomer Francesco Capuano di Manfredonia, this awesome little crater is 60 kilometers wide and has a very strong west wall!

Now let's head for Beta Capricorni and drop south-southeast to have a look at a pair of doubles—Rho and Pi (Figures 10.24 and 10.25).

Northernmost Rho is a multiple system slightly less than 100 light-years away, with each discernable member also being a spectroscopic double. Separated by about an eighth of a light-year, look for a 5th magnitude yellow–white giant with a very close 9th magnitude companion. Further south is Pi, a triple star system which has a traditional name—Okul. Located around 670 light-years away, look for a bright blue–white 5th magnitude primary that is also a spectroscopic double—and its much easier C component, which is around magnitude 8.

Figure 10.23.
Capuanus (Credit—Wes Higgins).

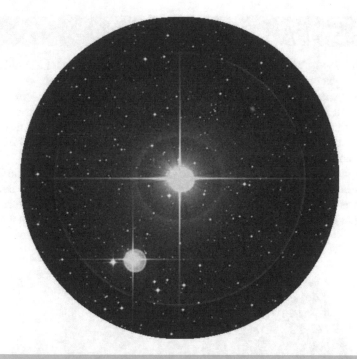

Figure 10.24. Rho Capricorni (Credit—Palomar Observatory, courtesy of Caltech).

Figure 10.25. Pi Capricorni (Credit—Palomar Observatory, courtesy of Caltech).

Friday, October 10

Today in 1846, William Lassell was busy at his scope as he made a new discovery—Neptune's moon Triton (Figure 10.26). Although our everyday equipment can't "see" Triton, we can still have a look at Neptune which is also hanging out in tonight's study constellation of Capricornus less than a degree south of the Moon. Try checking astronomy periodicals or many great online sites for accurate locator charts. For some lucky astronomers, it will be an occultation event!

Tonight we'll let Gassendi be our guide as we head north to examine the ruins of crater Letronne (Figure 10.27). Sitting on a broad peninsula on the south edge of Oceanus Procellarum, this class V crater once spanned 118 kilometers. Thanks to the lava flows which formed Procellarum, virtually the entire northern third

Figure 10.26.
William Lassell (widely used public image).

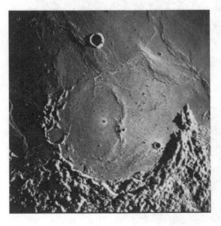

Figure 10.27. Apollo 16 image of Letronne (Credit—NASA).

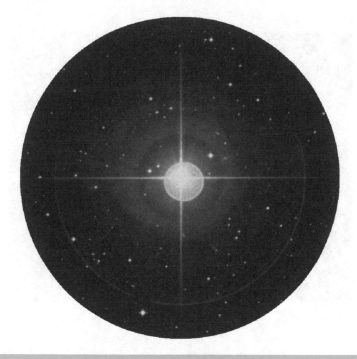

Figure 10.28. Zeta Capricorni (Credit—Palomar Observatory, courtesy of Caltech).

of the crater was submerged, leaving the remaining scant walls to rise no more than 1000 meters above the surface. While this might seem shallow, it's as high as El Capitan in Yosemite.

Although tonight's bright skies will make our next target a little difficult to find visually, look around four fingerwidths southwest of Delta Capricorni (RA 21 26 40 Dec −22 24 40) for Zeta...

Also known as 34 Capricorni, Zeta is a unique binary system. Located about 398 light-years from Earth, the primary star is a yellow supergiant with some very unusual properties—it's the warmest, most luminous barium star known (Figure 10.28). But that's not all, because the B component is a white dwarf almost identical in size to our own Sun!

Saturday, October 11

If you journey to the Moon tonight, you might return to the southern quadrant along the terminator to have a look at 227 kilometer diameter crater Schickard (Figure 10.29). Seen on the oblique, this great crater's floor is so humped in the middle that you could stand there and not see the crater walls! Be sure to note Schickard for your lunar challenge studies.

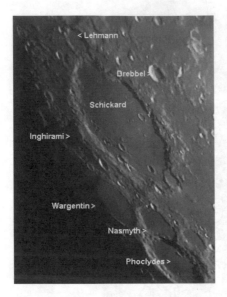

< Lehmann

Drebbel >

Schickard

Inghirami >

Wargentin >

Nasmyth >

Phoclydes >

Figure 10.29.
Schickard region
(Credit—Roger Warner,
annotation by Tammy
Plotner).

Figure 10.30. Fomalhaut (Credit—Palomar Observatory, courtesy of Caltech).

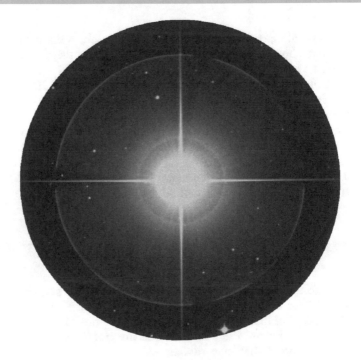

After having looked at the Moon, take the time out to view a bright southern star—Fomalhaut (RA 22 57 39 Dec −29 37 20). Also known as "The Lonely One," Alpha Piscis Austrini seems to sit in a rather empty area in the southern skies, some 23 light-years away (Figure 10.30). At magnitude 1, this main sequence A3 giant is the southernmost visible star of its type for Northern Hemisphere viewers, and is the 18th brightest star in the sky. The Lonely One is about twice the diameter of our own Sun, but 14 times more luminous! Just a little visual aid is all that it takes to reveal its optical companion...

Sunday, October 12

Today in 1891, the Astronomical Society of France was established. Exactly one year later in 1892, astronomer great E. E. Barnard was hard at work using the new tool of photography and became the first to discover a comet—1892 V—in this way! But Barnard's main photographic interest was in capturing details of the Milky Way. Just as soon as skies are dark again, we'll have a look at more of Barnard's work.

Do you like looking at things which are considered dubious? Then tonight let's start on the lunar surface and peek at a ray system whose origins are uncertain. You'll find the bright ring of Bessel almost in the center of Mare Serenitatis, but the ray system is splashed all over it (Figure 10.31). Did they come from Menelaus on the mare's edge? Or from as far south as Tycho? Next time the

Figure 10.31. Bessel Rays (Credit—Ricardo Borba).

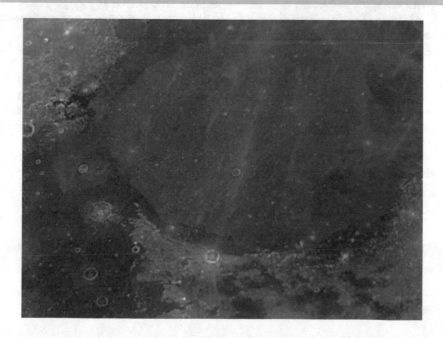

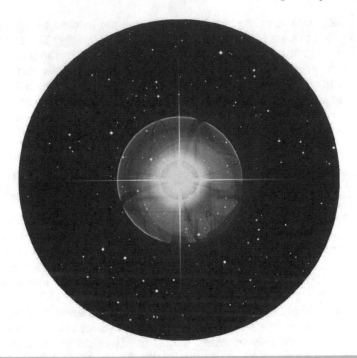

Figure 10.32. Beta Gruis (Credit—Palomar Observatory, courtesy of Caltech).

terminator passes over this region, look closely. Do you see the rays now—or just a complicated system of dorsa?

With tonight's bright skies, it will be difficult to practice any astronomy—or will it? Try re-locating Fomalhaut and drop about a handspan south-southwest into Grus to pick up bright star Beta (RA 22 42 40 Dec –46 53 04).

Around 170 light-years from Planet Earth, Beta is the 59th brightest star in the sky and the 2nd brightest star not to have a proper name (Figure 10.32). It's an M-type supergiant, but one that is also slightly irregular—changing by about a third of a magnitude in approximately 37 days. Well evolved, Beta is on its way to becoming a Mira type and is only the size of the orbit of Venus. Its loss of mass could mean it has a dead carbon–oxygen core, and studies at infrared wavelengths point to a shell waiting to be expelled.

In the telescope, you will see Beta also has a visual companion to the south. Although it is unrelated to Beta itself, modern interferometry suggests there may be a true companion star which has yet to be resolved. No matter how you view it, you'll like Beta for its rich color! Remember its position...

Monday, October 13

Today marks the founding of the British Interplanetary Society in 1933. "From imagination to reality," the BIS is the world's oldest established organization devoted solely to supporting and promoting astronautics and the exploration of space.

Tonight we'll make them proud as we have a look at a discovery in the "Crane." Although the constellation of Grus is a bit difficult for mid-northern latitudes, return to Beta Gruis and hop to the southeast with binoculars as we take on "Triple Tau." Easily caught with a little optical aid, you will see a descending line of three nearly equal-magnitude stars, each of a different spectral type. Now focus your attention, and your telescopes, on the southernmost of the trio—Tau 1 (RA 22 53 37 Dec –48 35 53).

At 6th magnitude, this G-type dwarf certainly isn't anything to write home about, but it's certainly not far away (Figure 10.33). A little less than 100 light-years distant, Tau 1 is only slightly larger than our own Sol, and within just a few degrees of its temperature. If you think about that, then you might know where this is going...

Toward an extra-solar planet!

Orbiting its parent star in slightly less than 4 years is a planet about 1.5 times the size of Jupiter. (No Pluto here, folks!) In an eccentric orbit, the planet circles at an average distance about equal to that of our own asteroid belt. Researchers began studying Tau 1 in 1998, but it wasn't until 2002 that this exoplanet was confirmed as the hundredth to be found, and a hundred more external planets were known by 2006!

Figure 10.33. Tau 1 Gruis (Credit—Palomar Observatory, courtesy of Caltech).

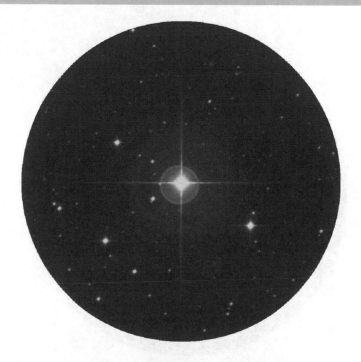

Tuesday, October 14

Tonight is Full Moon. Known at times as the "Hunter's Moon" or the "Blood Moon," its name came from a time when hunters would stalk the fields by Luna's cold light in search of prey before the winter season began. Pick a place at sunset to watch it rise—a place having a stationary point with which you can gauge its progress. Make note of the time when the first rim appears and then watch how quickly it gains altitude! How long does it take before it rises above your marker?

Tonight let's take a look at opposite poles as we begin in the far north with Alpha Ursa Minoris (RA 02 32 49 Dec +89 15 50). Known as Polaris, it is without a doubt the most famous of all stars in the sky (Figure 10.34). While it ranks only 49th in brightness, its fame comes from its use as a navigation aid. Although it is within a degree of true north, time itself will correct the position until it reaches near perfection in the year 2102.

When you aim binoculars toward Polaris, you will happily discover a circle of stars around it, known as the "Engagement Ring;" and Alpha is the sparkling diamond. In a telescope you will get an even better treat as its binary nature becomes clear! In 2006, the Hubble Space Telescope resolved a very close companion to Polaris A, so it is really a triple star system!

Figure 10.34. Polaris (Credit—Palomar Observatory, courtesy of Caltech).

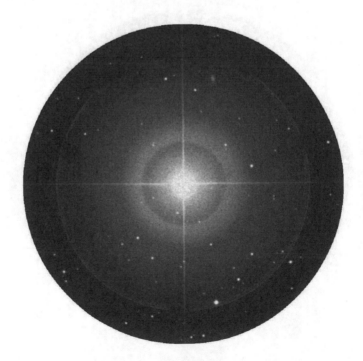

Figure 10.35. Lambda Centauri (Credit—Palomar Observatory, courtesy of Caltech).

But what about the south? Viewers in the Southern Hemisphere can never see Polaris—is there a matching star for the south? The answer is yes. Sigma Octantis. But at magnitude 5, it doesn't make a very good unaided-eye guide.

Ancient navigators found better success with the constellation Crux—better known as the Southern Cross—to guide them. Two bright stars of the Southern Cross, Gacrux and Acrux, are oriented north–south and point across the pole to brilliant Archenar. Splitting the distance between Gacrux and Archenar puts you within two degrees of the rather desolate south pole of the sky. Southern Hemisphere observers wishing to see a double star comparable to Polaris in appearance should choose Lambda Centauri (RA 11 35 46 Dec –63 01 11). The difference in magnitude between its components, and also their separation, is about the same as for its northern counterpart (Figure 10.35)!

Wednesday, October 15

Today in 1963 marked the first detection of an interstellar molecule. This discovery was made by Sander Weinreb (with Barrett, Meeks, and Henry) on the MIT Millstone Hill 84 foot dish. The discovery was made possible by new correlation receiver technology, and picked up the hydroxyl radical in an absorption band. By using the radio galaxy Cas A as a background continuum source, the

Figure 10.36. 19 Capricorni (Credit—Palomar Observatory, courtesy of Caltech).

detection occurred at 1667.46 megahertz and again at 1665.34 megahertz. By the dawn of 2000, nearly 200 different interstellar molecules had been identified, and many of these are classified as organic.

Tonight let's have a look at a radio source as we visit a pulsar located almost mid-way between Theta and Beta Capricorni—PSR 2045+16 (Figure 10.36).

While pulsars aren't truly visible objects, there is still something undeniably cool about locating the field in which a rotating neutron star is sending out staccato pulses of radio waves anywhere between 0.001 and 4.0 seconds apart. If you have bright star 19 in the binocular field, then you know you're in the right area for many radio sources, including many nearby quasars... Just imagine the possibilities!

Thursday, October 16

Tonight let's take the time before moonrise to have a peek at M72, just about a degree and a half west (RA 20 53.5 Dec –12 32) of previous study M73.

Originally found by Méchain on the night of August 29–30, 1780, this class IX globular cluster is one of the faintest and most remote of the Messiers, and Charles didn't catalog it until over a month after its discovery. At around

Figure 10.37. M72 (Credit—Palomar Observatory, courtesy of Caltech).

magnitude 9, this 53,000 light-year distant globular will be not much more than a faint round smudge in smaller aperture, but will take on a modicum of resolution in larger telescopes. Well beyond the galactic center and heading toward us at 255 kilometers per second, M72 is home to 42 variables and the average magnitude of its members is around 15 (Figure 10.37). While mid-sized scopes will pick up graininess in the texture of this globular, notice how evenly the light is distributed, with little evidence of a core region. Be sure to write down your observations!

Friday, October 17

With ample time to spare before the Moon rises, our destination for the evening is the mighty M2 (Figure 10.38). You'll find it located about three fingerwidths north-northeast of Beta Aquarii (RA 21 33 27 Dec –00 49 24).

At slightly dimmer than 6th magnitude, this outstanding globular cluster is just dim enough that it can't quite be viewed unaided, but even the smallest of binoculars will pick it out of a relatively starless field with ease. Holding a Class II designation, it was discovered by Maraldi on September 11, 1746 and rediscovered

Figure 10.38. M2 (Credit—Palomar Observatory, courtesy of Caltech).

independently by Messier exactly 14 years later. At a distance of roughly 37,500 light-years, it is estimated to contain 150,000 stars in the neighborhood.

Even a small telescope will reveal M2's rich and concentrated core region and slight ellipticity. Not bad for an 11 billion year old group of stars! As aperture increases, some of the brightest stars will begin to resolve, and in larger telescopes it will approach total resolution. You might well note a dark area in the northeastern section, and several more located throughout the splendid field. Feast your eyes on one of the finest in the skies!

Saturday, October 18

Today in 1959, the Soviet spacecraft Luna 3 began returning the first photographs of the Moon's far side. Also today—but in 1967—the Soviets again made history as Venera 4 became the first spacecraft to probe Venus' atmosphere. If you're up before dawn, be sure to have a look at brilliant Venus rising just ahead of the Sun!

Be thankful if we have dark skies this evening, as our target is usually around 10th magnitude—but not always! First set your sights toward 1 Aquarii and let's have a look (Figure 10.39)...

Figure 10.39. 1 Aquarii (Credit—Palomar Observatory, courtesy of Caltech).

Figure 10.40. AE Aquarii (center of image) (Credit—Palomar Observatory, courtesy of Caltech).

Star 1 (RA 20 39 24 Dec −00 29 11) is a beautiful disparate double! With a primary coming in about magnitude 5, and a well-spaced secondary holding an approximate magnitude of 11, this is a great exercise to get you warmed up and in the right field for what comes next. Now hop a bit east until your field lands on our next stop—a very strange variable!

In the center of the field you will find AE Aquarii (RA 20 40 09 Dec −00 52 15). At 10th magnitude, it's not exactly setting records for being the most conspicuous star in the field—but maybe it is (Figure 10.40)! On a very erratic schedule, AE can brighten by 200% in less than an hour and, roughly once a year, can actually jump two full magnitudes brighter. So what the heck is it?

AE is a prime example of a recurrent nova—a star with complex cycles which is probably similar to the SS Cygni type. Discovered in 1931, AE was later revealed to have a close binary companion of about the same mass, but in a different stage of evolution. While this is not so strange in itself, the two orbit each other in about 10 hours! This means the pair might actually touch one another during their passes, causing rather "explosive" results!

Sunday, October 19

Today we celebrate in 1910 the birth of one of the greatest astrophysicists of modern times—Subrahmanyan Chandrasekhar (Figure 10.41). His genius led him to an amazing discovery: not all stars end their lives as white dwarfs or neutron stars, those below a certain mass must undergo further collapse. This mass remains known as the "Chandrasekhar limit." He left a rich legacy of research in cosmology, astrophysics, and mathematics—the Chandra X-ray Observatory was named in his honor.

Figure 10.41.
Subrahmanyan
Chandrasekhar (widely
used public image).

Now let's get some practice in Capricornus as tonight we'll take on a more challenging target with confidence. Locate the centermost bright star in the northern half of the constellation—Theta—because we're headed for the "Saturn Nebula."

Three fingerwidths north of Theta you will see dimmer Nu, and only one fingerwidth west is NGC 7009 (RA 21 04 10 Dec −11 21 48). This wonderful blue planetary is around 8th magnitude and achievable in small scopes and large binoculars (Figure 10.42). NGC 7009 was the first discovery of Sir William Herschel on September 7, 1782—the night he started his sky survey—and he cataloged it as H IV.1. Sir William's original notes describe it as a "very bright nearly round planetary, not well-defined disk."

When viewed by Lord Rosse in the 1840s, he gave it the nickname Saturn Nebula, and it is considered one of the nine Struve rare celestial objects. Also known as Bennett 127 and Caldwell 55, it is generally believed to be around 2400 light-years away—but not so far that it doesn't make almost every list as an all time great!

Even at moderate magnification, you will see the elliptical shape which gave rise to its moniker. With larger scopes, those "ring-like" projections become even clearer as the 11th magnitude central star appears. No matter which aperture you choose, this challenging object is well worth the hunt. You can do it!

Figure 10.42. NGC 7009: The Saturn Nebula (Credit—Palomar Observatory, courtesy of Caltech).

Monday, October 20

Now we are slipping into the stream of Comet Halley and thus into one of the finest meteor showers of the year. If skies are clear tonight, this would be the perfect chance to begin your observations of the Orionid meteor shower. But go to bed early...because the best action happens just before dawn, despite the Moon! We'll learn more about this great shower tomorrow...

Tonight let us take out binoculars or a telescope using the widest possible field of view, and have a look at Barnard's passion as we view two regions in the westering Aquila—the "Double Dark Nebula."

Just northeast of Altair is bright star Gamma Aquilae, and about a fingerwidth west is a pair of Barnard discoveries: B 142 and B 143—two glorious absences of stars known as interstellar dust clouds (Figures 10.43 and 10.44). B 143 (RA 19 40 42 Dec +10 57 00) is no more than a half degree in size and will simply look like a blank area shaped like a horseshoe, with its extensions pointing toward the west. Just south is B 142 (RA 19 39 41 Dec +10 31 00), an elongated comma shape, which seems to underline its companion.

Located anywhere from 1000 to 3000 light-years away, these non-luminous clouds of gas and dust are a very fine example of Barnard's obsession. Do not be upset if you don't see them on your first attempt—for the chances are, if you are seeing "nothing?", you are looking in the right place!

Figure 10.43. Barnard 142 (Credit—Palomar Observatory, courtesy of Caltech).

Figure 10.44. Barnard 143 (Credit—Palomar Observatory, courtesy of Caltech).

Tuesday, October 21

Be sure to be outdoors before dawn to enjoy one of the year's most reliable meteor showers. The offspring of Comet Halley will grace the early morning hours as they return once again as the Orionid meteor shower. This dependable shower produces an average of 10–20 meteors per hour at maximum and the best activity begins before local midnight on the 20th, and reaches its height as Orion stands high to the south at about 2 hours before local dawn on the 21st.

Although Comet Halley has long since departed our solar system, the debris left from its trail still remain scattered along Earth's orbital path, allowing us to predict when this meteor shower will occur (Figure 10.45). We first enter the "stream" at the beginning of October and do not leave it until the beginning of November, making your chances of "catching a falling star" even greater! These meteors are very fast and, although they are faint, it is still possible to see an occasional fireball leave a persistent trail.

For best success, try to get away from city lights—even though we can't escape the Moon. Facing south-southeast, simply relax and enjoy the stars of the winter Milky Way. The radiant, or apparent point of origin, for this shower will be near the red giant Alpha Orionis (Betelguese), but meteors may occur from any point in the sky. You will make your meteor watching experience much more comfortable if you take along a lawn chair, a blanket and a thermos of your favorite beverage.

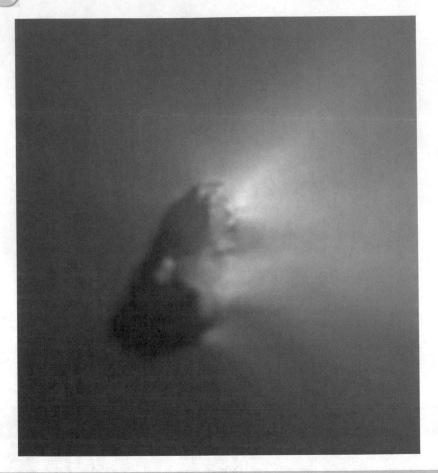

Figure 10.45. Comet Halley's nucleus (Credit—NASA).

Clouded out? Don't despair. You don't always need your eyes or perfect weather to meteor watch. By tuning an FM radio to the lowest frequency that does not receive a clear signal, you can practice radio meteor listening! An outdoor FM antenna pointed at the zenith and connected to your receiver will increase your chances, but it's not necessary. Simply turn up the static and listen. Those hums, whistles, beeps, bongs, and occasional snatches of signals are our own radio signals being reflected off the meteor's ion trail!

Pretty cool, huh?

Wednesday, October 22

Something very special happened today in 2136 BC. There was a solar eclipse, and for the first time it was seen and recorded by Chinese astronomers. And

Figure 10.46. Karl Jansky and his antenna (widely used public image).

probably a very good thing because in those days the royal astronomers were executed for failure to predict! Today is also the birthday of Karl Jansky (Figure 10.46). Born in 1905, Jansky was an American physicist and electrical engineer. One of his pioneer discoveries was non-Earth-based radio waves at 20.5 megahertz, a detection he made while investigating noise sources during 1931 and 1932.

In 1975, the Soviet probe Venera 9 was busy sending Earth the very first look at Venus' surface. If you are up before dawn this morning, why not take a moment to have a look at Venus yourself... Can you tell what phase it is in through the telescope?

Tonight let's head to the eastern portion of Capricornus and start by identifying Zeta about a fistwidth southwest of the eastern corner star—Delta. Now look southeast about two fingerwidths to identify 5th magnitude star 41. About one half degree west is our target globular for the evening, M30 (RA 21 40 22 Dec −23 10 44).

At near magnitude 8, this class V globular cluster is well suited to even binoculars, and becomes spectacular in a telescope (Figure 10.47). Originally discovered by Messier in August 1764, and resolved by William Herschel in 1783, M30's most attractive features include the several branches of stars which seem to radiate from its concentrated core region. Estimated to be about 26,000 light-years away, you'll find it fairly well resolved in large aperture, but take time to really look. The dense central region may have already undergone core collapse—yet as close as these stars are, very few have collided to form X-ray binaries. For the smaller scope, notice how well M30's red giants resolve, and be sure to mark your notes!

Figure 10.47. M30 (Credit—Palomar Observatory, courtesy of Caltech).

Thursday, October 23

If you're up before dawn this morning, be sure to step outside to catch the pleasing pairing of the Moon and Regulus less than two degrees apart!

Tonight it's time for a telescopic challenge—a compact galaxy group. You'll find it less than half a degree southeast of the stellar pair 4 and 5 Aquarii (RA 20 52 26 Dec –05 46 19).

Known as Hickson 88, this grouping of four faint spiral galaxies is estimated to be about 240 million light-years away and is by no means an easy object—yet the galactic cores can just be glimpsed with mid-sized scopes from a very dark site (Figure 10.48). Requiring around 12.5" in aperture to study in detail, you'll find the brightest of the group to be northernmost NGC 6978 and NGC 6977. While little detail can be seen in the average large backyard scope, NGC 6978 shows some evidence of being a barred spiral, while NGC 6977 shows the even appearance of a face-on. Further south, NGC 6976 is much smaller and considerably fainter. It is usually caught while averting and studying the neighborhood. The southernmost galaxy is NGC 6975, whose slender, edge-on appearance makes it much harder to catch.

Although these four galaxies seem to be in close proximity to one another, no current data suggests any interaction between them. While such a faint galaxy grouping is not for everyone, it's a challenge worthy of seasoned astronomer with a large scope! Enjoy...

Figure 10.48. Hickson 88 (Credit—Palomar Observatory, courtesy of Caltech).

Friday, October 24

Today we remember the launch of Deep Space 1 from Cape Canaveral in 1998 (Figure 10.49). Its primary mission was extremely successful, testing a dozen advanced, high-risk technologies. During its extended mission, Deep Space 1 headed for Comet Borrelly and sent back the best images from a comet up to that time. The mission continued to test new techniques until it was finally retired after three fantastic years of service on December 18, 2001.

Tonight in 1851, William Lassell—a busy astronomer —was at the eyepiece as he discovered Uranus' moons Ariel and Umbriel. Although the equipment he used is far beyond backyard equipment (Figure 10.50), we can have a look at that distant world, as we find Uranus about 25 degrees north-northwest of Fomalhaut.

While Uranus' small, blue–green disc isn't exactly the most exciting thing to see in a small telescope or binoculars, the very fact we are looking at a planet that's over 18 times further from the Sun than we are is pretty impressive (Figure 10.51)! Usually holding close to magnitude 6, we watch as the tilted planet orbits our nearest star once every 84 years. Its atmosphere is composed of hydrogen, helium, and methane, yet pressure causes about a third of this distant planet to behave as a liquid. Larger telescopes may be able to discern a few of Uranus' moons, for Titania (the brightest) is around magnitude 14.

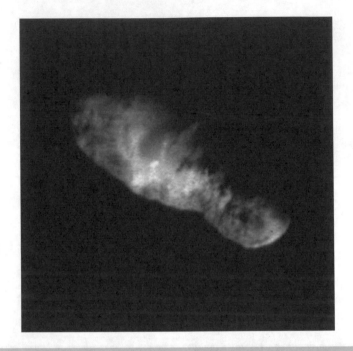

Figure 10.49. Deep Space 1 image of Comet Borrelly (Credit—NASA).

Figure 10.50. Rendition of Lassell's Telescope (widely used public image).

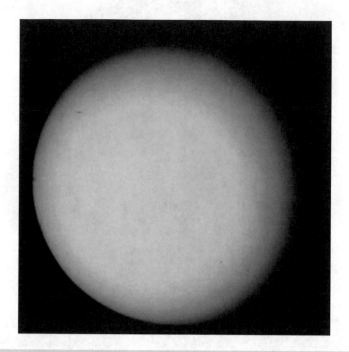

Figure 10.51. True color image of Uranus (Credit—NASA).

Now let's head toward the southwest corner star of the Great Square of Pegasus—Alpha. Our goal will be 11th magnitude NGC 7479, located about three degrees south (RA 23 04 56 Dec +12 19 23).

Discovered by Sir William Herschel in 1784 and cataloged as H I.55, this barred spiral galaxy can be spotted in average telescopes and comes to beautiful life with larger aperture (Figure 10.52). Also known as Caldwell 44 on Sir Patrick Moore's observing list, what makes this galaxy special is its delicate "S" shape. Smaller scopes will easily see the central bar structure of this 105 million light-year distant island universe and, as aperture increases, the western arm will become more dominant. This arm itself is a wonderful mystery—containing more mass than it should and having a turbulent structure. It is believed that a minor merger may have occurred at one time, yet no evidence of a companion galaxy can be found.

On July 27, 1990 a supernova occurred near NGC 7479's nucleus and reached a magnitude of 16. When observed in the radio band, there is a polarized jet near the bright nucleus that is unlike any other structure known. If at first you do not see a great deal of detail, relax... Allow your mind and eye time to look carefully. Even with telescopes as small as 8–10 inches, the structure can easily be seen. The central bar becomes "clumpy" and this well-studied Seyfert galaxy is home to an abundance of molecular gas and is actively forming stars. Enjoy the incredible NGC 7479...

Figure 10.52. NGC 7479 (Credit—Palomar Observatory, courtesy of Caltech).

Saturday, October 25

And who was watching the planets in 1671? None other than Giovanni Cassini—because he'd just discovered Saturn's moon Iapetus. If you're up before dawn this morning, have a look at Saturn for yourself as it poses less than five degrees away from the Moon. Iapetus usually holds around a magnitude of 12, and orbits well outside of bright Titan's path.

Today is the birthday of Henry Norris Russell (Figure 10.53). Born in 1877, Russell was the American leader in establishing the modern field of astrophysics. As the namesake for the American Astronomical Society's highest award (for lifetime contributions to the field), Russell is the "R" in H–R diagrams, along with Hertzsprung. This work was first used in a 1914 paper, published by Russell. Tonight let's have a look at a star that resides right in the middle of the H–R diagram as we have a look at Beta Aquarii (RA 21 31 33 Dec –05 34 16).

Named Sadal Suud ("Luck of Lucks"), this spectral-type G star is around 1030 light-years distant from our solar system and shines 5800 times brighter than our own Sun (Figure 10.54). The main sequence beauty also has two 11th magnitude optical companions. The one closest to Sadal Suud was discovered by John Herschel in 1828, while the further star was reported by S. W. Burnham in 1879.

Figure 10.53. Henry Norris Russell (widely used public image).

Figure 10.54. Beta Aquarii (Credit—Palomar Observatory, courtesy of Caltech).

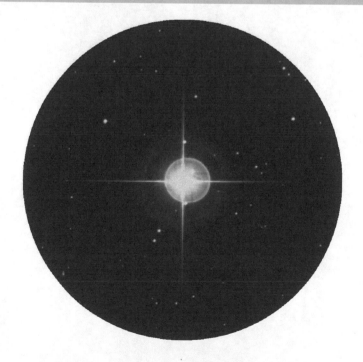

Sunday, October 26

If you're up early, be sure to look for Venus and Antares making a close pairing in the predawn sky!

For observers all over the world, it's almost time for Halloween! For the next five days, let's take a look at some of the "spookiest" objects in the night sky...

This evening we are once again going to study a single star, which will help you become acquainted with the constellation of Perseus. Its formal name is Beta Persei (RA 03 08 10 Dec +40 57 20) and it is the most famous of all eclipsing variable stars (Figure 10.55). Tonight let's identify Algol and learn all about the "Demon Star."

Ancient history has given this star many names. Associated with the mythological figure Perseus, Beta was considered to be the head of Medusa the Gorgon, and was known to the Hebrews as Rosh ha Satan or "Satan's Head." Seventeenth century maps labeled Beta as Caput Larvae, or the "Specter's Head," but it is from the Arabic culture that the star was formally named. They knew it as Al Ra's al Ghul, or the "Demon's Head," and we now know it as Algol. Because these medieval astronomers and astrologers associated Algol with danger and misfortune, we are led to believe that Beta's strange visual variable properties had been noted throughout history.

Figure 10.55. Beta Persei: Algol (Credit—Palomar Observatory, courtesy of Caltech).

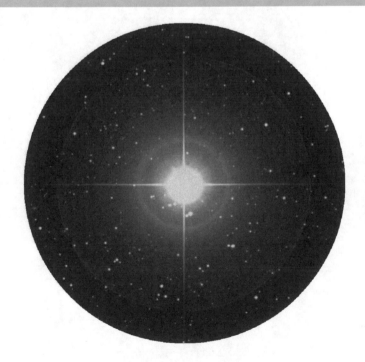

Italian astronomer Geminiano Montanari was the first to record that Algol occasionally "faded," and its regular timing was cataloged in 1782 by John Goodricke, who surmised that it was being partially eclipsed by a dark companion orbiting it. Thus was born the theory of the eclipsing binary, which was proved spectroscopically for Algol in 1889 by H. C. Vogel. Ninety three light-years away, Algol is the nearest eclipsing binary, and is treasured by the amateur astronomer because it requires no special equipment to easily follow its stages. Normally Beta Persei holds a magnitude of 2.1, but approximately every 3 days it dims to magnitude 3.4 and gradually brightens again. The entire eclipse only lasts about 10 hours!

Although Algol is known to have two additional spectroscopic companions, the true beauty of watching this variable star is not telescopic—but visual. The constellation of Perseus is well placed this month for most observers, and appears like a glittering chain of stars that lie between Cassiopeia and Andromeda. To help further assist you, re-locate last week's study star, Gamma Andromedae (Almach), east of Algol. Almach's visual brightness is about the same as Algol's at maximum.

Monday, October 27

Now we need a jack-o-lantern...

Asteroid Vesta is considered to be a minor planet since its approximate diameter is 525 kilometers (326 miles), making it slightly smaller in size than the state of Arizona. Vesta was discovered on March 29, 1807 by Heinrich Olbers and it was the fourth such "minor planet" to be identified (Figure 10.56). Olbers' discovery was fairly easy because Vesta is the only asteroid bright enough at times to be seen unaided from Earth. Why? Orbiting the Sun every 3.6 years and rotating on its axis for 5.24 hours, Vesta has an albedo (or surface reflectivity) of 42%. Although it is about 350 million kilometers away, pumpkin-shaped Vesta is the brightest asteroid in our solar system because it has a unique geological surface. Spectroscopic studies show it to be basaltic, which means lava once flowed on the surface. (Very interesting, since most asteroids were once thought to just be rocky fragments leftover from our forming solar system!)

Studies by the Hubble telescope have confirmed this, as well as shown a large meteor impact crater which exposed Vesta's olivine mantle. Debris from Vesta's collision then set sail away from the parent asteroid. Some of the debris remained within the asteroid belt near Vesta, and became asteroids themselves with the same spectral pyroxene signature. But some of the debris escaped the asteroid belt through the "Kirkwood Gap" created by Jupiter's gravitational pull. This allowed these small fragments to be kicked into orbits that would eventually bring them "down to Earth." Did one make it? Of course! In 1960 a piece of Vesta fell to Earth and was recovered in Australia. Thanks to Vesta's unique properties, the meteorite was definitely identified as coming from our third largest asteroid. Now, that we've learned about Vesta, let's talk about what we can see from our own backyards.

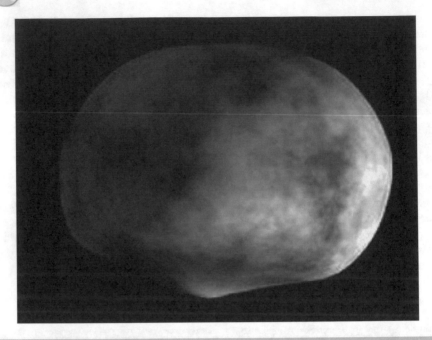

Figure 10.56. Asteroid Vesta (Credit—NASA).

As you can discern from the image, even the Hubble Space Telescope doesn't give incredible views of this bright asteroid. What we will be able to see in our telescopes and binoculars will closely resemble a roughly magnitude seven "star." Tonight you can find Vesta about seven degrees west of Alpha Ceti (RA 02 35 00 Dec +03 46 00). Vesta will be at opposition in just 3 days, and is now in retrograde, so you will be able to watch it slowly move away from Alpha Ceti for the rest of the year.

Of course, the approximate coordinates given above are only accurate for a short time, so I strongly encourage you to use a good planetarium program to print accurate locator charts, or visit an online resource such as the IAU Minor Planet Center for more details. When you locate the proper stars and the asteroid's probable location, mark physically on the map Vesta's position. Keeping the same map, return to the area a night or two later and see how Vesta has moved since your original mark. Since Vesta will stay located in the same area for awhile, your observations need not be on consecutive nights, but once you learn how to observe an asteroid and watch it move—you'll be back for more!

Tuesday, October 28

Today in 1971, Great Britain launched its first satellite—Prospero.

One of the scariest movies in recent times was the "Ring"... Let's find one! Tonight's dark-sky object is a difficult one for northern observers and is truly a challenge. Around a handspan south of Zeta Aquarii and just a bit west of finderscope star Upsilon (RA 22 51 48 Dec –20 36 31) is a remarkably large area of nebulosity that is very well suited to large binoculars, rich-field telescopes and wide-field eyepieces. Are you ready to walk into the "Helix?"

Known as NGC 7293, this faint planetary nebula's "ring" structure is around half the size of the Full Moon. While its total magnitude (6.5) and large size should indicate it would be an easy find, the Helix is anything but easy because of its low surface brightness (Figure 10.57). Binoculars will show it as a large, round, hazy spot, while small telescopes with good seeing conditions will have a chance to outshine larger ones by using lower-power eyepieces to pick up the braided ring structure.

Figure 10.57. NGC 7293: The Helix Nebula (Credit—R. Jay GaBany).

As one of the very closest of planetary nebulae, NGC 7393 is very similar in structure to the more famous ring, M57. It is a spherical shell of gas lighted by an extremely hot, tiny central star that's only around 2% of our own Sun's diameter—yet exceeds Sol in surface temperature by over 100,000 Kelvin. Can you resolve it? Best of luck!

Wednesday, October 29

On this night in 1749, the French astronomer Le Gentil was at the eyepiece of an 18' focal length telescope. His object of choice was the Andromeda Galaxy (Figure 10.58), which he believed to be a nebula. Little did he know at the time that his descriptive notes also included M32, a satellite galaxy of M31. It was the first small galaxy discovered, and it would be another 175 years before these were recognized as such by Edwin Hubble.

Tonight take the time to view the Andromeda Galaxy for yourself. Located just about a degree west of Nu Andromedae, this "ghost" set against the starry night was known as far back as 905 ad, and was referred to as the "Little Cloud." Located about 2.2 million light-years from our solar system, this expansive member of our Local Galaxy Group has delighted observers of all ages throughout the years. No

Figure 10.58. The Andromeda Galaxy (Credit—John Chumack).

matter if you view with just your eyes, a pair of binoculars or a large telescope, M31 still remains one of the most spectacular galaxies in the night.

"Boo" tiful...

Thursday, October 30

Today in 1981 Venera 13 was launched on its way toward Venus—did you catch the bright planet before dawn?

What Halloween celebration would be complete without a black cat? Tonight let's cruise Draco (the Dragon) in search of the "Cat's Eye"...

Located about halfway between Delta and Zeta Draconis is one of the brightest planetary nebulae in the night—8.8 magnitude NGC 6543 (Figure 10.59). Around 3000 light-years away, it was one of the first planetaries to be studied spectroscopically, and the resulting emission lines proved the phenomenon was actually a shell of gas emitted from a dying star. Our own Sun awaits a similar fate.

Figure 10.59. NGC 6543: The Cat's Eye (Credit—Hubble Heritage Team/NASA).

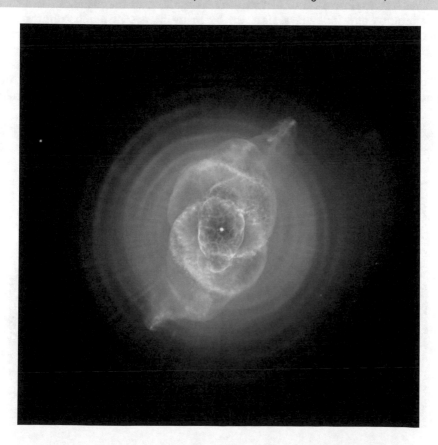

While a small telescope will never reveal NGC 6543 as gloriously as a Hubble image, you can expect (even in a small telescope or binoculars) to make out a small, blue–green, glowing object. But a large aperture telescope and good sky conditions are needed to reveal some of the braided structure seen within. No matter how you view it, the Cat's Eye belongs to the list of spooky objects!

Friday, October 31

Happy Halloween! Many cultures around the world celebrate this day with a custom known as "Trick or Treat." Tonight instead of tricking your little ghouls and goblins, why not treat them to a sweet view through your telescope or binoculars?

So far we've collected a demon, a pumpkin, a galactic ghost, and the eye of the cat... And what Halloween would be complete without a witch! Easily found from a modestly dark site with the unaided eye, the Pleiades can be spotted well above the northeastern horizon within a couple of hours of nightfall (Figure 10.60). To average skies, many of the seven bright components will resolve easily without the use of optical aid, but to telescopes and binoculars?... M45 is stunning!

First let's explore a bit of history. The recognition of the Pleiades dates back to antiquity, and its stars are known by many names in many cultures. The Greeks and Romans referred to them as the "Starry Seven," the "Net of Stars," "The Seven Virgins," "The Daughters of Pleione," and even "The Children of Atlas." The Egyptians referred to them as "The Stars of Athyr;" the Germans as "Siebengestiren" (the Seven Stars); the Russians as "Baba" after Baba Yaga, the witch who flew through the skies on her fiery broom. The Japanese call them

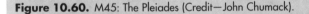

Figure 10.60. M45: The Pleiades (Credit—John Chumack).

"Subaru;" Norsemen saw them as packs of dogs; and the Tongans as "Matarii" (the Little Eyes). American Indians viewed the Pleiades as seven maidens placed high upon a tower to protect them from the claws of giant bears, and even Tolkien immortalized the star group in *The Hobbit* as "Remmirath." The Pleiades were even mentioned in the Bible! So, you see, no matter where we look in our "starry" history, this cluster of seven bright stars has been part of it. But let's have some Halloween fun!

The date of the Pleiades culmination (its highest point in the sky) has been celebrated throughout its rich history, being marked with various festivals and ancient rites—but there is one particular rite that really fits this occasion! What could be more spooky on this date than to imagine a group of Druids celebrating the Pleiades' midnight "high" with Black Sabbath? This night of "unholy revelry" is still observed in the modern world as "All Hallow's Eve," more commonly called Halloween. Although the actual date of the Pleiades' midnight culmination is now on November 21 instead of October 31, why break with tradition? Thanks to its nebulous regions, M45 looks wonderfully like a "ghost" haunting the starry skies.

Treat yourself and your loved ones to the "scariest" object in the night. Binoculars give an incredible view of the entire region, revealing far more stars than are visible with the naked eye. Small telescopes at lowest power will enjoy M45's rich, icy-blue stars and fog-like nebulosity. Larger telescopes and higher power reveal many pairs of double stars buried within its silver folds. No matter what you choose, the Pleiades definitely rocks!

CHAPTER ELEVEN

November, 2008

Saturday, November 1

On this day in 1977, Charles Kowal made a wild discovery—Chiron (Figure 11.1). This was the first sighting of one of the multitude of tiny, icy bodies inhabiting the outer reaches of our solar system. Collectively known as Centaurs, they reside in unstable orbits between Jupiter and Neptune and are almost certainly "refugees" from the Kuiper Belt.

This evening have a look at the lunar surface and the southeast shoreline of Mare Crisium for Agarum Promontorium (Figure 11.2). To a small telescope it will look like a bright peninsula extending northward across the dark plain of Crisium's interior, eventually disappearing beneath the ancient lava flow. Small crater Fahrenheit can be spotted at high power to the west of Agarum, and it is just southeast of there that Luna 24 landed. If you continue south of Agarum along the shoreline of Crisium, you will encounter 15 kilometer high Mons Usov. To its west is a gentle rille known as Dorsum Termier—where the remains of the Luna 15 mission lie. Can you spot 23 kilometer wide Shapely further south?

Tonight let's use our eyes, binoculars, or even a telescope at lowest power to have a look at two objects cataloged by Sir William Herschel on this night. You may know them as the "Double Cluster"...

Properly designated as NGC 869 and NGC 884 (Figures 11.3 and 11.4), this pair has definitely got to go into history as "How come Messier missed them?" Probably already known in pre-historic times, and first cataloged by Hipparchus (ah! that's why!), they are easily seen by the eye as a hazy patch in the Milky Way

Figure 11.1. Chiron (Credit—NASA).

Figure 11.2. The 3 day Moon (Credit— Ricardo Borba).

Figure 11.3. NGC 869 (Credit—Palomar Observatory, courtesy of Caltech).

Figure 11.4. NGC 884 (Credit—Palomar Observatory, courtesy of Caltech).

between Cassiopeia and Andromeda. Located a little more than 7000 light-years away, and a few hundred light-years apart, they are both very young as clusters go—but they differ radically from each other in age: NGC 884 is about 5.6 million years old, while NGC 889 is only 3.2 million.

Both clusters are listed on many "Best Objects" lists, so be sure to congratulate yourself for noting them as Caldwell 14!

Sunday, November 2

Today celebrates the birth of an astronomy legend—Harlow Shapely (Figure 11.5). Born in 1885, the American-born Shapley paved the way in determining distances to stars, clusters, and the center of our Milky Way galaxy. Among his many achievements, Shapely was also the Harvard College Observatory director for many years. Today in 1917 also represents the night first light was seen through the Mt. Wilson 100" telescope.

If you didn't spot diminutive crater Shapley last night, try again! There's an annotated map here to help guide you to both the crater and the Luna landing sites (Figure 11.6).

Of course, Shapley spent his fair share of time on the Hooker telescope as well. A particular point of his studies were globular clusters, their distances, and the relationship they have to the halo structure of our galaxy. Tonight let's look at a very unusual little globular located about a fistwidth south-southeast of Beta Ceti, and just a couple of degrees north-northwest of Alpha Sculptoris (RA 00 52 47 Dec −26 34 43). Its name is NGC 288...

Discovered by William Herschel on October 27, 1785, and cataloged by him as H VI.20, this class X globular cluster blew apart scientific thinking in the late 1980s when a study of perimeter globulars showed it to be older than its X-class peers by three billion years—thanks to the color-magnitude diagrams of

Figure 11.5. Harlow Shapley (widely used public image).

Figure 11.6. Crater Shapely (Credit—Greg Konkel, annotation by Tammy Plotner).

Figure 11.7. NGC 288 (Credit—Palomar Observatory, courtesy of Caltech).

Hertzsprung and Russell. NGC 288 is currently thought to be about 11 billion years old (Figure 11.7).

By identifying both its blue and red branches, it was shown that many of NGC 288's stars are being stripped away by tidal forces and contributing to the formation of the Milky Way's halo structure. In 1997, three additional variable stars were discovered in this cluster.

At magnitude 8, this small globular is easy for southern observers, but faint for northern ones. If you are using binoculars, be sure to look for the equally bright spiral galaxy NGC 253 to the globular's north.

Monday, November 3

On this day in 1955, one of the few documented cases of a person being hit by a meteorite occurred. What are the odds on that?

In 1957 the Russian space program launched its first "live" astronaut into space—Laika (Figure 11.8). Carried on board Sputnik 2, our canine hero was the first living creature to reach orbit. The speedily developed Sputnik 2 was designed with sensors to transmit the ambient pressure, breathing patterns, and heartbeat of its passenger, and it also had a television camera on board to monitor its occupant. The craft also studied ultraviolet and X-ray radiation to further assess the impact of space flight upon live occupants. Unfortunately, the technology of

Figure 11.8. Laika (widely used public image).

the time offered no way to return Laika to Earth, so she perished in space. On April 14, 1958 Laika and Sputnik 2 returned to Earth in a fiery re-entry after 2570 orbits.

This evening on the Moon we will be returning to familiar features Theophilus, Cyrillus, and Catharina (Figure 11.9). Why not take the time to really power up on them and look closely? Curving away just to the southwest of Catharina, on the terminator, is another lunar challenge feature: Rupes Altai, or the Altai Scarp. Look for smaller craters beginning to emerge, such as Kant to the northwest, Ibn-Rushd just northwest of Cyrillus, and Tacitus to the west.

Since we've got the scope out, let's go have another look at the galaxy we spied last night!

Discovered by Caroline Herschel on September 23, 1783, NGC 253 (RA 00 47 33 Dec –25 17 18) is the brightest member of a concentration of galaxies known as the Sculptor Group, and is the brightest galaxy outside our own local group (Figure 11.10). Cataloged as both H V.1 and Bennett 4, this 7th magnitude beauty is also known as Caldwell 65, and due to both its brightness and oblique angle, it is often called the "Silver Dollar Galaxy." As part of the Saguaro Astronomy Club's "SAC 110" best NGCs, you can even spot this one if you don't live in the Southern Hemisphere. At around 10 million light-years away, this very dusty, star-forming Seyfert galaxy rocks in even a modest telescope!

Figure 11.9. Theophilus, Cyrillus, Catharina, and the Altai Scarp (Credit—Ricardo Borba).

Figure 11.10. NGC 253: The Silver Dollar Galaxy (Credit—R. Jay GaBany).

Tuesday, November 4

This morning will be the peak of the Southern Taurid meteor shower. Already making headlines around the world for producing fireballs, the Taurids will be best visible in the early morning hours with no Moon to interfere. The radiant for this shower is, of course, the constellation of Taurus and red giant Aldebaran, but did you know the Taurids are divided into two streams?

It is surmised that the original parent comet shattered as it passed our Sun 20,000–30,000 years ago. The larger "chunk" continued orbiting, and is known as periodic comet Encke. The remaining debris field turned into smaller asteroids, meteors, and other fragments. These occasionally pass through our atmosphere, creating the astounding "fireballs" known as bolides. Although the fall rate for this particular shower is rather low at seven per hour, these slow-traveling meteors (27 kilometers per second) are usually very bright, and appear to almost "trundle" across the sky. With the chances high all week of seeing a bolide, this makes a bit of quiet contemplation under the stars well worth a morning walk!

Tonight your lunar journey is to locate the ruins of crater Descartes northwest of Theophilus. What's left of this ancient crater still takes up a 48 kilometer diameter residence on the lunar surface. Look to its north for the deep well of Dolland—a much younger impact which drops 1580 meters down. Apollo 16 touched down just to the northeast of Dolland (Figure 11.11). Can you imagine the view as the astronauts flew over?

For unaided eye or binocular observers—or for those who just want something a bit different tonight—have a look at 19 Piscium (RA 23 46 23 Dec +03 29 12). You'll find it as the easternmost star in the small circlet just south of the Great Square of Pegasus. Also known as TX, it's a cool giant star which varies slightly

Figure 11.11. Descartes and Dolland (Credit—Wes Higgins).

Figure 11.12. 19 Piscium (Credit—Palomar Observatory, courtesy of Caltech).

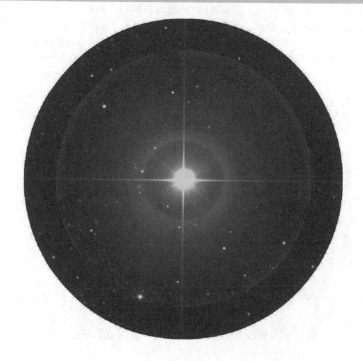

in magnitude on an irregular basis (Figure 11.12). You'll find this carbon star quite delightful because of its strong red color. Located between 400 and 1000 light-years away, it rivals even R Leporis' crimson beauty.

Wednesday, November 5

Today in 1906, a man named Fred Whipple was born (Figure 11.13). If that name doesn't ring a bell for you—it should. Thanks to Whipple's work, we have a clearer understanding of the orbital mechanics of comets and their relation to meteoroid streams. Not only that, he also founded the SAO observatory in Arizona, discovered six comets, made invaluable contributions to research in the upper atmosphere, and was the first to call a comet a "dirty snowball." His guess about the outgassing properties of comets was proved true when the first flyby of Comet Halley was made!

To honor Whipple a bit, let's have a look at a beautiful optical pair (possibly a multiple system), as we journey to the southernmost star in the Circlet—Kappa Piscium (RA 23 26 55 Dec +01 15 20).

Easily split even in binoculars, this lovely green and violet combination of stars may have once belonged to the Pleiades group. Magnitude 5 Kappa A is a chromium star which rotates completely in around 48 hours (Figure 11.14). It shows unusual iron lines in its spectrum, and also displays lines of uranium and osmium—and possibly those of a very rare element known as holmium. Both the uranium and the osmium contents could be the result of a supernova explosion in a nearby star. Enjoy this colorful pair tonight!

Closer to home, let's get a rough idea of how large craters really are on the Moon, by taking a look at Sacrobosco (Figure 11.15)—it's just west of the Theophilus–Cyrillus–Catharina trio. When you've located it, power up—because we're going to compare it to a terrestrial feature. The choice? Arecibo! As huge

Figure 11.13. Fred Whipple (widely used public image).

Figure 11.14. Kappa Piscium (Credit—Palomar Observatory, courtesy of Caltech).

Figure 11.15. Detail view of Sacrobosco (Credit—Wes Higgins, annotation by Tammy Plotner).

as the 1000 foot Arecibo dish looks, it would take 91 of them lined up end to end to span Sacrobosco's largest interior crater (A). Can you spot the dot in the A crater that's about the same size as Arecibo?

And can you imagine the possibilities if we could use a lunar crater to house an even larger radio telescope?!

Thursday, November 6

Tonight let's head toward the lunar surface as we have a look at a series of Lunar Club challenges you may not have logged yet. Just slightly below central, and toward lunar south, look for a series of rings which grow smaller as they progress. Once again, these are Ptolemaeus, Alphonsus, and Arzachel (Figure 11.16). But focus your attention on the largest of these, and in particular the small crater caught on its northern edge.

Named for Sir William himself, crater Herschel spans 41 kilometers and drops to a depth of 3770 kilometers below the surface. While you're journeying, look for small Ammonius caught in Ptolemaeus' interior. Further south, see if you can catch Alphonsus' bright central peak. Ranger 9's remains lie just northeast of there!

Figure 11.16. Ptolemaeus, Alphonsus, and Arzachel (Credit—Greg Konkel).

Friday, November 7

Today in 1996, the Mars Global Surveyor left on its journey. Just 30 years beforehand on this same day, Lunar Orbiter 2 was launched. Tonight let's launch our way toward the Moon as we begin our observing evening with a look at a far northern crater—J. Herschel (Figure 11.17).

Residing on the mid-northern edge of Mare Frigoris, this huge, shallow, old crater spans 156 kilometers and bear the scars of the years. Look for the deeper and younger crater Horrebow on the southwestern wall—for it has obliterated another, older wall crater.

Ready to aim for a bullseye? Then head for the bright, reddish star Aldebaran (Figure 11.18). Set your eyes, scopes, or binoculars there and let's look into the "eye" of the Bull.

Known to the Arabs as Al Dabaran, or "the Follower," Alpha Tauri got its name because it appears to follow the Pleiades across the sky. In Latin it was called Stella Dominatrix, yet the Olde English knew it as Oculus Tauri, or very literally the "eye of Taurus." No matter which source of ancient astronomical lore we explore, there are references to Aldebaran.

As the 13th brightest star in the sky, it almost appears from Earth to be a member of the V-shaped Hyades star cluster, but this association is merely coincidental, since it is about twice as close to us as the cluster is. In reality,

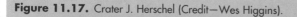

Figure 11.17. Crater J. Herschel (Credit—Wes Higgins).

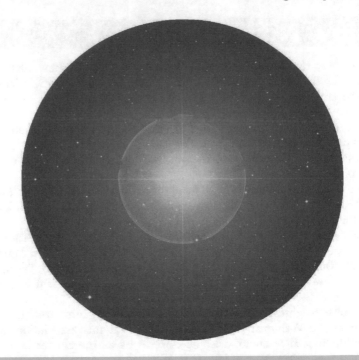

Figure 11.18. Aldebaran (Credit—Palomar Observatory, courtesy of Caltech).

Aldebaran is on the small end as far as K5 stars go, and like many other orange giants, it could possibly be a variable. Aldebaran is also known to have five close companions, but they are faint and very difficult to observe with backyard equipment. At a distance of approximately 68 light-years, Alpha is "only" about 40 times larger than our own Sun and approximately 125 times brighter. To try to grasp such a size, think of it as being about the same size as Earth's orbit! Because of its position along the ecliptic, Aldebaran is one of the very few stars of first magnitude that can be occulted by the Moon.

Saturday, November 8

Even if you use only binoculars tonight, you can't miss the beautiful C-shape of Sinus Iridum as it comes into view on the lunar surface (Figure 11.19). As we have learned, the mountains ringing it are called the Juras, and the crater punctuating them is named Bianchini. Do you remember what the bright tips of the opening into the "Bay of Rainbows" are called? That's right: Promontorium LaPlace to the northeast and Promontorium Heraclides to the southwest. Now take a good look at Heraclides... Just south of here is where Luna 17 landed, leaving the Lunokhod rover to explore!

Figure 11.19. Sinus Iridum (Credit—Wes Higgins).

Figure 11.20. Edmund Halley (widely used public image).

Born on this day in 1656, the great Edmund Halley made his mark on history as he became best known for determining the orbital period of the comet which bears his name (Figure 11.20). English scientist Halley had multiple talents, however, and in 1718 discovered that what were then referred to as "fixed stars," actually displayed (proper) motion! If it were not for Halley, Sir Isaac Newton may never have published his now famous work on the laws of gravity and motion.

Now turn your eyes or binoculars just west of bright Aldebaran and have a look at the Hyades Star Cluster (Figure 11.21). As noted yesterday, Aldebaran appears to be part of this large, V-shaped group, but is not an actual member. The Hyades cluster is one of the nearest galactic clusters, and it is roughly 130 light-years away at its center. This moving group of stars is drifting slowly away toward Orion, and in another 50 million years will require a telescope to view!

Figure 11.21. The Hyades Star Cluster (Credit—NASA).

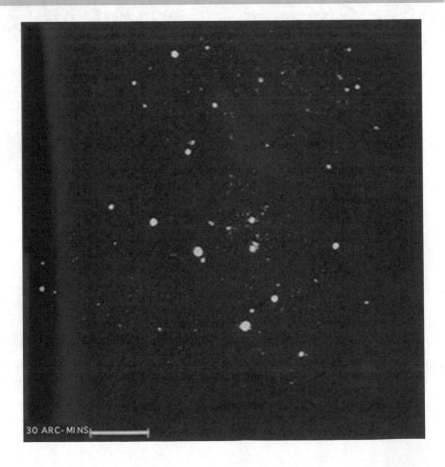

30 ARC-MINS

Sunday, November 9

Today is the birthdate of Carl Sagan (Figure 11.22). Born in 1934, Sagan was an American planetologist, exobiologist, popularizer of science and astronomy, and novelist. His influential work and enthusiasm inspired us all. If Carl were with us tonight, he would encourage amateurs at every level of astronomical ability! So let us honor his memory by beginning with an optical pairing of stars known as Zeta and Chi Ceti (Figure 11.23), a little more than a fistwidth northeast of bright Beta. Now have a look with binoculars or small scopes because you'll find that each has its own optical companion!

Now drop south-southwest less than a fistwidth to have a look at something so unusual that you can't help but be charmed—the UV Ceti System (RA 01 39 01 Dec –17 57 01).

What exactly is it? Also known as L 726-8, you are looking at two of the smallest and faintest stars known (Figure 11.24). This dwarf red binary system is the sixth nearest star to our solar system and resides right around 9 light-years away. While you are going to need at least an intermediate-size scope to pick up these near 13th magnitude points of light, don't stop observing right after you locate it. The fainter member of the two is what is known as a "Luyten's Flare Star" (hence the "L" in its name). Although it doesn't have a predictable timetable, this seemingly uninteresting star can jump two magnitudes in less than 60 seconds and drop back to "normal" within minutes—the cycle repeating possibly 2 or 3 times every 24 hours. A most incredible incident was recorded in 1952 when UV jumped from magnitude 12.3 to 6.8 in just 20 seconds!

No matter what you choose to look at tonight, as Sagan would say: "We are all star stuff."

Figure 11.22. Carl Sagan (widely used public image).

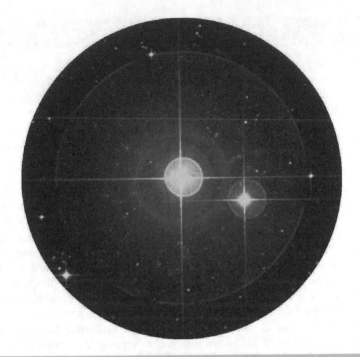

Figure 11.23. Chi Ceti (Credit—Palomar Observatory, courtesy of Caltech).

Figure 11.24. UV Ceti System (Credit—Palomar Observatory, courtesy of Caltech).

Monday, November 10

Tonight the gibbous Moon will dominate the sky. If you haven't had a chance to log some features like Copernicus, Gassendi, Tycho, and Plato, be sure to pick them up before the glare overpowers them. While you're there, be sure to look for "the Man in the Moon!"

Almost everyone is familiar with the legend of Cassiopeia and how the Queen came to be bound to her chair, destined for an eternity to turn over and over in the sky...but did you know that the constellation Cassiopeia holds a wealth of double stars and galactic clusters? Seasoned sky watchers have long been familiar with this constellation's many delights, but let's remember that not everyone knows them all, and tonight let's begin our exploration of Cassiopeia with two of its primary stars (Figure 11.25).

Looking much like a flattened "W," its southernmost bright star is Alpha. Also known as Schedar, this magnitude 2.2 spectral-type K star was once suspected of being a variable, but no changes have been detected in modern times. Binoculars will reveal its orange–yellow coloring, but a telescope is needed to bring out its unique features. In 1781, Sir William Herschel discovered a 9th magnitude companion, and our modern optics easily separate the blue–white component at a distance of 63". A second, even fainter, companion at 38" is mentioned in lists of double stars, and even a third at 14th magnitude was spotted

Figure 11.25. Alpha Cassiopeiae (Credit—Palomar Observatory, courtesy of Caltech).

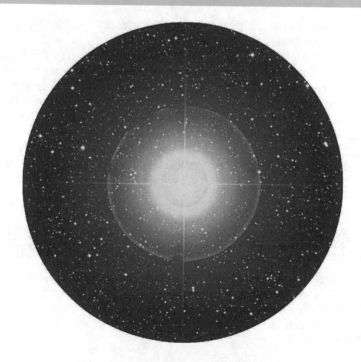

Figure 11.26. Eta Cassiopeiae (Credit—Palomar Observatory, courtesy of Caltech).

by S. W. Burnham in 1889. All three stars are optical companions only, but make 150–200 light-year distant Schedar a delight to view!

Just north of Alpha is the next destination for tonight—Eta Cassiopeiae (Figure 11.26). Discovered by Herschel in August of 1779, Eta is one of the most well-known binary stars. The 3.5 magnitude primary star is of spectral-type G, meaning it has a yellowish color much like our own Sun. It is about 10% larger than Sol and about 25% brighter. The 7.5 magnitude secondary (or B star) is very definitely a K type: metal poor and distinctively red. In comparison, it is half the mass of our Sun, crammed into about a quarter of its volume, and is about 25 times dimmer. In the eyepiece, the B star will angle off to the northwest, providing a wonderful and colorful look at one of the season's finest!

Tuesday, November 11

A true observer was born on this day in 1875. His name was Vesto Slipher (Figure 11.27), who spent some very quality time with the 60" and 100" telescopes on Mt. Wilson. Slipher was the first to photograph galaxy spectra and measure their redshifts, which led to the discovery of the expansion of the universe by Edwin Hubble.

Figure 11.27. Vesto Slipher (widely used public image).

As our observing year draws to a close, let's take another look at a feature you might have missed—Wargentin. Located in the southwest quadrant on the terminator, just south of the larger crater Schickard, we return again because Wargentin is one of the Moon's curiosities (Figure 11.28). You can capture it in

Figure 11.28. Schickard and Wargentin (Credit—Roger Warner).

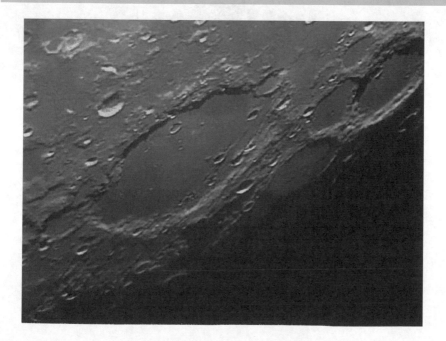

binoculars, but it's best seen through a telescope at high power—really take a look at what was once a normal small crater! Unlike most craters, Wargentin's walls were solid—able to contain the lava which eventually filled it to a height of 84 meters above the lunar surface.

While at first you might not notice, compare it to nearby Nasmyth and Phocylides. While both of these craters go below the surface, they also contain interior strikes—Wargentin has none! Except for a gentle, unnamed rille across its elevated surface, Wargentin is smooth.

On this night in 1572, the incomparable Tycho Brahe set out to record a bright new star. Today we realize he was looking at a supernova! "Visible" now as a supernova remnant only at very long wavelengths in the constellation of Cassiopeia, if you are good with your finderscope, you can still view it as a 7th magnitude star. Using Gamma, Alpha, and Beta as your visual starting point, use binoculars to locate Kappa just north of this trio. Small Kappa will also be part of a configuration of stars which will look much like the three of our starting point, only much dimmer. From Kappa, you will see a line of stars heading northwest. The very first in this series of 7th magnitude stars is SN 1572 (RA 00 25 08 Dec +64 09 55) (Figure 11.29). According to Tycho's report, as taken from Burnham's *Celestial Handbook*:

On the 11th day of November in the evening after sunset, I was contemplating the stars in a clear sky. I noticed that a new and unusual star, surpassing the other stars in brilliancy, was shining almost directly above msy head; and since I had, from boyhood, known all

Figure 11.29. Field of SN 1572 (Credit—Palomar Observatory, courtesy of Caltech).

the stars of the heavens perfectly, it was quite evident to me that there had never been any star in that place of the sky, even the smallest, to say nothing of a star so conspicuous and bright as this. I was so astonished of this sight that I was not ashamed to doubt the trustworthyness of my own eyes. But when I observed that others, on having the place pointed out to them, could see that there was really a star there, I had no further doubts. A miracle indeed, one that has never been previously seen before our time, in any age since the beginning of the world.

So bright was the event that it rivaled Jupiter at the time, and soon even surpassed Venus—being visible during the day for nearly 2 weeks. It had faded by the end of November, slowly changing color to red, and then passed out of visibility almost 16 months later. We'll be forever glad it wasn't cloudy at the time, for the event inspired Tycho Brahe to dedicate his life to astronomy... And who'd blame him?!

Wednesday, November 12

Wouldn't we all have loved to have been there in 1949 when the first scientific observations were made with the Palomar 5 meter (200 inch) telescope? Or to have seen what Voyager 1 saw as it made its closest approach to Saturn on this date in 1980? To watch Space Shuttle Columbia launch in 1981? Or even better, to have been around in 1833—the night of the Great Leonid Meteor Shower! But this is here and now, so let's make our own mark on the night sky as we view the Moon.

First let's explore tonight's lunar feature—Galileo (Figure 11.30). It is a challenge for binoculars to spot this feature, but telescopes of any size capable of higher power will find it easily on the terminator in the west-northwest section of the Moon. Set in the smooth sands of Oceanus Procellarum, Galileo is a very tiny, eye-shaped crater and has a soft rille that accompanies it. It was named for the very man who first viewed and contemplated the Moon through a telescope. No matter what lunar resource you choose to follow, all agree that giving such an insignificant crater a great name like Galileo was unthinkable! For those of you familiar with some of the outstanding lunar features, read any good account of Galileo's life and just look at how many spectacular craters were named for people he supported! We cannot change lunar cartography, but we can remember Galileo's many accomplishments each time we view this crater...

Now let's continue our stellar studies with the centralmost star in the lazy "W" of Cassiopeia—Gamma (RA 00 56 42 Dec +60 43 00).

At the beginning of the 20th century, the light from Gamma appeared to be steady, but in the mid-1930s it took an unexpected jump in brightness. In less than two years it rose by a magnitude! Then, just as unexpectedly, it dropped back down in roughly the same amount of time—a performance it repeated some 40 years later!

Gamma Cassiopeiae isn't quite a giant and is still fairly young on the evolutionary scale (Figure 11.31). Spectral studies show violent changes and variations in the star's structure. After its first recorded episode, Gamma ejected a shell

Figure 11.30. A Galileo Moon (Credit—Roger Warner).

Figure 11.31. Gamma Cassiopeiae (Credit—Palomar Observatory, courtesy of Caltech).

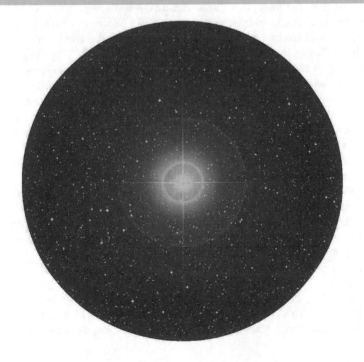

of gas which expanded its size by over 200%—yet it doesn't appear to be a candidate for a nova event.

The best estimate now is that Gamma is around 100 light-years away and approaching us in a very slow rate. If conditions are good, you might be able to telescopically pick up its disparate 11th magnitude visual companion, discovered by S. W. Burnham in 1888. It shares the same proper motion—but doesn't orbit this unusual variable star. For those who like a challenge, visit Gamma again on a dark night! Its shell left two bright (and difficult!) nebulae, IC 59 and IC 63.

Thursday, November 13

Today in 1990 Carolyn Porco was appointed leader of the imaging team for the Cassini mission to Saturn (Figure 11.32). Carolyn's career as a planetary scientist is unsurpassed, and she is an expert on planetary ring systems. For all of you who look at Saturn's rings with wonder, be sure to send your best to Carolyn—her undying love of astronomy began with observations just like yours!

Today is also the birthday of James Clerk Maxwell (Figure 11.33). Born in 1831, Maxwell was a leading English theoretician on electromagnetism and the nature of light. Tonight let's take a journey of 150 light-years as we honor Maxwell's theories of electricity and magnetism and take a look at a star that is in nuclear decay—Alpha Ceti (RA 03 02 19 Dec +04 14 10).

Its name is Menkar, and this second magnitude orange giant is slowly using up its nuclear fuel and gaining mass (Figure 11.34). According to Maxwell's theories of the electromagnetic and weak nuclear forces, W bosons must be produced

Figure 11.32. Carolyn Porco (widely used public image).

Figure 11.33. James Clerk Maxwell (widely used public image).

Figure 11.34. Alpha Ceti: Menkar (Credit—Palomar Observatory, courtesy of Caltech).

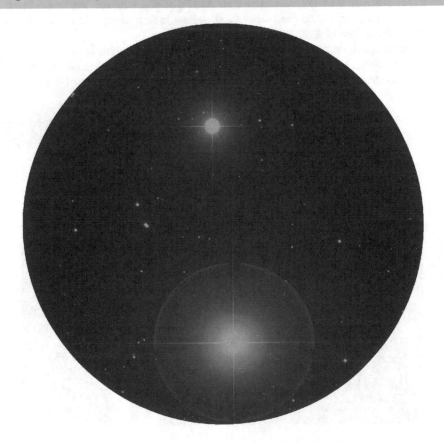

in such circumstances—this was an extremely advanced line of thinking for the time. Without getting deep into the physics, simply enjoy reddish Alpha for the beauty that it is. Even small telescopes will reveal its 5th magnitude optical partner 93 Ceti to the north. It's only another 350 light-years further away! You'll be glad you took the time to look this one up, because the wide separation and color contrast of the pair make this tribute to Maxwell worth your time!

Tonight is the Full "Frost Moon," and there is little doubt about how its name came to be! For those of you interested in viewing lunar features tonight, libration could be favorable for studying a collection of shallow, dark craters known as Mare Australe. Located on the southeastern limb, this large binocular and telescopic object is well-worth looking out for...because it's a challenge which isn't always visible.

Friday, November 14

This date in history marks the discovery of what we now refer to as a "Trans-Neptunian Object"—Sedna. In 2003 Brown, Trujillo and Rabinowitz went into the books for having observed the most distant natural solar system body to date. The rethinking of what it means to be a planet that this discovery inspired would eventually spell the end to Pluto's reign as our ninth planet!

Also on this day in 1971, Mariner 9 became the first space probe to orbit Mars. Can you still spot the faint Mars at sunset?

While Cassiopeia is in prime position for most northern observers, let's return tonight for some additional studies. Starting with Delta, let's hop to the northeast corner of our "flattened W" and identify 520 light-year distant Epsilon. For larger telescopes only, it will be a challenge to find the 12" diameter, magnitude 13.5, planetary nebula known as I.1747 in the same field as magnitude 3.3 Epsilon!

Using both Delta and Epsilon as our "guide stars," let's draw an imaginary line between the pair extending from southwest to northeast, continuing it the same distance until you stop at visible Iota (RA 02 29 03 Dec +67 24 08). Now go to the eyepiece...

As a quadruple system, Iota will require a telescope and a night of steady seeing to split its three visible components (Figure 11.35). Approximately 160 light-years away, this challenging system will show little or no color to smaller telescopes, but to large aperture the primary may appear slightly yellow and the companion stars a faint blue. At high magnification, the 8.2 magnitude C star will easily break away from the 4.5 magnitude primary, 7.2" to the east-southeast. But look closely at that primary: hugging in very close (2.3") to the west-southwest and looking like a bump on its side is the B star!

Dropping back to the lowest of powers, place Iota at the southwest edge of the eyepiece. It's time to study two incredibly interesting stars that should appear in the same field of view to the northeast. When both of these stars are at their maximum, they are easily the brightest stars in the field. Their names are SU (southernmost) and RZ (northernmost) Cassiopeiae (Figures 11.36 and 11.37), and each is unique! SU (RA 02 51 58 Dec +68 53 18) is a pulsing Cepheid variable located about 1000 light-years away, and will show a distinctive red coloration.

Figure 11.35. Iota Cassiopeiae (Credit—Palomar Observatory, courtesy of Caltech).

Figure 11.36. SU Cassiopeiae (Credit—Palomar Observatory, courtesy of Caltech).

Figure 11.37. RZ Cassiopeiae (Credit—Palomar Observatory, courtesy of Caltech).

RZ (RA 02 48 55 Dec +69 38 03) is a rapidly eclipsing binary which can change from magnitude 6.4 to magnitude 7.8 in less than 2 hours. Wow!

Saturday, November 15

On this day in 1990, Phil Harrington's first book *Touring the Universe through Binoculars* was released, making the author a household name in the astronomy world. Since that time, Phil has published seven additional books, given countless lectures, is a contributing author to well-known astronomy periodicals, and presents technical training at Brookhaven National Laboratory. His achievements are many, and we salute you Phil!

Above all, today we mark a very special birthday: on this day in 1738 my personal hero William Herschel was born (Figure 11.38). Among this British astronomer and musician's many accomplishments, Herschel was credited with the discovery of the planet Uranus in 1781; detecting the motion of the Sun in the Milky Way in 1785; finding Castor's binary companion in 1804—and he was the first to record infrared radiation. Herschel was well known as the discoverer of many clusters, nebulae, and galaxies. This came through his countless nights studying the sky and writing catalogs whose information we still use today. Just look at how many we've logged this year! Tonight let's look toward Cassiopeia as we remember this great astronomer...

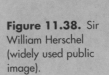

Figure 11.38. Sir William Herschel (widely used public image).

Although Herschel discovered many of the famous "400" objects in Cassiopeia just two days after his birthday in 1787, we only have a short time before the Moon rises, so let's set our sights on the area between Delta and Epsilon and have a look at three of them: NGC 654, NGC 663, and NGC 659.

Figure 11.39. NGC 654 (Credit—Palomar Observatory, courtesy of Caltech).

Figure 11.40. NGC 663 (Credit—Palomar Observatory, courtesy of Caltech).

At magnitude 6.5, NGC 654 (RA 01 44 00 Dec +61 53 00) is achievable in binoculars (Figure 11.39), but shows as nothing more than a hazy spot bordered by the resolvable star HD 10494. Yet, set a telescope its way and watch this diminutive beauty resolve. It is a very young open cluster which has been extensively studied spectroscopically. Oddly enough, it did not cease production of low-mass stars after the larger ones formed, and shows distinct polarization. Enclosed in a shell of interstellar matter, almost all of NGC 654's stars have reached main sequence and two have been identified as detached binaries.

Now shift your attention to NGC 663 (RA 01 46 12 Dec +61 14 00) (Figure 11.40). At magnitude 7, it is also viewable as a faint glow in binoculars—but is best in a telescope. With an age of about nine million years, this cluster contains the largest concentration of Be-type stars known: such stars show strong emission lines in hydrogen. While this might be considered "normal" for a B-type star, the mystery behind Be-types is that their emissions can simply end at any time—only to resume later. This could be in a matter of days, or it could be decades—but these odd stars may very well be victims of rapid rotation, high magnetic activity (similar to flares), or even interactions with a companion.

Time to head toward the faintest of the three—NGC 659 (RA 01 44 24 Dec +60 40 00). At magnitude 8, it is still within the reach of larger binoculars and will be fully resolved with a mid-sized telescope (Figure 11.41). Studied as recently as 2001, this looser collection contains seven newly discovered variables—three

Figure 11.41. NGC 659 (Credit—Palomar Observatory, courtesy of Caltech).

of which are Be stars. But give credit where credit is due! For as avid as Sir William was about observing, he had an equally avid observing partner: his sister Caroline. This time it was her call, as she is credited with the discovery of this particular open cluster—four years before her brother added it to his list in 1787!

Sunday, November 16

Today in 1974, there was a party at Arecibo, Puerto Rico, as the new surface of the giant 1000 foot radio telescope was dedicated (Figure 11.42). At this time, a quick radio message was released in the direction of the globular cluster M13.

Tonight let's take advantage of early dark and venture further into Cassiopeia. Returning to Gamma, we will move toward the southeast and identify Delta. Also known as Ruchbah, this long-term and very slightly variable star is about 45 light-years away, but we are going to use it as our marker as we head just one degree northeast and discover M103 (RA 01 33 24 Dec +60 39 00).

As the last object in the original Messier catalog, M103 (NGC 581) was actually credited to Méchain in 1781 (Figure 11.43). Easily spotted in binoculars and small scopes, this rich open cluster is around magnitude 7, making it a prime study object. About 8000 light-years away and spanning approximately 15 light-years, M103 offers up superb stars in a variety of magnitudes and colors, with a notable red in the south and a pleasing yellow and blue double to the northwest.

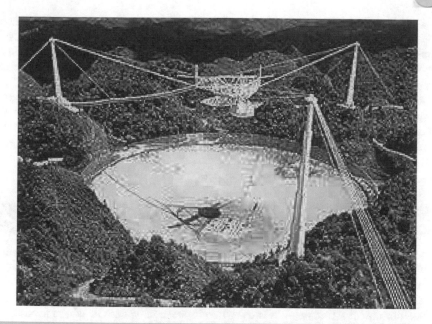

Figure 11.42. Arecibo (widely used image).

Figure 11.43. M103 (Credit—Palomar Observatory, courtesy of Caltech).

Keep watch for shooting stars tonight, because the annual Leonid meteor shower is underway. For those of you seeking a definitive date and time, it isn't always possible. The meteor shower itself belongs to the debris shed by comet 55/P Tempel-Tuttle as it passes our Sun in its 33.2 year orbit. Although it was once assumed it would simply be about 33 years between the heaviest "showers," we later came to realize the debris formed a cloud which lagged behind the comet and dispersed irregularly. With each successive pass of Tempel-Tuttle, new filaments of debris are left in space along with the old ones, creating different "streams" the orbiting Earth passes through at varying times, which makes blanket predictions unreliable at best.

Monday, November 17

If you didn't stay up late, then get up early this morning to catch the Leonids (Figure 11.44). Each year during November, we pass through the filaments of its debris—both old and new ones—and the chances of impacting a particular stream from any one particular year of Tempel-Tuttle's orbit becomes a matter of mathematical estimates. We know when it passed... We know where it passed... But will we encounter it and to what degree?

Figure 11.44. Leonid meteor shower (Credit—NASA).

Traditional dates for the peak of the Leonid meteor shower occur as early as the morning of November 17 and as late as November 19, but what about this year? On November 8, 2005 the Earth passed through an ancient stream shed in 1001. Predictions ran high for viewers in Asia, but the actual event resulted in a dud. There is no doubt that we crossed through that stream, but its probability of dissipation was impossible to calculate. Debris trails left by the comet in subsequent years look promising, but we simply don't know.

We may never know precisely where and when the Leonids might strike, but we do know that a good time to look for this activity is well before dawn on November 17, 18, and 19. With the Moon blocking the way, it will be difficult this year, but wait until the radiant constellation of Leo rises and the chances are good of spotting one of the offspring of periodic comet Tempel-Tuttle. Remember to dress warmly and provide for your viewing comfort!

On this day in 1970, the long-running Soviet mission Luna 17 successfully landed on the Moon (Figure 11.45). Its Lunokhod 1 rover became the first wheeled vehicle on the Moon. Lunokhod was designed to function for three lunar days but actually operated for eleven. Transmissions from Lunokhod officially stopped on October 4, 1971, the anniversary of Sputnik 1. Lunokhod had traversed 10,540 meters, transmitted more than 20,000 television pictures, over 200 television panoramas, and performed more than 500 lunar soil tests. *Spaseba!*

Figure 11.45. Luna 17 (Credit—NASA).

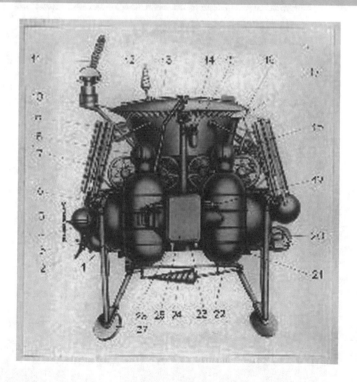

Tonight while skies are fairly dark, also be sure to keep watch for members of the Pegasid meteor shower—the radiant is roughly near the Great Square. This stream is at its peak tonight and endures from mid-October until late November, and used to be quite spectacular.

Tuesday, November 18

If you got clouded out of the Leonids yesterday morning, there is no harm in trying again. The meteor stream varies, and your chances are still quite good of catching one of these bright meteors.

Thanks to the later rise of the Moon, let's return again to Cassiopeia and start at the centralmost bright star, Gamma. Four degrees southeast is our marker for this starhop, Phi Cassiopeiae. By aiming binoculars or telescopes at this star, it is very easy to locate an interesting open cluster, NGC 457 (RA 01 19 40 Dec +58 17 18), because they will be in the same field of view (Figure 11.46).

This bright and splendid galactic cluster has received a variety of names over the years because of its uncanny resemblance to a figure. Some call it an "Angel," others see it as the "Zuni Thunderbird." I've heard it called the "Owl" and the "Dragonfly," but perhaps my favorite is the "E.T. Cluster." As you view it, you can see why! Bright Phi and HD 7902 appear like "eyes" in the dark, and the

Figure 11.46. NGC 457 (Credit—Palomar Observatory, courtesy of Caltech).

dozens of stars that make up the "body" appear like outstretched "arms" or "wings." (For *E.T.* fans? Check out the red "heart" in the center.)

All this is very fanciful, but what is NGC 457 really? Both Phi and HD 7902 may not be true members of the cluster. If 5th magnitude Phi were actually part of this grouping, it would have to be at a distance of approximately 9300 light-years, making it the most intrinsically luminous star in the sky, far outshining even Rigel! To get a rough of idea of what this means, if we were to view our own Sun from this far away, it would be no more than magnitude 17.5. The fainter members of NGC 457 comprise a relatively young cluster which spans about 30 light-years. Most of the stars are only about 10 million years old, yet there is an 8.6 magnitude red supergiant at its heart. No matter what you call it, NGC 457 is an entertaining and bright cluster that you will find yourself returning to again and again. Enjoy!

Wednesday, November 19

Once again utilizing early darkness, let's go back to Cassiopeia. Remembering Alpha's position as the westernmost star, go there with your finderscope or binoculars and locate bright Sigma and Rho (each has a dimmer companion). They will appear to the southwest of Alpha. It is between these two stars that you will find NGC 7789 (RA 23 57 24 Dec +56 42 30).

Figure 11.47. NGC 7789 (Credit—Palomar Observatory, courtesy of Caltech).

Absolutely one of the finest examples of a rich open cluster which border on a loose globular, NGC 7789 has a population of about 1000 stars and spans a mind-boggling 40 light-years (Figure 11.47). At well over a billion years old, the stars in this 5000 light-year distant galactic cluster have already evolved into red giants or supergiants. Discovered by Caroline Herschel in the 18th century, this huge cloud of stars has an average magnitude of 10, making it a great large binocular object, a superb small telescope target, and a total fantasy of resolution for larger instruments.

Now head to the westernmost of the bright stars—Beta. Also known as Caph, Beta Cassiopeiae is approximately 45 light-years away and is known to be a rapid variable. Viewers with larger telescopes are challenged to find the 14th magnitude optical companion to Caph at about 23" in separation. Tonight, using our previous study stars Alpha and Beta, we are going to learn to locate a Messier object with ease! By drawing an imaginary line between Alpha and Beta, we extend that line the same distance beyond Beta and find M52 (RA 23 24 48 Dec +61 35 00).

Found on September 7, 1774 by Charles Messier, this magnitude 7 galactic cluster is easily seen in both binoculars and small telescopes (Figure 11.48). Comprised of roughly 200 members, it is roughly 3000 light-years distant and spans approximately 10–15 light-years. Containing stars of several different magnitudes, larger telescopes will easily perceive blue components as well as

Figure 11.48. M52 (Credit—Palomar Observatory, courtesy of Caltech).

orange and yellow ones. Also known as NGC 7654, M52 is a young, very compressed cluster which is approximately as old as the Pleiades.

For those with large telescopes wanting a challenge, try spotting a faint patch of nebulosity just 36' to the southwest. This is NGC 7635, more commonly known as the "Bubble Nebula." Best of luck!

Thursday, November 20

Today celebrates the birth of another significant astronomer—Edwin Hubble (Figure 11.49). Born in 1889, Hubble became the first American astronomer to identify Cepheid variables in M31—which in turn established the extragalactic nature of the spiral nebulae. Continuing with the work of Carl Wirtz, and using Vesto Slipher's redshifts, Hubble could then calculate the velocity–distance relation for galaxies. This has become known as Hubble's Law and demonstrates the expansion of our Universe.

Tonight we're going to head just a little more than a fistwidth west of the westernmost bright star in Cassiopeia to have a look at Delta Cephei (RA 22 29 10 Dec +58 24 54). This is the most famous of all variable stars and the granddaddy of all Cepheids (Figure 11.50). Discovered in 1784 by John Goodricke, its changes in magnitude are not due to a revolving companion—but rather the pulsations of the star itself.

Ranging over almost a full magnitude in 5 days, 8 hours, and 48 minutes precisely, Delta's changes can easily be followed by comparing it to nearby Zeta and Epsilon. Upon reaching its dimmest point, it will brighten rapidly in a period of about 36 hours—yet take 4 days to slowly dim again. Take time out of your busy night to watch Delta change and change again. It's only 1000 light-years away, and doesn't even require a telescope! (But even binoculars will show its optical companion.)

Figure 11.49. Edwin Hubble (widely used public image).

Figure 11.50. Delta Cephei (Credit—Palomar Observatory, courtesy of Caltech).

Friday, November 21

Tonight we will haunt Cassiopeia one last time—with studies for the seasoned observer. Our first challenge of the evening will be to return to Gamma where we will locate two patches of nebulosity in the same field of view. IC 59 and IC 63 are challenging because of the bright influence of the star, but by moving the star to the edge of the field of view you may be able to locate these two splendid small nebulae. If you do not have success with this pair, why not move on to Alpha? About 1.5 degrees due east, you will find a small collection of finderscope stars that mark the area of NGC 281 (RA 00 52 25 Dec +56 33 54). This distinctive cloud of stars and ghostly nebulae make this NGC object a fine challenge (Figure 11.51)!

The last things we will study are two small elliptical galaxies that are achievable in mid-sized scopes. Locate Omicron Cassiopeiae about seven degrees north of M31, and discover a close galactic pair that is associated with the Andromeda group—NGC 185 (RA 00 38 57 Dec +48 20 14) and NGC 147 (RA 00 33 11 Dec +48 30 24) (Figures 11.52 and 11.53).

The constellation of Cassiopeia contains many more fine star clusters and nebulae—and even more galaxies. For the casual observer, simply tracing over the rich star fields with binoculars is a true pleasure, because there are many

Figure 11.51. NGC 281 (Credit—Palomar Observatory, courtesy of Caltech).

Figure 11.52. NGC 185 (Credit—Palomar Observatory, courtesy of Caltech).

Figure 11.53. NGC 147 (Credit—Palomar Observatory, courtesy of Caltech).

bright asterisms best enjoyed at low power. And scopists will return year after year to "rock with the Queen." Enjoy its many challenging treasures tonight!

Saturday, November 22

Tonight let's have a look at one of the most elusive Messiers of all as we head about two fingerwidths northeast of Eta Piscium in search of M74 (RA 01 36 Dec +15 47).

Discovered at the end of September 1780 by Méchain, M74 is a real challenge to smaller backyard telescopes—even at magnitude 9 (Figure 11.54). This near perfect presentation of a face-on spiral galaxy has low surface brightness, and it takes really optimal conditions to spot much more than its central region. Located 30–40 million light-years away, M74 is roughly the size of the Milky Way, yet has no central bar. Its tightly wound spiral arms contain clusters of young blue stars and traces of nebulous star-forming regions that can be seen in photos, yet little more than some vague concentrations in structure are all that can be noted visually...even in a large scope. But if sky conditions are great, even a small telescope can see details! Add the slightest bit of light pollution and even the biggest scopes will have problems locating it.

Don't be disappointed if all you see is a bright nucleus surrounded by a small hazy glow—just try again another time. Who knows what might happen?

Figure 11.54. M74 (Credit—R. Jay GaBany).

A supernova was discovered in 2002 by a returning amateur, and again in 2003 by an observer in the Southern Hemisphere. When it comes to M74, this is the very best time of the year to try with a smaller scope!

Sunday, November 23

Tonight in 1885, the very first photograph of a meteor shower was taken. Also, the weather satellite TIROS II was launched on this day in 1960 (Figure 11.55). Carried to orbit by a three-stage Delta rocket, the "Television Infrared Observation Satellite" was about the size of a barrel—it successfully tested experimental television techniques and infrared equipment. Operating for 376 days, Tiros II sent back thousands of pictures of Earth's cloud cover and was successful in its experiments to control the orientation of the satellite's spin. Coincidentally, a similar mission—Meteosat 1—became the first satellite put into orbit by the European Space Agency, in 1977 on this day. Where is all this leading? Why not try observing satellites on your own? Thanks to wonderful online tools from NASA, you can be alerted by e-mail whenever a bright satellite makes a pass over your specific area, or you can use other available tools to predict passes. It's fun and doesn't require any special equipment!

Tonight let's test our starhopping and observing talents by starting first with a beautiful double—Gamma Arietis. Now look about a fistwidth east-southeast for dim little Pi. When you have Pi centered, move about half a degree southwest for an alternative catalog study—DoDz 1 (Figure 11.56).

While you might find this little, sparkling, double handful of stars of little interest—think twice before you hop on. While DoDz studies are far more

Figure 11.55. TIROS II (archival image).

scattered and less populous that most galactic clusters, it doesn't make them less interesting. What you are looking at are basically the fossils of a once active and more concentrated region of stars. As the cluster itself has matured, the lower mass members have been stripped away and have gone off to join the general

Figure 11.56. DoDz 1 (Credit—Palomar Observatory, courtesy of Caltech).

population. Known as a "dissolving cluster," DoDz 1 is all that's left of a far grander collection. Very ancient...yet still very beautiful!

Monday, November 24

Tonight let's head less than a degree south-southeast of Delta Ceti (RA 02 43 40 Dec –00 00 48) to have a look at a galaxy grouping that features the magnificent M77 (Figure 11.57).

Discovered on October 29, 1780 by Pierre Méchain, Messier cataloged it as Number 77 around six weeks later, classifying it as a "nebulous cluster"—an accurate description for a small telescope. Not until 1850 when Lord Rosse uncovered its spiral nature did we begin to view it as the grand structure seen in modern telescopes.

Around 47 million light-years away, larger instruments will reveal the wide spiral arms which the older stars call home, and the concentrated core region where gigantic gas clouds move rapidly and new stars are being formed—a core which contains such a massive energy source that it emits a spectrum of radio waves. After decades of study, the highly active nucleus of this Seyfert galaxy is known to have a mass equaling 10 million suns. The nucleus has a five light-year wide disc rotating around it which includes intense star-forming regions. This is

Figure 11.57. M77 (Credit—Palomar Observatory, courtesy of Caltech).

Figure 11.58. NGC 1055 (Credit—Palomar Observatory, courtesy of Caltech).

Figure 11.59. NGC 1073 (Credit—Palomar Observatory, courtesy of Caltech).

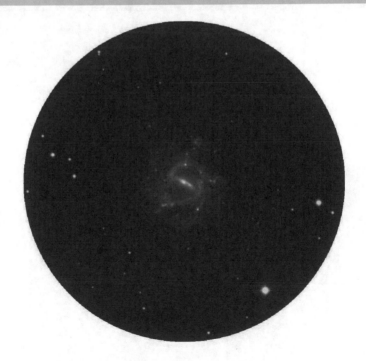

one of the brightest known, and was cataloged by Arp as number 37 on his list of peculiar galaxies.

While even binoculars can spot the core, and modest scopes can reveal M77's glory, larger telescopes will also spy 10th magnitude, edge-on NGC 1055 (RA 02 41 45 Dec +00 26 32) about half a degree north-northwest. And you may also spot 11th magnitude, face-on NGC 1073 (RA 02 43 40 Dec +01 22 34) about a degree north-northeast of M77. Enjoy them tonight (Figures 11.58 and 11.59)!

Tuesday, November 25

Tonight let's journey back to the area around M77—because we've got more to explore!

Let's start with Delta Ceti and head north about a degree for NGC 1032 (RA 02 39 23 Dec +01 05 37). Discovered in 1783 by Sir William Herschel and cataloged as H II.5, this 13th magnitude edge-on galaxy isn't for the smaller scope, but that doesn't mean it's uninteresting (Figure 11.60). Possessing a bright core region and an almost stellar nucleus, this superb galaxy was home to a supernova event in 2005!

Now have a look at M77 again and head less than two degrees east for a pair of north–south-oriented galaxies—NGC 1090 and NGC 1087 (RA 02 46 33 Dec –00

Figure 11.60. NGC 1032 (Credit—Palomar Observatory, courtesy of Caltech).

Figure 11.61.
NGC 1090/1087
(Credit—Palomar
Observatory, courtesy
of Caltech).

14 49). At around 120 million light-years distance, northern NGC 1090 (H II.465) is also a supernova host candidate, with events being reported in both 1962 and 1971. At close to magnitude 13, this barred spiral isn't easy, but it can be spotted with aversion and a mid-sized telescope.

About 15' south is NGC 1090/1087 (Figure 11.61). Although the pair seem quite close, no interaction between them has been detected. At magnitude 11, smaller scopes stand a much better chance of picking out 1087's faint, round glow...while large scopes will get a sense of tightly-wound spiral arms around Herschel II.466. Its barred structure is quite curious—far smaller than is common in this type of spiral, but still a grand star-forming region. A region that hosted a supernova in 1995!

Check out this active group tonight...

Wednesday, November 26

Today in 1965 marked the launch of the first French satellite—Asterix 1. Today is also the ninth anniversary of the discovery of the meteorites SaU 005 and 008—the "Mars Meteorites." These meteorites are known to be of Martian origin

Figure 11.62. Abell 194 (Credit—Palomar Observatory, courtesy of Caltech).

because of gases preserved in the glassy material of their interiors. They were hurled into space some 600,000 years ago when a probable asteroid impact on Mars tossed them high enough to escape the planet's gravity. They were captured by our own gravity these many thousands of years later. Besides SaUs 005 and 008, at least 30 other meteorites found on Earth have been positively determined to be of Martian origin from their chemical compositions.

Are you ready to take on a real "big scope" challenge tonight? Then set sail with the largest of telescopes to investigate a galaxy cluster known as Abell 194 (RA 01 26 01 Dec –01 22 02) (Figure 11.62). Over 100 galaxies have been found in this area, and most of them are about 265 million light-years away; the brightest being NGC 547. The pairing of NGCs 547 and 545 may be interacting with the elliptical NGC 541. Other viewable members include NGCs 548, 543, 535, 530, 519, 538, and 557, as well as the far more southern 564, 560, and 558—the last of these is just north of another optical double.

Thursday, November 27

Today is the birthday of Anders Celsius—born in 1701 (Figure 11.63). While you might easily recognize the name Celsius in connection with temperature, you might not know about the contributions which Anders made to astronomy three centuries ago. Born to a Swedish family of mathematicians and astronomers, one

Figure 11.63.
Anders Celsius (widely
used image).

of his first achievements came when he participated in an effort to determine
the true shape of Earth. He was also the very first scientist to recognize the
connection between magnetism and the aurora, and by age 39 he had become
the director of an observatory.

Figure 11.64. IC 1613 (through blue filter) (Credit—Palomar Observatory, courtesy of Caltech).

Celsius also developed the first instrument for measuring the brightness of starlight. Ever resourceful, he already possessed tools to measure position and motion—but had nothing with which to gauge magnitude. His idea was so simple it was downright elegant: he simply blocked the light with identical glass plates until the star disappeared. The brighter the star the more plates it took!

While Celsius died far too young, we can honor his memory on this New Moon night by having a look at one of the members of our own Local Group—IC 1613 (Figure 11.64). You'll find it on the Cetus–Pisces border, due south across the sky from another "neighbor"—the Andromeda Galaxy.

Discovered just over 100 years ago by Max Wolf, IC 1613 is booked at magnitude 11, but has very low surface brightness. Although very difficult for a small telescope, it has been spotted in large binoculars, and nearly comes to full resolution with large telescopes. It is very irregular, barred, and even has a detached star cloud. Very similar to the Magellanic Clouds, it's full of Cepheid variables and its distance has been estimated to be between 1.8 and 2.5 million light-years.

If you have a chance to study it, please look for a brighter concentration of stars at the northeastern edge. It's a wonderful star-forming region spanning 1000 light-years, and is filled with gas and dust—with the brightest of these stars very similar to our own Rigel!

Friday, November 28

Tonight in 1659, Christian Huygens was busy at the eyepiece—but he wasn't studying Saturn. This was the first time any astronomer had seen dark markings on Mars!

If Huygens and Herschel were alive to enjoy today's new technology, you could bet they'd have a big backyard scope aimed about four degrees east of the Zeta-Chi pairing in Cetus to have a look at Hickson Compact Galaxy Group 16 (RA 02 09 31 Dec –10 08 59).

Consisting of four faint, small, galaxies designated as NGCs 835, 833, 838, and 839 clustered around a 9th magnitude star, these aren't for average equipment—but are a true challenge for a seasoned observer (Figure 11.65). Groups of galaxies such as Hickson 16 are believed to be some of the very oldest things in our Universe—this particular assemblage has a reputation for having an extremely large amount of starburst activity. It is also close enough for scientists to study. Its members were all discovered and cataloged by Sir William Herschel on this very night—223 years ago! The northernmost, NGC 833, is known as H II.482, roughly of magnitude 13, followed by NGC 835 (H II.483) which holds a magnitude of 12. Next in line is NGC 838 (H II.484) at close to magnitude 13, followed by southernmost NGC 839 (H II.485) at magnitude 13. Not easy... But this beautiful crescent of four is worth the effort. If Herschel could do it, so can you!

Figure 11.65. Hickson 16 (Credit—Palomar Observatory, courtesy of Caltech).

Saturday, November 29

Today in 1961 Enos the Chimp launched into fame! His story is a long and colorful one—but Enos was a true astronaut (Figure 11.66). Selected to make the first American orbital animal flight only three days before the launch, he flew into space on board Mercury-Atlas 5 and completed his first orbit in just under 90 minutes. Although Enos was scheduled to complete three orbits, he was brought back due to "attitude difficulties." But whose? Malfunctions caused the chimp to be repeatedly shocked when performing the correct maneuvers, but Enos continued to perform flawlessly—and was said to run and jump enthusiastically on board the recovery ship, shaking the hands of the crew.

Although he died a year later from an unrelated disease which could not be cured at the time, Enos the chimp remains one of our most enduring space heroes. Tonight let's monkey around with a planetary nebula as we'll take a look at 8th magnitude NGC 246 about five degrees north of Beta Ceti (Figure 11.67).

On the large side as planetaries go, this variegated shell of gas envelops a dying star about 1600 light-years away. Once upon a time, the star was much like our Sun, but over thousands of years its atmosphere expanded, interacting with the gas and dust in the interstellar medium to create what you see tonight. With the outer shell visible to even small telescopes, larger aperture can spot the fainter

Figure 11.66. Enos (and handler) (Credit—NASA).

Figure 11.67. NGC 246 (south is up) (Credit—Palomar Observatory, courtesy of Caltech).

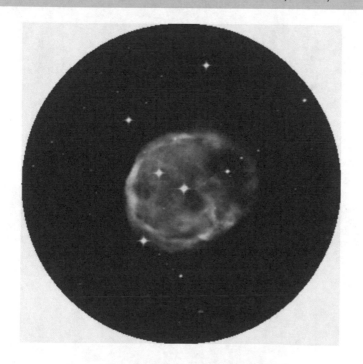

member of the binary at the heart of this planetary...a star well on it's way to becoming a white dwarf.

Discovered by Sir William Herschel, it is often referred to as the "Skull" nebula, but perhaps tonight you'll see the smiling face of Enos forever leaving its mark on space!

Sunday, November 30

Just as a curiosity, on this day in 1954, Elizabeth Hodges was struck by a five kilogram meteorite in Alabama. Duck! If you're out at sunset tonight, you'll be struck by the beauty of the slender crescent Moon illuminated by Earthshine. Not far away, look for the pairing of Venus and Jupiter, because things are going to get a lot cozier as the last month of the year begins!

Before the Moon once again interferes with study, last take one last look at Cetus before we move on. Our last target is a beauty—one which can be seen with larger binoculars, is easy with a small telescope, and becomes breathtaking with large aperture. Set your sights on bright Beta and drop less than three degrees south-southeast for NGC 247 (RA 00 47 09 Dec −20 45 38)...

As one of the largest members of a group of galaxies located around our galactic south pole, NGC 247 seems to be standing still in space—at a distance of 6–8 million light-years (Figure 11.68). At its core is a near stellar-sized nucleus—a

Figure 11.68. NGC 247 (Credit—Palomar Observatory, courtesy of Caltech).

bright, central mass of stars which dominates its patchy looking structure. Look closely at its northern edge for NGC 247 sports a huge area of dark, obscuring dust—or what may just be an empty space between its clouds of stars. Note a bright star caught on its southern flank.

While you may find this low surface brightness galaxy a bit difficult unless you stick with the most minimal of magnification, you can congratulate yourself for capturing not only another Herschel "400" object, but Bennett 3 as well!

CHAPTER TWELVE

December, 2008

Monday, December 1

Born today in 1811 was Benjamin (Don Benito) Wilson. He was the namesake of Mt. Wilson, California—home to what were once the largest telescopes in the world—the 60" Hale and the 100" Hooker. Later, three solar telescopes were added on the mountain—two of which are still in use—as well as the CHARA array and active interferometers. It was here that Edwin Hubble first realized that the "spiral nebulae" were actually distant galaxies, and discovered Cepheid variables in them (Figure 12.1). As we approach the end of our SkyWatching year together, let us stand as the last of the light leaves the sky and admire one of the most beautiful conjunctions of the year.

Look no further than the ecliptic plane to the west where Venus, Jupiter, and the tender crescent Moon have gathered to present a picturesque view of the workings of our solar system. Perhaps at your time, Venus is about to be occulted—or will merely sit at the edge of lunar limb showing the wonderful perspective from which we are able to look out into what is beyond. Picture in your mind the positions of the Earth, Sun, Moon, Venus, and Jupiter. Enjoy this splendid sight tonight!

When the Moon has set, let's head far north for one of the oldest galactic clusters in our visible sky—NGC 188 (Figure 12.2).

Hovering near Polaris (RA 00 48 26 Dec +85 15 21), this circumpolar open cluster also goes by other names: Collinder 1 and Melotte 2. Discovered by John Herschel on November 3, 1831, this 8th magnitude collection of faint stars will

Figure 12.1. Edwin Hubble working at the Hooker Telescope (widely used public image).

require a telescope to resolve its 120 members. At one time, it was believed to be as old as 24 billion years, later revised down to 12 billion but it is now considered to be around 5 billion years old. No matter how old it may truly be, it was around long before us and will endure long after we are gone. NGC 188 is one of the time-honored great studies, and is also number one on the Caldwell list!

Figure 12.2. NGC 188 (Credit—Palomar Observatory, courtesy of Caltech).

Tuesday, December 2

Today in 1934, the largest mirror in telescope history began its life when the blank for the 200″ telescope was cast in Corning, New York. We'll take a look in the weeks ahead at the important role the 200″ scope played in astronomy...

Let's begin our observations tonight with a lunar feature, as we turn our telescopes toward the southern limb of the Moon to seek out the previously studied double crater Boussingault (Figure 12.3). As we know, this ancient formation spans at least 131 kilometers across the lunar surface, yet we have no clear idea of how deep it may be. Look further south, this may be your last chance this year to glimpse crater Hale as well. Named for George Ellery Hale, director of Mt. Wilson Observatory (and namesake of the above-mentioned 200″ telescope), this on-the-edge feature will never be fully disclosed to our terrestrial telescopes. At the present time we know it is far smaller than Boussingault with an 84 kilometer span, yet—like its predecessor—we know little more about it.

Now that you've viewed a challenging set of craters, would you like to have a look at a challenging double star? All you have to do is locate Theta Aurigae on the east side of the pentagonal shape of this constellation (Figure 12.4).

Located about 110 light-years away, 2.7 magnitude Theta is a four-star system, whose members range in magnitude from 2.7 to 10.7. Suited even to a small telescope, the brightest member—Theta B—is itself a binary at magnitude 7.2, and was first recorded by Otto Struve in 1871. The pair moves quite slowly, and

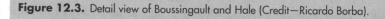

Figure 12.3. Detail view of Boussingault and Hale (Credit—Ricardo Borba).

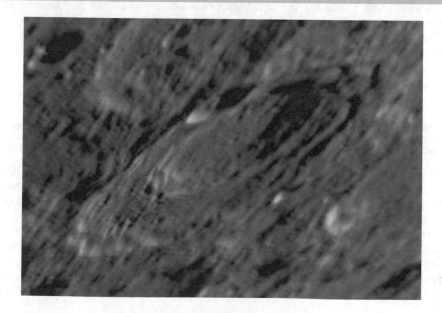

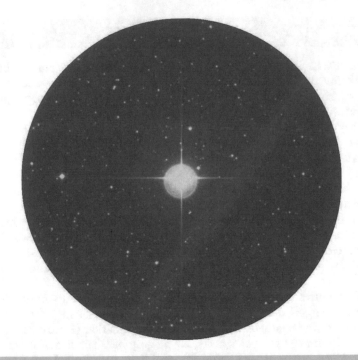

Figure 12.4. Theta Aurigae (Credit—Palomar Observatory, courtesy of Caltech).

may take as long as 800 years to orbit at their separation of about 110 AU. The furthest member of this system was also noted by Struve as far back as 1852, but it is not a true member—the separation only occurring thanks to Theta's own proper motion.

While you are there, be sure to note Theta's unusual color. While it will appear "white," look closely at the diffraction caused by our own atmosphere which acts much like a prism... You'll notice a lot more purple and blue around this star than many others of the same spectral type. Why? Theta is a silicon star!

Wednesday, December 3

Today in 1971, the Soviet Mars 3 became the first spacecraft to make a soft landing on the red planet, and two years later on this same date the Pioneer 10 mission became the first to fly by Jupiter. One year later on this same date? Pioneer 11 did the same thing!

Our focus tonight is on this date 36 years ago. With only four days to go before the last manned lunar mission headed skyward, I wonder how Eugene Cernan felt to know he was about to be the last man of the 20th century to set foot on the Moon. Somehow I'd like to believe Cernan knew the importance of his role in history: on December 14, 1972 he called his last step in Taurus-Littrow Valley the "end of the beginning." As we reach the end of our observing year, let this be only the beginning for you as we look to that distant orb to seek out the Apollo 17 landing area, using our photographic map (Figure 12.5).

You have learned so much over the last 11 months ! Even if the terminator has not progressed as far as the illustration shows, you should know the approximate location of Posidonius on the surface and recognize Mare Crisium and the Taurus Mountains to its east, as well as the small, gray expanse of Sinus Amoris between them. Littrow is on its western shore, and although it is rather small with a 31 kilometer diameter, Mons Vitruvius will shine like a beacon to the south. Enjoy your Moon walk!

Now for the "star" of tonight's show, Beta Andromedae (Figure 12.6). More commonly known as Mirach, you'll find this irregular, slightly variable star at home 199 light-years away. Notable even to the eye as slightly red in color, this dying star's core may be mainly helium or carbon now—soon Beta will become a dense white dwarf. For now, it is fairly small, no more than the size of Mercury's orbit from our own Sun. If you have steady skies, power up. This is also a study of E. E. Barnard, who in 1898 discovered a 14th magnitude companion star over 800 times fainter than Sol and 60,000 times fainter than Mirach!

Return again to Beta on a good dark night and see if you can spot "The Ghost of Mirach" in the same field, and no more than 7' to the northwest. This is the small elliptical galaxy NGC 404—not an error!

Figure 12.5. Apollo 17 landing area (Credit—Ricardo Borba, annotation by Tammy Plotner).

Apollo 17 -->

Figure 12.6. Beta Andromedae (Credit—John Chumack).

Thursday, December 4

Today in 1978, the Pioneer Venus Orbiter became the first spacecraft to orbit Venus. And in 1996, the Mars Pathfinder mission was launched. Tonight launch yourself toward the Moon as we have a look at southern crater Maurolycus (Figure 12.7). Although we have visited it before, look again! At an overall diameter of 114 kilometers, this double impact crater sinks below the surface to a depth of 4730 meters and displays a wonderful multiple mountain-peaked center. If you have not collected Gemma Frisius for your studies, you will find it just north of this grand crater, looking much like a "paw print" at low power.

If your lunar curiosity is still sparking, then head north from Maurolycus to the emerging Sinus Medii to view the incredible Rima Hyginus (Figure 12.8). This tremendous crack in the smooth lava of the "Bay in the Middle" stretches for 220 kilometers. If this is indeed a volcanic artifact, it's precisely the same length as the lahar flows affecting the rivers around the Mt. Saint Helens volcano!

Now let's travel 398 light-years away as we have a look at AR Aurigae—the centermost star in a brilliant collection (Figure 12.9). It is about one-third the distance from southern Beta to northern Alpha (Capella). AR is an eclipsing binary which consists of two main sequence white dwarf stars. About every

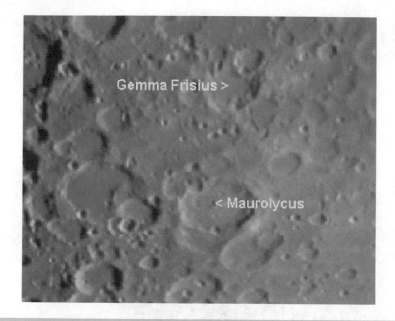

Figure 12.7. Maurolycus and Gemma Frisius (Credit—Greg Konkel, annotation by Tammy Plotner).

Figure 12.8. Rima Hyginus (Credit—Wes Higgins).

Figure 12.9. AR Aurigae (Credit—Palomar Observatory, courtesy of Caltech).

4.1 days, this pair will have a slight magnitude drop. While both are chemically peculiar, neither fills its Roche Lobe—meaning they are not stripping material from each other to cause these unusual abundances. Recent studies have shown the possibility of a third, unseen companion! But even binoculars will see that AR resides in a great field of stars and is worth a little of your time...

Friday, December 5

With only 20 days left until the holiday, astronomers have recently discovered a unique feature on the lunar surface. While accepted for many years as a natural feature of selenography, modern photography coupled with today's high-powered telescopes have discovered an area near the lunar North Pole being used as a runway by a man in a red suit piloting an unusual spacecraft (Figure 12.10). Be sure to spark the imaginations in your young viewers as you show them the Alpine Valley!

    Tonight your stellar destination is K-type star, 51 Andromedae (RA 01 37 59 Dec +48 37 41). You'll find it as the northernmost star in the A-shape which forms the constellation—it is considered to be the Lady's foot (Figure 12.11). Located 174 light-years away, star 51's claim to fame is being one of the few well-evolved

Figure 12.10.
Santa's Landing Strip
(Credit—Wes Higgins).

Figure 12.11. 51 Andromedae (Credit—Palomar Observatory, courtesy of Caltech).

stars for which we know the exact parallax. While this is interesting enough in itself, the true beauty of this region is simply the field which accompanies this 3.5 magnitude star. Go tonight and enjoy it in binoculars or in low power scopes!

Saturday, December 6

Tonight there are craters galore to explore: Plato, Aristotle, Eudoxus, Archimedes... But let's head to the deep, deep, south as we go out on the limb for Klaproth and Casatus (Figure 12.12). Differing by only eight kilometers in width, this pair of extreme features is well worth all the magnification skies will allow!

With deep sky studies improbable for the next few days, why don't we try taking a look at another interesting variable star? RT (star 48) Aurigae is a bright Cepheid that is located roughly halfway between Epsilon Geminorum and Theta Aurigae. This perfect example of a pulsating star follows a precise timetable of 3.728 days, and varies by close to one full magnitude (Figure 12.13).

Located 1600 light-years away, RT was discovered in 1905 by T. H. Astbury of the British Astronomical Association. Like all Cepheids, it expands and contracts rhythmically—for reasons science is not completely sure of. Yet we do know that it takes about 1.5 days for it to expand to its largest and brightest, and then 2.5 days for it to contract, cool, and dim.

Figure 12.12.
Klaproth and Casatus
(Credit—Wes Higgins).

Figure 12.13. RT Aurigae (Credit—Palomar Observatory, courtesy of Caltech).

Sunday, December 7

Today is the birthday of Gerard Kuiper. Born in 1905, Kuiper was a Dutch-born American planetary scientist who discovered moons of both Uranus and Neptune. He was the first to know that Titan had an atmosphere, and he studied the origins of comets and the solar system.

Tonight on the south shore of the emerging Mare Nubium, look for ancient craters Pitatus and Hesiodus right on the terminator (Figure 12.14). During this phase, something wonderful can happen! If you are at the right place at the right time, sunlight will shine briefly through a break in Hesiodus' wall and cast an incredible ray across the lunar surface! If you don't catch it, you can still enjoy one of the few concentric craters on the Moon.

When you are done with your lunar observations, turn the scope toward lovely Gamma Andromedae (RA 02 03 53 Dec +42 19 47).

Visible to the unaided eye and known as Almach, this 2nd magnitude K-type star is perhaps one of the most beautiful of all double stars for a small telescope (Figure 12.15). Believed to have been discovered in 1788 by J. T. Mayer, one of the reasons this particular star is considered extraordinary is its color contrast. The primary star is a warm, golden yellow, while the 5th magnitude secondary is notably green. But that's not all...

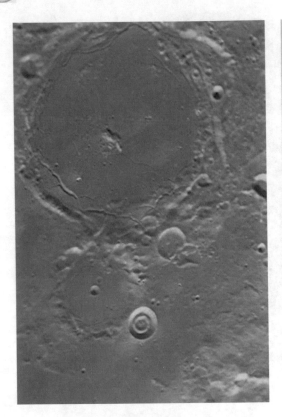

Figure 12.14. Pitatus and Hesiodus (Credit—Wes Higgins).

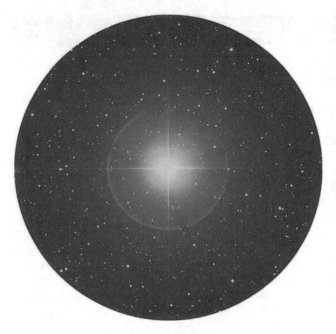

Figure 12.15. Gamma Andromedae: Almach (Credit—Palomar Observatory, courtesy of Caltech).

In 1842, Otto Struve noticed the secondary was itself a binary star—its secondary is only a magnitude less bright, and quite blue. This pair has a highly elliptical orbit of about 61 years. While they last reached maximum separation in 1982, even in 2008 they can be split easily with larger optics. But that isn't all either! The third component is also a spectroscopic binary which has a rotational period of just under three days! Be sure to catch the quadruple Almach system... It may be 260 light-years away, but tonight you'll find it as close as your telescope!

Monday, December 8

While the mighty Copernicus on the terminator will draw the eye like no other crater tonight, it's time to pick up another study which you may not have logged—crater Davy. You will find it just west of the large ring of Ptolemaeus on the northeastern edge of Mare Nubium. It will appear as a small, bright ring, with the large crater Davy A on its southern border. Now skip across the gray sands of Nubium further west to take a look at a peninsula-like series of uplands dominated by crater Guericke (Figure 12.16). Named for Dutch physicist Otto von Guericke, this 58 kilometer diameter crater has all but eroded away. Look for a break in its eastern wall, and notice how lava flow has eradicated the northern wall.

Figure 12.16. Guericke (Credit—Wes Higgins).

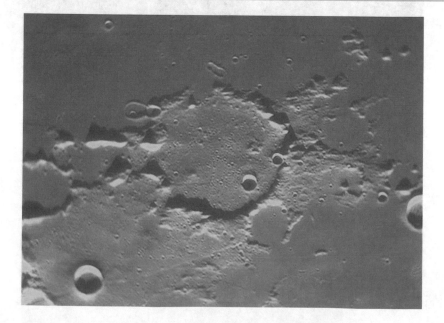

Tonight in history (1908) marked "first light" for the 60" Hale Telescope at Mt. Wilson Observatory (Figure 12.17). Not only was it the largest telescope of the time, but it ended up being one of the most productive of all. Almost 100 years later, the 60" Hale is still in service as a public outreach instrument. If we could use the 60" tonight to study, where would we go? My choice would be Omicron 2 Eridani, located roughly a handspan west of Rigel. As the southernmost of the Omicron pair, it is sometimes known as 40 Eridani, and you'll find it to be an interesting multiple star system that's very worthy of your time and backyard scope (Figure 12.18).

Discovered by William Herschel in 1783, this 16 light-year distant system is the eighth nearest of the unaided visible stars. Well spaced from the primary, the companion star is also a double for high powers, and will reveal a red dwarf discovered by Otto Struve. Now look closely at the 9th magnitude B star. This is the only white dwarf that can be considered "easy" for the average telescope. Its diameter is only about twice the size of Earth and its mass is about the same as our Sun's. Power up and locate the 11th magnitude companion...for it's one of the least massive stars known! This white dwarf may be the smallest stellar object visible in an amateur telescope—and spotting it would be comparable to spotting a tennis ball...on the Moon!

Figure 12.17. The 60 inch Hale Telescope (Credit—Palomar Observatory, courtesy of Caltech).

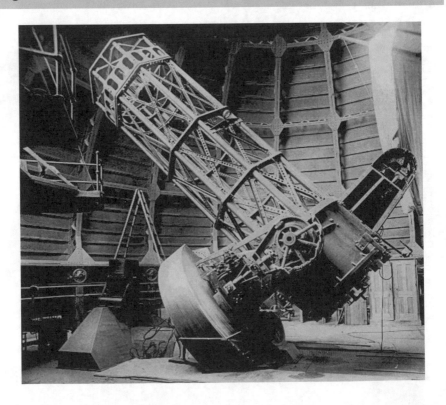

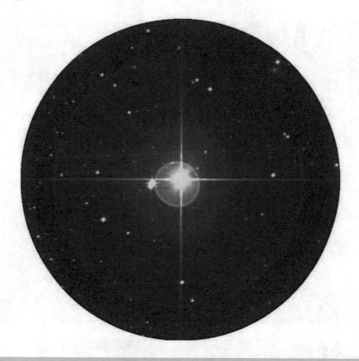

Figure 12.18. 40 Eridani (Credit—Palomar Observatory, courtesy of Caltech).

Tuesday, December 9

Southern Hemisphere viewers, you're in luck! This is the maximum of the Puppid-Velid meteor shower. With an average fall rate of about 10 per hour, this particular shower could also be visible to those Northerners far enough south to see the constellation of Puppis. Very little is known about this shower except that the streams and radiants are very tightly bound together. Since studies of the Puppid-Velids are just beginning, why not take the opportunity to watch? Viewing will be possible all night long—although most of the meteors are faint, this one is known to produce an occasional fireball.

Since we're favoring the south tonight, let's journey to the southern edge of the Moon as we pick up one of the last of our lunar studies—Longomontanus (Figure 12.19). Named for the Danish Astronomer Christian Longomontanus (an assistant to Tycho Brahe), this wonderland of details stretches for about 145 kilometers across the surface. Look for a great collection of interior craters along its northwest interior wall and note how it has eradicated a much older crater which still shows an edge to the east.

Now, for apparently no good reason, let's head for Alpha Persei (Mirfak). While there's nothing particularly interesting about this 570 light-year distant star, what is incredible is the field in which it resides! Take a look at lowest power with a rich field telescope or binoculars and be prepared to be blown away...

Figure 12.19. Longomontanus (Credit—Wes Higgins).

Figure 12.20. Partial field of Melotte 20 (Credit—Palomar Observatory, courtesy of Caltech).

This is the Alpha Persei moving group—a fantastic field of main sequence stars that contains just over a hundred members. Even though it will take 90,000 years before any perceptible change is seen in this bright collection, they are happily moving along together at a pace of about 16 kilometers per second toward Beta Tauri! Enjoy this fine group also known as Melotte 20 (RA 03 25 00 Dec +49 07 00) (Figure 12.20)...

Wednesday, December 10

If you are out stargazing until the morning hours, look for the peak of the Monocerid meteor shower. Its fall rate is around one per hour and its radiant point is near Gemini.

When the sky darkens early tonight, let's turn our eyes toward the Moon. Ah, yes... Is there any more beautiful crater on the Moon than graceful Gassendi? While we have visited it before, take the time to power up and enjoy its features (Figure 12.21). Look for the rimae which crisscross the shallow floor, and the strong A crater which mars its northern wall. How many of its interior features can you resolve?

While we're out, let's have a look at one of the best-known double stars in the night—Gamma Arietis (RA 01 53 31 Dec +19 17 37). Also known as Mesarthim, this combined magnitude 4 beauty was unintentionally discovered in 1664 by Robert Hooke, who was following a comet at the time (Figure 12.22). Hooke,

Figure 12.21. Gassendi region (Credit—Greg Konkel, annotation by Tammy Plotner).

Figure 12.22. Gamma Arietis: Mesarthim (Credit—Palomar Observatory, courtesy of Caltech).

who was one of the first people to note the phenomenon of double stars, wrote of his discovery:

Of this kind, the most remarkable is the star in the left horn of Aries, which, whilst I was observing the comet which appeared in the year 1664, and followed till he passed by this star, I took notice that it consisted of two small stars very near together; a like instance to which I have else met within all the heavens.

While no visual change has been spotted in the almost 344 years since then, there has been a slight difference detected in the components' radial velocities. Roughly 160 light-years away, you'll enjoy this almost matched-magnitude pair of white stars—but look carefully: in 1878, S. W. Burnham found a third star nearby that might not be a physical member, but is also a double!

Thursday, December 11

On this date in 1863, Annie Jump Cannon was born (Figure 12.23). She was an American astronomer who created the modern system for classifying stars by their spectra. Tonight let's ignore the Moon and celebrate this achievement. Come along with me and have a look at some very specific stars with unusual visual spectral qualities! Let's grab a star chart, brush up on our Greek letters, and start first with Mu Cephei.

Figure 12.23. Annie Jump Cannon (widely used public image).

Nicknamed the "Garnet Star," this is perhaps one of the reddest stars visible to the unaided eye. At around 1200 light-years away, this spectral-type M2 star will show a delightful blue/purple "flash." If you still don't perceive color, try comparing Mu to its bright neighbor Alpha, a spectral type A7, or "white," star. Perhaps you'd like something a bit more off the beaten path? Then head for S

Figure 12.24. Aldebaran (Credit—John Chumack).

Cephei, about halfway between Kappa and Gamma toward the pole. Its intense shade of red makes this magnitude 10 star an incredibly worthwhile hunt.

To see an example of a B spectrum star, look no further than the Pleiades... All the components are blue–white. Want to taste an "orange?" Then look again at Aldebaran, or Alpha Tauri, and say hello to the K spectrum star (Figure 12.24). Now that I have your curiosity aroused, would you like to see what our own Sun would look like? Then choose Alpha Aurigae, better known as Capella, and discover a spectral class G star that's "only" 160 times brighter than the one that holds our solar system together!

Still no luck in seeing color? Don't worry. It does take a bit of practice! The cones in our eyes are the color receptors and when we go out in the dark, the color-blind rods take over. By intensifying the starlight with either a telescope or binoculars, we can usually excite the cones in our dark-adapted eyes to pick up on color.

Tonight is also the peak of the Sigma Hydrid meteor stream. Its radiant is near the head of the Serpent and the fall rate is about 12 per hour—but these are fast!

Friday, December 12

Today we honor the birth of S. W. Burnham (Figure 12.25). Born in 1838, this American astronomer spent 50 years of his life surveying the night sky for double stars. Although at the time it was believed that all visual binaries had

Figure 12.25.
S. W. Burnham
(historical image).

Figure 12.26. OSCAR 1 (archival image).

been accounted for, Burnham's work was eventually published as the General Catalogue of 1290 Double Stars. His keen eye and diligent study opened the doors for him at observatories such as Yerkes and Lick. His lifetime count of binaries discovered eventually reached 1340. He was also the very first to observe what would eventually be termed a "Herbig-Haro object," and he discovered 6 NGC and 21 IC objects.

Today in 1961, OSCAR 1 was launched (Figure 12.26). The project started in 1960, the name stands for Orbital Satellite Carrying Amateur Radio. OSCAR 1 operated in orbit for 22 days, transmitting a signal in Morse Code—the simple greeting "Hi." The success of the mission helped to promote interest in amateur radio which continues to this day!

Tonight it's the "Full Moon before Yule." Not only that, but the Moon is at perigee—its closest point to the Earth. While you might hear a tall tale or two about it being brighter than normal since it is also close to solstice, judge for yourself! Is it truly brighter? Or just an illusion? While you're out, turn the telescope Selene's way and let's scan the surface. On the eastern limb we

Figure 12.27. Furnerius (Credit—Roger Warner).

Figure 12.28. Lambda Tauri (Credit—Palomar Observatory, courtesy of Caltech).

see the bright splash ray patterns surrounding ancient Furnerius—yet the rays themselves emanate from the much younger crater Furnerius A (Figure 12.27). All over the visible side, we see small points light up: a testament to the Moon's violent past written in its scarred lines. Take a look now at the western limb...for the sunrise is about to advance around it.

Now let's take a visual journey about a fistwidth west-southwest of brilliant Aldebaran to take a look at Lambda Tauri (RA 04 00 40 Dec +12 29 25). Although it has no proper name, it is one of the very brightest of eclipsing variable stars, and was one of the first to be identified as such, in 1848 (Figure 12.28). Orbiting about 13 million kilometers away from the primary star is its spectroscopic companion—so close that we can only distinguish the two stars by the changes which take place about every 4 days. Keep an eye on Lambda and watch as it drops sharply by almost a magnitude one night, and recovers less than 24 hours later!

Saturday, December 13

Today in 1920, the first stellar diameter was measured by Francis Pease with an interferometer at Mt. Wilson. His target? Betelgeuse!

Tonight will bring one of the most hauntingly beautiful and mysterious displays of celestial fireworks all year—the Geminid meteor shower (Figure 12.29). First noted in 1862 by Robert P. Greg in England, and B. V. Marsh and Prof. Alex C. Twining of the United States in independent studies, the annual appearance of the Geminid stream was weak initially, producing no more than a few per hour, but it has grown in intensity during the last century and a half. By 1877, astronomers had realized this was a new annual shower—producing about 14 meteors per hour. At the turn of the last century, the rate had increased to over 20, and by the 1930s up to 70 per hour. Only 10 years ago observers recorded an outstanding 110 per hour during a moonless night... But unfortunately it's not moonless tonight.

So why are the Geminids such a mystery? Most meteor showers are historic—documented and recorded for hundreds of years—and we know them as originating with cometary debris. But when astronomers began looking for the Geminids' parent comet, they found none. It wasn't until October 11, 1983 that Simon Green and John K. Davies, using data from NASA's Infrared Astronomical Satellite, detected an object (confirmed the next night by Charles Kowal) that matched the orbit of the Geminid meteoroid stream. But this was no comet, it was an asteroid—in fact, a 17th magnitude asteroid which is transiting tonight around 02:00 UT at roughly RA 02 and Dec +31!

Originally designated as 1983 TB, but later renamed 3200 Phaethon, this apparently rocky solar system member has a highly elliptical orbit that places it within 0.15 AU of the Sun about every year and a half. But asteroids can't fragment like a comet—can they? The original hypothesis was that since Phaethon's orbit passes through the asteroid belt, it may have collided with one or more asteroids, creating rocky debris. This sounded good, but the more we studied the more we realized the meteoroid "path" occurred when Phaethon neared the Sun. So now our asteroid is behaving like a comet, yet it doesn't develop a tail.

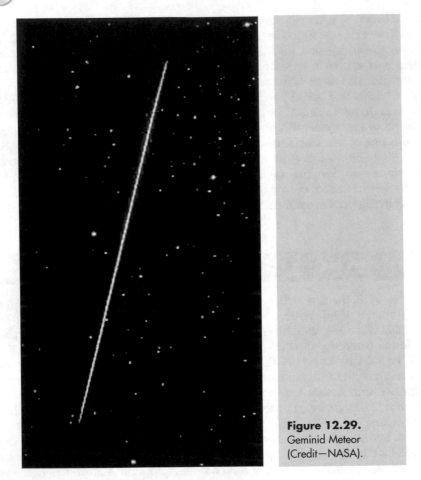

Figure 12.29.
Geminid Meteor
(Credit—NASA).

So what exactly is this "thing?" Well, we do know that 3200 Phaethon orbits like a comet, yet has the spectral signature of an asteroid. By studying photographs of the meteor showers, scientists have determined that the meteors are denser than cometary material, yet not as dense as asteroid fragments. This leads them to believe Phaethon is probably an extinct comet which has gathered a thick layer of interplanetary dust during its travels, yet retains the ice-like nucleus. Until we are able to take physical samples of this "mystery," we may never fully understand what Phaethon is, but we can fully appreciate the annual display it produces!

Thanks to the wide path of the stream, folks all over the world get an opportunity to enjoy the show. The traditional peak time is tonight as soon as the constellation of Gemini appears, around mid-evening. The radiant for the shower is near the bright star Castor, but meteors can originate from many points in the sky. From around 2 AM tonight until dawn (when our local sky window is aimed directly into the stream) it is possible to see about one "shooting star" every 30 seconds. The most successful of observing nights are ones where you are comfortable, so be sure to use a reclining chair or pad on the ground while looking up. Although the rising Moon will greatly interfere, please get away from light sources when possible—it will triple the amount of meteors you see. Enjoy the incredible and mysterious Geminids!

Sunday, December 14

Today was a very busy day in astronomy history. Tycho Brahe was born in 1546 (Figure 12.30). Brahe was a Danish pre-telescopic astronomer who established the first modern observatory in 1582 and gave Kepler his first job in the field. And in 1962, the Mariner 2 spacecraft made a flyby of Venus and became the first successful interplanetary probe.

The Moon will rise a little later this evening, but we're going to run ahead of it tonight and enjoy some studies in Auriga! Looking roughly like a pentagon in shape, start by identifying the brightest of these stars—Alpha Aurigae (Capella). Due south of it is the second brightest star, Beta (Menkalinan). After aiming binoculars at Beta, go north about one-third the distance between the two and enjoy all the stars!

You will note two very conspicuous clusters of stars in this area, and so did Le Gentil in 1749. Binoculars will show them both in the same field, as will telescopes using lowest power. The dimmest of these clusters is M38 (RA 05 28 43 Dec +35 51 18), and it will appear vaguely cruciform in shape (Figure 12.31). At roughly 4200 light-years away, larger aperture will be needed to resolve the 100 or so fainter members. About 2.5 degrees to the southeast you will see the much brighter M36 (RA 05 36 12 Dec +34 08 24). More easily resolved in binoculars and small scopes, this "jewel box" galactic cluster is quite young—and about 100 light-years closer (Figure 12.32)!

Figure 12.30. Tycho Brahe (Credit—NASA).

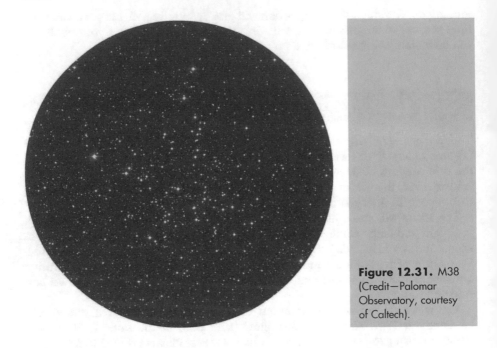

Figure 12.31. M38 (Credit—Palomar Observatory, courtesy of Caltech).

Figure 12.32. M36 (Credit—Palomar Observatory, courtesy of Caltech).

Monday, December 15

Today in 1970, the Soviet spacecraft Venera 7 made a successful soft landing on Venus, and so went into the history books as the first craft to land on another planet. You can catch Venus yourself just after sunset. It's the brightest "star" to the southwest!

How about something a little more suited to the mid-sized scope tonight? Set your sights on Alpha Fornacis and let's head about three fingerwidths northeast (RA 03 33 15 Dec –25 52 18) for NGC 1360 (Figure 12.33). In a 6" telescope, you'll find the 11th magnitude spectroscopic double star in the center of this planetary nebula to be very easy—but be sure to avert because the nebula itself is very elongated. Like most of my favorite things, this planetary is a rule-breaker since it doesn't have an obvious shell structure. But why? Rather than believe it is not a true planetary, studies have shown that it could quite possibly be a very highly evolved one—an evolution which has allowed its gases to begin to mix with the interstellar medium. Although faint and diffuse for northern observers, those in the south will recognize this as Bennett 15!

If you're out later tonight, take the time to look at the Moon, even if you just use binoculars. Very often we only study lunar features as the Moon waxes—later hours keep us from observing as it wanes. But the journey is quite worthwhile, since it allows us to see old features in a new light—literally! Recognizable craters

Figure 12.33. NGC 1360 (Credit—Palomar Observatory, courtesy of Caltech).

Figure 12.34. Theophilus, Cyrillus, and Catharina (Credit—Greg Konkel).

such as Theophilus, Cyrillus, and Catharina change little (Figure 12.34), but look at the shock waves in Mare Nectaris, or how much more prominent the Altai Scarp appears!

Tuesday, December 16

Tonight we celebrate the birthday of Edward Emerson (E. E.) Barnard (Figure 12.35). Born in 1857, Barnard was an American observational astronomer and an absolute legend. He led a very colorful life, and his sharp skills have led to a multitude of discoveries. His life was a very fascinating one: Barnard was often known to simply set the scope on one point in the sky and just watch for new objects as the field moved! Tonight let's take a look at a bright star which is an example of Barnard's touch as we begin our explorations with Beta Aurigae.

First identified as a spectroscopic binary by A. Maury in 1890, Beta itself is part of a moving group of stars that includes Sirius, and is an Algol-type variable. While you won't see changes as dramatic as those of the Demon Star, it has a precise drop of 0.09 in magnitude every 3.96 days. This system contains nearly identical stars which are more than 2.5 times the size of our Sun, but they orbit each other at a distance of less than 0.1 AU! While Menkalinan's 10th magnitude optical companion was first spotted by Sir William Herschel in 1783, only E. E. Barnard noticed the 14th magnitude true tertiary to this incredible multiple system (Figure 12.36)!

Figure 12.35.
E. E. Barnard at the
36 inch scope
(Credit—UCO/Lick
Observatory (archival
image)).

Now let's have a look at a bright open cluster known by many names: Herschel VII.32, Melotte 12, Collinder 23, and NGC 752 (Figure 12.37). You'll find it three fingerwidths south (RA 01 57 41 Dec +37 47 06) of Gamma Andromedae...

Under dark skies, this 5.7 magnitude cluster can just be spotted with the unaided eye, is revealed in the smallest of binoculars, and can be completely

Figure 12.36. Beta Aurigae: Menkalinan (Credit—Palomar Observatory, courtesy of Caltech).

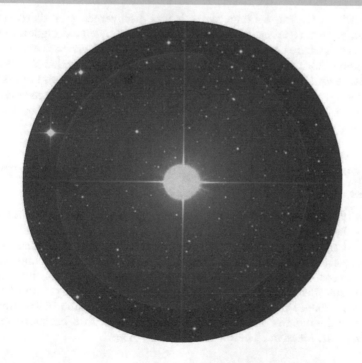

Figure 12.37. NGC 752 (Credit—Palomar Observatory, courtesy of Caltech).

resolved with a telescope. Chances are it was discovered by Hodierna over 350 years ago, but it was not cataloged until Sir William gave it a designation in 1786. But give credit where credit is due... For it was Caroline Herschel who observed it on September 28, 1783! Containing literally scores of stars, galactic cluster NGC 752 could be well over a billion years old, strung out in chains and knots in an X pattern over a rich field. Take a close look at the southern edge for orange star 56: while it is a true binary star, the companion you see is merely optical. Enjoy this unsung symphony of stars tonight!

Wednesday, December 17

Today in 1965, David Levy began his telescopic comet search (Figure 12.38). Since that time, Levy has found 8 visual and 13 photographic comets, credit for which was shared with the Shoemakers. His career is a very illustrious one, including authoring seven books and cataloging more than 300 deep sky objects during his sky sweeps. Along with his wife Wendee and Tom Glinos, he discovered over 150 asteroids, including the first Martian Trojan. With yet another comet discovery in 2006, David has now taken third place all-time in the number of comets found—most of his discoveries were made with a backyard telescope! Move over, Mr. Messier... There's a new kid in town!

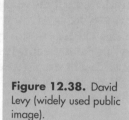

Figure 12.38. David Levy (widely used public image).

Tonight let's set our sights about halfway between Theta and Beta Aurigae. Our study object will be the open cluster M37 (Figure 12.39). Apparently discovered by Messier himself in 1764, this galactic cluster will appear almost nebula-like to binoculars and very small telescopes—but will come to perfect resolution with larger instruments.

Figure 12.39. M37 (Credit—NOAO/AURA/NSF).

About 4700 light-years away, and spanning a massive 25 light-years, M37 is often billed as the finest of the three Aurigan open clusters for bigger scopes. Offering beautiful resolvability, this one contains around 150 members down to magnitude 12, and has a total population in excess of 500.

What makes it unique? As you view, you will note the presence of several red giants. For the most part, open clusters are comprised of stars that are all about the same age, but the brightest star in M37 appears orange in color and not blue! So what exactly is going on in here? Apparently some of these big, bright stars have evolved much faster—consuming their fuel at an incredible rate. Other stars in this cluster are still quite young on a cosmic scale, yet they all left the "nursery" at the same time! In theory, this allows us to judge the relative age of open clusters. For example, M36 is around 30 million years old and M38 about 40, but the presence of the red giants in M37 puts its estimated age at 150 million years! Just awesome...

Thursday, December 18

So where has Sir William Herschel been lately? Rest assured that one of the most prolific observers of the cosmos had never ceased to continue to explore, discover, and document some of the finest deep space objects and was doing so

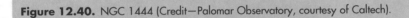

Figure 12.40. NGC 1444 (Credit—Palomar Observatory, courtesy of Caltech).

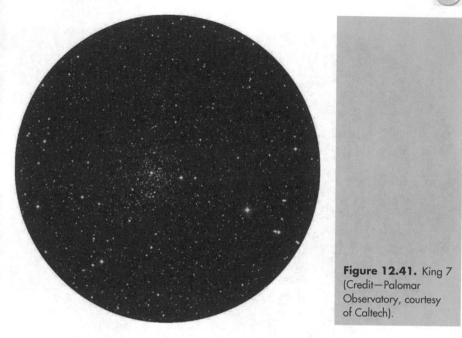

Figure 12.41. King 7 (Credit—Palomar Observatory, courtesy of Caltech).

almost every single night of the year. Tonight let's start with a Herschel discovery made on this date (1788) as we take a look in northern Perseus, about a fistwidth northeast of Alpha, for NGC 1444 (RA 03 49 24 Dec +52 40 00).

Well known as a source of radio emissions, NGC 1444 holds a rough cumulative magnitude of 6.5 and will show as a small compression of stars around SAO 24248 in binoculars—but use a scope if you can (Figure 12.40)! Even modest aperture and magnification will reveal a delightful chain of stars in an S-pattern around this Herschel "400" object.

If you'd like to explore something a little more off the beaten path (and an object not discovered by Herschel), head about a degree and a half southwest for King 7 (RA 03 59 00 Dec +51 48 00). Very rich and compressed, this alternative catalog study is slightly larger than tonight's Herschel, and is definitely more set apart from the surrounding starfields (Figure 12.41). Studied by (and named for) Ivan R. King, this intermediate-to-old open cluster seems to be very relaxed in its evolutionary state. Be sure to power up, because King 7 is definitely a stellar region you won't want to miss!

Friday, December 19

Tonight let's familiarize ourselves with the vague constellation of Fornax. Its three brightest stars form a shallow V just south of the Cetus/Eridanus border and span less than a handwidth of sky. Although it's on the low side for northern observers, there is a wealth of sky objects in this area.

Figure 12.42.
Alpha Fornacis
(Credit—Palomar
Observatory, courtesy
of Caltech).

Figure 12.43. NGC 1049 (Credit—Palomar Observatory, courtesy of Caltech).

Try having a look at the easternmost star—40 light-year distant Alpha Fornacis (Figure 12.42). At magnitude 4, it is not easy, but what you'll find there is quite beautiful. For binoculars, you'll see a delightful cluster of stars around this long-term binary—but telescopes will enjoy it as a great golden double star! First measured by John Herschel in 1835, the distance between the pair has narrowed and widened over the last 172 years, and it is suspected its orbital period may be 314 years. While the 7th magnitude secondary can be spotted with a small scope, watch out because it is a variable which may drop by as much as a full magnitude!

For larger telescopes, set sail for Beta Fornacis and head three degrees southwest (RA 02 39 42 Dec –34 16 08) for a real curiosity—NGC 1049 (Figure 12.43).

At magnitude 13, this globular cluster is a challenge for even large scopes—with a good reason. It isn't in our galaxy. This cluster is a member of the Fornax Dwarf Galaxy—a one degree span that's so large it was difficult to recognize as extra-galactic—or at least it was until the great Harlow Shapely figured it out! NGC 1049 was discovered and cataloged by John Herschel in 1847, only to be reclassified as Hodge 3 (by Paul Hodge) in a 1961 study of the system's five globular clusters. Since that time, yet another globular has been discovered in the Fornax Dwarf! Good luck...

Saturday, December 20

Tonight is the peak of the Delta Arietid meteor shower. While most showers are best after midnight, this is an early evening shower which must be viewed before the radiant sets. The fall rate is modest—about 12 per hour... And there's no Moon!

Today marks the founding of Mt. Wilson Solar Observatory. It officially opened its doors in 1904. We also celebrate the birth of Walter S. Adams on this date (Figure 12.44). Born in 1876, Adams was the astronomer at Mt. Wilson who revealed the nature of Sirius B, the first known white dwarf star. Sirius B was first seen by Alvan Clark in 1862, and recently the Hubble Space Telescope precisely measured the mass of B for the first time. When Sirius is well risen tonight, why not have a go at spotting the B star for yourself?

Until then, let's pretend the skies are still as dark as they were on Mt. Wilson in Adams' time as we aim our binoculars and telescopes toward one of the most elusive galaxies of all—M33 (Figure 12.45).

Located about one-third the distance between Alpha Trianguli and Beta Andromedae (RA 01 33 51 Dec +30 39 37), this member of our Local Group was probably first seen by Hodierna, but was recovered independently by Messier some 110 years later. Right on the edge of visibility unaided, M33 spans about four Moon-widths of sky, making it a beautiful binocular object and a prime view in a low-power telescope.

Smaller than both the Milky Way and the Andromeda Galaxy, the Triangulum is about average in size, but is anything but average to study. So impressed was Herschel that he gave it its own designation of H V.17—after having already cataloged one of its bright star-forming regions as H III.150! In 1926, Hubble

Figure 12.44. Walter Adams (Credit—Yerkes Observatory, University of Chicago).

also studied M33 at Mt. Wilson with the Hooker telescope during his work with Cepheid variables. Larger telescopes often "can't see" M33 with good reason—it overfills the field of view—but what a view! Not only did Herschel discover a region much like our own Orion Nebula, but the entire galaxy contains many NGC and IC objects (even globular clusters) reachable with a larger scope.

Figure 12.45. M33 (Credit—John Chumack).

Although M33 might be 3 million light-years away, tonight it's as close as your own dark-sky site...

Sunday, December 21

Today marks Winter Solstice—for the Northern Hemisphere, the shortest day and the longest night of the year—the point when the Sun is furthest south. Now is a wonderful time to demonstrate for yourself our own movements by choosing a "solstice marker." Anything from a fence post to a stick in the ground will suffice! Simply measure the shadow when the Sun reaches the zenith and repeat your experiment in the weeks ahead and watch as the shadow grows shorter...and the days grow longer (Figure 12.46)!

Tonight let's go north for a mid-sized scope challenge about two fingerwidths east-northeast of the beautiful double star Gamma Andromedae (RA 02 22 32 Dec +43 20 45). At 12th magnitude, NGC 891 is a perfect example of a spiral galaxy seen edge-on (Figure 12.47). To the mid-sized scope, it will appear as a pencil-slim scratch of light, but larger telescopes will be able to make out a fine, dark, dustlane upon aversion. Discovered by Caroline Herschel in 1783, NGC 891 contained a magnitude 14 supernova recorded on August 21, 1986. Often considered a "missed Messier," you can add this one to your Caldwell list as number 23!

For more advanced observers, let's take a look at a galaxy cluster—Abell 347— located almost directly between Gamma Andromedae and M34 (Figure 12.48). Here you will find a grouping of at least a dozen galaxies that can be fitted into a wide field view. Let's tour a few...

© V. Rumyantsev

Figure 12.46.
Analemma
(Credit—Vasilij
Rumyantsev (Crimean
Astrophysical
Observatory)/NASA).

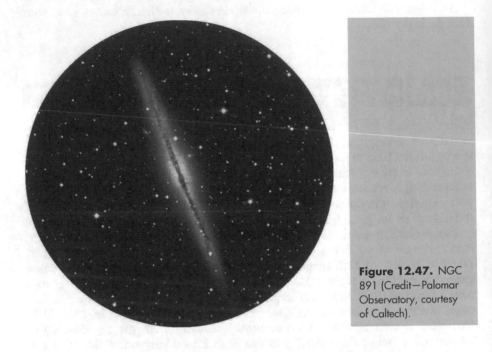

Figure 12.47. NGC 891 (Credit—Palomar Observatory, courtesy of Caltech).

Figure 12.48. Abell 347 (Credit—Palomar Observatory, courtesy of Caltech).

The brightest and largest is NGC 910, a round elliptical with a concentrated nucleus. To the northwest you can catch faint, edge-on NGC 898. NGC 912 is northeast of NGC 910, and you'll find it quite faint and very small. NGC 911 to the north is slightly brighter, rounder, and has a substantial core region. NGC 909 further north is fainter, yet similar in appearance. Fainter yet is the more northern NGC 906, which shows as nothing more than a round contrast change. Northeast is NGC 914, which appears almost as a stellar point with a very small haze around it. To the southeast is NGC 923 which is just barely visible with wide aversion as a round contrast change. Enjoy this Abell quest!

Monday, December 22

Up early? Fantastic! In the predawn hours of this morning, I have a treat for you—the Ursid meteor shower! Cruising around the Sun every 13.5 years, Comet 8P/Tuttle sheds a little skin. Although it never passes inside Earth's orbit, some six years after its closest approach we do pass through its debris stream. Not so unusual? Then think again, because it takes as much as six centuries before a particular meteoroid trail is affected enough by Jupiter's gravitation to deflect its stream into our atmosphere.

Figure 12.49. AE Aurigae: The Flaming Star (Credit—Palomar Observatory, courtesy of Caltech).

With little interference from the Moon, this circumpolar meteor shower could see activity of up to 12 per hour in the hours before dawn. By keeping watch on the constellation of Ursa Major, you might just spot one of these slow-moving, 600 year old travelers on their path only halfway between us and Selene!

Tonight let's go to our maps west of M36 and M38 to identify AE Aurigae (RA 05 16 19 Dec +34 18 44). As an unusual variable, AE is normally around 6th magnitude and resides approximately 1600 light-years distant (Figure 12.49). The beauty in this region is not particularly the star itself but the faint nebula in which it resides. Known as IC 405, this is an area consisting mostly of dust, and having very little gas. What makes this view so entertaining is that we are looking at a "runaway" star.

It is believed that AE originated in the M42 region in Orion. Cruising along at a very respectable speed of 130 kilometers per second, AE flew the "stellar nest" some 2.7 million years ago! Although IC 405 is not directly related to AE, there is evidence within the nebula of areas which have been cleared of their dust by the rapid northward motion of the star. AE's hot, blue illumination and high-energy photons fuel what little gas is contained within the region, and its light reflects off the surrounding dust as well—because of this AE is known as "The Flaming Star." Although we cannot "see" with our eyes like a photograph, together this pair makes an outstanding view for the small backyard telescope.

Tuesday, December 23

Tonight in 1672, astronomer Giovanni Cassini discovered Saturn's moon Rhea. Although you will have to wait until a little later in the evening to catch the ringed planet at its best (Figure 12.50), why not try your hand at finding Rhea as well? A well-collimated scope as small as 4.5" is perfectly capable of seeing Tethys, Rhea, and Dione as they orbit very near the edge of the ring system. All it requires is steady skies and a little magnification!

With only two days to go before the holiday and dark skies on our hands, what present would you ask for? My choice would be the Fornax Galaxy Cluster (Figure 12.51)!

Containing around 20 galaxies brighter than 13th magnitude in a roughly one degree field (RA 03 36 00 Dec -35 30 00), here is where a galaxy hunter's paradise begins! Telescopes will love NGC 1365 at the heart of the cluster proper. This great barred spiral gives an awesome view in even the smallest of scopes. As you slide north, you will encounter a host of galaxies: NGCs 1386, 1389, 1404, 1387, 1399, 1379, 1374, 1381, and 1380. There are galaxies everywhere! But if you lose track? Remember the brightest of these are two ellipticals—1399 and 1404. Have fun!

About a degree and a half north of Tau Fornacis (RA 03 35 57 Dec -35 30 28) is the large, bright and round spiral NGC 1398. A little more than a degree west-northwest is the easy ring of the planetary nebula NGC 1360. Look for the concentrated core and dark dustlane of NGC 1371 a degree north-northeast—or the round NGC 1385 which accompanies it. Why not visit Bennett 10 or Caldwell 67, as we take a look at NGC 1097 about six degrees west-southwest of Alpha? This one is bright enough to be caught with binoculars!

Figure 12.50. Saturn (Credit—JPL/NASA).

Figure 12.51. A 60 arcminute view of the Fornax Cluster centered on NGC 1387 (Credit—Palomar Observatory, courtesy of Caltech).

Wednesday, December 24

If you received a new telescope or binoculars as a present (and opened it early!), this would be a great opportunity to locate an easy Messier object—M34 (RA 02 42 06 Dec +42 46 00). If you remember our previous study stars Almach and Algol, you're halfway there. Draw an imaginary line between them and look with your binoculars or finder scope just a shade north of the line's center.

In binoculars, M34 will show around a dozen fainter stars clustered together, and perhaps a dozen more scattered around the field (Figure 12.52). Small telescopes at low power will appreciate M34 for its resolvability and the distinctive orange star in the center. Larger aperture scopes will need to stay at lowest power to appreciate the 18 light-year span of this 100 million year old cluster, but take the time to power up and study. You will find many challenging doubles inside!

Today in 1968, Apollo 8 became the first manned spacecraft to orbit the Moon. Until this date, no one had seen with their own eyes what lay beyond. Frank Borman, James Lovell, and William Anders were to become the first to directly view the "dark side"—and so would be the first to witness Earthrise over the Moon (Figure 12.53). If you are up just before dawn this morning, step outside and let your mind take flight to the distant orb. With the terminator almost near the western limb, all that's left is a slender crescent lit by brilliant

Figure 12.52. M34 (Credit—Palomar Observatory, courtesy of Caltech).

Figure 12.53. Earthrise (Credit—Apollo 8/NASA).

earthshine...truly the "Old Moon in the New Moon's Arms." What courage it took for these brave men to journey so far from hearth and home!

And from the crew of Apollo 8, we close, with good night, good luck, a Merry Christmas, and God bless all of you, all of you on the good Earth.

–Astronaut Frank Borman

Thursday, December 25

Wishing you all the very best for the Christmas season! Like a present, Sir Isaac Newton was born on this day in 1642 (Figure 12.54). Newton was the British "inventor" of calculus and a huge amount of what we now consider classical physics. Even young children are aware of his simple laws of motion and gravity. It wasn't until the time of Einstein that things changed!

In keeping with the season, tonight's astronomical object is a celebration of both starlight and asterism. Located 10 degrees east of Betelgeuse (RA 06 41 00 Dec +09 53 00) is NGC 2264 (Figure 12.55). Also known as the "Christmas Tree Cluster," this asterism of approximately 20 bright stars and over 100 fainter ones is embroiled in a faint nebula—a delightful Christmas tree shape adorned with stars! Can you also spot the "Cone Nebula?"

The very brightest of these stars is 5th magnitude S Monocerotis which will show clearly in the finderscope, and will be seen as a double at magnification.

Figure 12.54. Sir Isaac Newton (widely used public image).

Figure 12.55. NGC 2264: The Christmas Tree Cluster and Cone Nebula (Credit—R. Jay GaBany).

Steady skies will reveal that the "star" at the top of our "tree" is also a visual double. Many of the stars will also appear to have companions, as well as tints of silver and gold. The visual effect of this splendid open cluster and surrounding nebulosity is well worth the challenge it presents. After having noted it in a different category earlier, Sir William cataloged the Christmas Tree Cluster as H V.27 on the 26th of December 223 years ago. He wrote: "Some pretty bright stars [near S Monocerotis] are involved in much extremely faint nebulosity, which loses itself imperceptibly." Match your perceptions to Sir William's tonight!

Happy Holidays!

Friday, December 26

Sir William Herschel stopped exploring because of the holidays? Never! I'm even beginning to believe the master also never had a Moon or a cloudy night. So what was he into (besides yesterday's study NGC 2264) on this night in 1785? Let's find out—beginning with binoculars a little less than a fistwidth northeast of Aldebaran for a triple treat: two clusters within a cluster. Their designations are NGC 1746, 1758, and 1750 (Figure 12.56).

Figure 12.56. Sixty arcminute view centered on NGC 1746 (Credit—Palomar Observatory, courtesy of Caltech).

Located near the galactic anti-center in the direction of the Taurus dark clouds (RA 05 03 48 Dec +23 46 00), Dreyer was the first besides Herschel to believe this trio were physically overlapping star clusters. Studied photometrically, the neighboring Pleiades and Hyades clearly show as foreground objects while our "questionable clusters" appear reddened to different degrees. Of course, like many disputed regions, the larger, sparser, NGC 1746 may not be considered a cluster by some books—even though the two interior collections of stars show marked distance differences.

No matter how you view it, enjoy this large collection for yourself. NGC 1746 shows as a widely scattered field with two areas of compression to binoculars, while even a small telescope will resolve southern NGC 1750 (a Herschel "400" object) with its prominent double star. The smaller collection—NGC 1758—will be just to its northeast. Until the proper motion of this trio is properly studied by proper equipment, you can still consider it another good call on Herschel's part, and a real triple treat!

Saturday, December 27

Born today in 1571 was Johannes Kepler—a Danish astronomer and assistant to Tycho Brahe (Figure 12.57). Kepler used Brahe's copious notes of Mars' positions to help formulate his three laws of planetary motion. These laws are still applicable today. If you're up before dawn this morning, you can see them in action as Mars has returned low on the eastern horizon!

Figure 12.57.
Johannes Kepler (widely used public image).

Tonight is New Moon and there is a vast array of things we could choose to look at. I am a galaxy hunter at heart, and nothing makes it beat just a little bit quicker than an edge-on. Tonight let's walk into the lair of the Dragon as we seek out the incredible NGC 5907 (Figure 12.58).

Located just a few degrees south of Iota Draconis (RA 15 15 53 Dec +56 19 43), this particular galaxy is worth staying up just a bit late to catch. Located about 40 million light-years away, 10th magnitude NGC 5907 contains far more than that which meets the casual eye. It's warped. Long believed to have been the prototype for non-interacting galaxies, things changed drastically when two companion dwarf galaxies were discovered. A faint, photographic ring structure revealed itself, exposing tidal disruption—the ellipsoid involving the nuclear region of the primary galaxy pulling apart the small spheroid. Also part of the picture is PGC 54419, another dwarf so close to the warp as to almost belonging to NGC 5907 itself!

In smaller scopes, prepare yourself to see nothing more than an averted vision scratch of light. The larger the aperture, the more there is revealed, as 5907 gains a bright and prominent nucleus. Although it doesn't look like the grand spiral we envision our own Milky Way to be, we are looking at it from a different angle. In this respect, it behaves much like our own microcosm—a living, interacting, member of a larger group and of a much, much larger Universe.

Figure 12.58. NGC 5907 (Credit—Palomar Observatory, courtesy of Caltech).

Sunday, December 28

Today we celebrate the birth of Arthur S. Eddington (Figure 12.59). Born in 1882, Eddington was a British theoretical astrophysicist whose work was fundamental to interpreting and explaining stellar nature. He also coined the phrase "expanding universe" to refer to the mutual recession of the galaxies. This idea eventually became known as "Hubble's Law," as the massive 200" telescope at Palomar Observatory played another important role when Eddington's work in this field was continued by Edwin Hubble. Tonight let us honor both the great minds as we take a look at a galaxy which is indeed receding from us—NGC 1300 (Figure 12.60).

Located about a fingerwidth north of Tau 4 Eridani (RA 03 19 41 Dec –19 24 40), this is probably the most incredible barred spiral you will ever encounter. At magnitude 10, it will require at least a 4.5" telescope in northern latitudes, but can probably be spotted with binoculars in the far south.

Seventy-five million light-years away, NGC 1300's central bar alone is larger than the Milky Way, and this galaxy has been intensively studied because the manner of its formation was so similar to our own. Although it is so distant, it is seen face-on: allowing us to see this formation without looking through the gas and dust which block our own Galaxy's center from view. Enjoy this one's fantastic structure!

Figure 12.59.
Arthur Eddington
(Credit—American
Institute of Physics Niels
Bohr Library).

Figure 12.60. NGC 1300 (Credit—Palomar Observatory, courtesy of Caltech).

Monday, December 29

Tonight will mark the return of the ultra-slim crescent Moon, fresh from its journey between the Earth and the Sun. If your horizon allows, be sure to look for it just as the very last rays of Sol have quit the sky. Joining it less than a degree to the south will be faint little Mercury. To the north, also less than a degree away, is nearly the last view of Jupiter you will get this year. It is possible there could be an occultation of one planet or the other (depending on your time and location), so please be sure to check with an organization such as IOTA for more precise predictions.

Tonight let's take another Herschel journey as we look in the area almost precisely between Alpha Persei and Alpha Aurigae for three open clusters discovered by Sir William on December 28, 1790.

I think the Master will forgive us for being a day late as we start a breath west of the finderscope optical double b Persei (HD 26961) for NGC 1545 (RA 04 20 54 Dec +50 15 00). While at first you might feel a little disappointed at such a small and loose collection, don't be! It's bright, has a host of mixed magnitudes, and color contrasts...and a secret (Figure 12.61). One of its members, Hoag 4, is also a variable star!

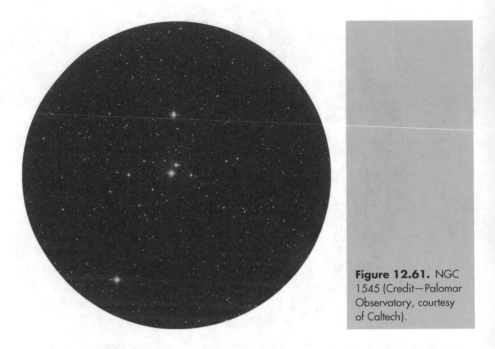

Figure 12.61. NGC 1545 (Credit—Palomar Observatory, courtesy of Caltech).

Figure 12.62. NGC 1528 (Credit—Palomar Observatory, courtesy of Caltech).

Figure 12.63. NGC 1513 (Credit—Palomar Observatory, courtesy of Caltech).

Now hop just a bit north-northwest for NGC 1528 (RA 04 15 24 Dec +51 13 00). A little larger, a little richer, and definitely more compressed, this open cluster is easily viewed in binoculars and is part of many challenge lists (Figure 12.62). Is it special? Not particularly. Color-magnitude diagrams show it to be a rather normal open cluster, but how many Herschel "400" objects are this easy?!

Now head west to Lambda Persei and fade south for NGC 1513 (RA 04 10 38 Dec +49 31 00). Ah, I saw you raise an eyebrow! While NGC 1513 isn't as bright as the others to the eye, it certainly is more interesting as an asterism, and is known as the "Double Ring Cluster" (Figure 12.63). Of its 111 possible members, 33 are confirmed to be genuine, and all of them will come to light with aperture. One circle is devoid of interior stars and the other is full of them!

Tuesday, December 30

As our year together nears its end, tonight let's skywalk for another northern gem, M76 (RA 01 42 19 Dec +51 34 31).

Located in western Perseus just slightly less than one degree north-northwest of Phi, M76 is often referred to as the "Little Dumbbell" (Figure 12.64). Originally discovered by Messier's assistant Méchain in September of 1780, Charles didn't get around to cataloging it for another six weeks. What a shame it took him

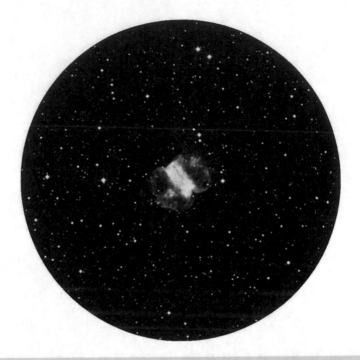

Figure 12.64. M76: The Little Dumbbell (Credit—Palomar Observatory, courtesy of Caltech).

Figure 12.65. Eta Carinae (Credit—Hubble Heritage Team/NASA).

so long to view this fine planetary nebula! Its central star is one of the hottest known, but its resemblance to M27 is what makes it so fascinating. Looking very much like a miniature of the much larger M27, M76 is rather faint at magnitude 11, but is quite achievable in scopes of 4.5" aperture or larger. It is small, but its irregular shape makes this planetary nebula a real charmer.

For our Southern Hemisphere friends, get thee out there and view Eta Carinae (Figure 12.65)! First recorded by Halley in 1677, even the great Sir John Herschel was at a loss to describe the true beauty and complexity of this nebular variable star. This "slow nova" is filled with all the wonders that we "Northerners" can only dream about...

Wednesday, December 31

Today is the birthday of Robert G. Aitken (Figure 12.66). Born in 1864, Aitken was a prolific American observer who discovered and catalogued more than 3100 double and binary stars. Just look at what a prolific observer *you* have become in just a year! You've learned lunar features, explored double and variable stars, and have located countless clusters, nebulae, and galaxies! Even if you've had cloudy nights, you might be pleasantly surprised at just how much you've done. Try celebrating this accomplishment and the end of the year in a unique way... Go observing!

In the hours before midnight, you could take a cosmic journey which spans millions of light-years. In the Northern Hemisphere, visit with the Andromeda Galaxy again—or the Small and Large Magellanic Clouds if you live in the South.

Figure 12.66. Robert Aitken (Credit—The Mary Lea Shane Archives/Lick Observatory).

Figure 12.67. NGC 2419 (Credit—Palomar Observatory, courtesy of Caltech).

Feast your eyes on vast and wondrous displays of stars like the Double Cluster in Perseus, or the Jewel Box—the Kappa Crucis star cluster. Rejoice in the birth of new stars by voyaging to the Orion Nebula...and remember the old by returning to the Crab Nebula. Take delight at peeking in on Saturn's rings or scanning the horizon for Mars before sunrise. Perhaps the ISS will make a pass over your area tonight, or maybe only a single star will shine through a cloudy sky. It may be something as spectacular as watching a meteor go down in a blaze of glory, or as quiet and contemplative as watching the earthshine-lit Moon sink below the horizon as night falls.

We've become Intergalactic Wanderers, you and I. On this night 220 years ago, Sir William Herschel discovered tonight's destination—a very fitting one for all of us who love nothing more than to contemplate the mysteries of the Cosmos. Located in the constellation of Lynx, but most easily found by hopping about seven degrees north of Castor, the globular cluster NGC 2419 is so distant that it was once classed as an extra-galactic object (Figure 12.67). Located about 182,000 light-years from Earth and 210,000 light-years from the galactic center, it is well outside the halo where almost all globulars reside. Even though you might think this object impossible for a smaller telescope, it's not. While you can't expect resolution, you can still admire the faint glow of the "Intergalactic Wanderer!"

Now take a moment to look up at the stars and think about all the billions of years they have been in the making and all the time it has taken for their light to reach us. Salute! Our observing year has been wonderful together...

May all your journeys be at light speed!

Glossary

A

Absolute Magnitude

The brightness any given star would have if viewed 10 parsecs from Earth

Absorption Nebula

A dark nebula; one as seen in silhouette

Accretion

An accumulation; dust or gas circling larger bodies such as stars, planets or moons; a disc of particles around an existing body

Albedo

The measure of reflectivity of an object; a value expressed in a ratio of light reflected as compared to incident light

Analemma

The painstaking photographic process of recording the Sun's position as seen from the same vantage point, at the same time of day, over a period of a year

Annular Eclipse

A variation of a solar eclipse which leaves a full or partial ring around the Sun

Aphelion

An object's furthest point from the Sun during its orbit

Aperture

Term applied to the size of the light gathering element in either a telescope or binoculars

Apparent Magnitude

Brightness factor of any object as seen visually

Apparition

A period of time which a planet, comet, etc. is visible

Arcminute

A unit of astronomical measurement; 1/60th of a degree;

Arcsecond

A unit of astronomical measurement; 1/60th of an arcminute; 1/360 of a degree;

Asterism

A recognizable pattern of stars

Asteroid

An orbiting minor planet; a small, rocky planetoid with a diameter of less than 1,000 kilometers

Asteroid Belt

A large number of minor planets which orbit between Mars and Jupiter

Astronomical Unit (AU)

The average distance from the Earth to the Sun (149,598,550 kilometers)

Astrometry

The precise measurement of an object's distance and motion

Atmosphere

The layer of gas which surrounds an astronomical body; e.g., a planet

Aurora

A magnetic, electric, and solar phenomenon causing a variety of light phenomena visible around the Earth

Averted Vision

An observing technique for increasing contrast of faint visual objects by using the more light sensitive area of the retina at its edges

B

Barnard Nebulae

A classification catalog of dark nebulae as studied by E. E. Barnard

Bayer Designation

The Greek letters assigned to the stars in a constellation, usually in descending order of brightness; the stellar catalog entries of Johannes Bayer

Belts

Dark bands of clouds which show the differences in temperate zones; e.g., Jupiter

Binary Star

A multiple star system whose members are gravitationally bound to each other

Blackbody

An area which absorbs all incident of radiation, regardless of wavelength

Black Hole

A region of space where gravity is so intense that nothing can escape—not even light

Bok Globule

An area of dark dust spanning no more than one light year

Brown Dwarf

A failed star; one with too little mass to ignite nuclear fusion; energy released is by gravitational contraction

Cassini Division

The wide separation between Saturn's A and B rings

Celestial Equator

The imaginary circle which resides above Earth's equator on the celestial sphere

Celestial Longitude

Measured in degrees hours and minutes along the ecliptic; e.g. right ascension

Celestial Pole

The point in the sky where Earth's poles intersect the celestial sphere

Celestial Sphere

The imaginary sphere around the Earth showing the positions of the celestial bodies

Cepheid Variable

A star which varies in brightness in the manner of Delta Cephei

Circumpolar

A star, constellation, or object which from the viewpoint of the observer is viewable all year

Comet

A rocky, icy body which orbits the solar system in an expanded ellipse not necessarily in the ecliptic plane

Conjunction

The alignment of two celestial bodies occurring when they reach the same celestial longitude; a close visual pairing

Constellation

One of 88 regions of the sky divided by borders; asterisms of bright stars which represent mythological figures

Cosmic ray

An extremely fast, an energetic particle

Cosmos

The Universe; derived from Greek meaning "everything"

Crescent

One of the phases of the Moon or the inner planets; descriptive of appearance

Culmination

The point where an object reaches its highest point in the visible sky

D

Danjon Scale

The scale by which the apparent darkness of a lunar eclipse is judged

Dark Adaptation

An observer's term meaning time to allow the eyes to adjust to lower light conditions

Dark Nebula

An area of un-illuminated gas or dust

Declination

The angular distance (in degrees) of a point either north or south of the celestial equator

Deep Sky Object

A slang term (DSO) given to an observed astronomical subject beyond our solar system

Deep Space

The region outside our solar system

Double Star

Two stars appearing close together in the sky; either a random optical alignment or a gravitationally interacting pair

Dustlane

A thin disc or line of light obscuring material located in another galaxy

Dwarf Star

A main sequence star too small to be classified as a giant or supergiant; e.g. our Sun is a yellow dwarf star

E

Earthshine

The visible portion of the Moon not lit by direct sunlight; the darker portion appearing grey or blue-grey

Eclipse

When one celestial body passes in front of another, from the viewpoint of an observer

Eclipsing Binary

A multiple star system where one stellar body passes in front of the other from our viewpoint

Ecliptic

The apparent path the Moon, Sun and planets take across the visible sky

Egress

Term given to the movement of an astronomical object when leaving a specified area

Elliptical

An elongated oval; e.g., an "elliptical" orbit; elliptical galaxy structure

Elongation

The angle between the inner planets and the Sun as seen from Earth's viewpoint

Emission Nebula

A region of gas illuminated by the energy absorbed from a nearby star

Equinox

The time when the Sun crosses the celestial equator denoting a change in seasons; e.g. Spring "Vernal", Fall "Autumnal"; times of equal hours of day and night

Eyepiece

The viewing lens of a telescope

Faint Fuzzy

Astronomical slang for an unresolved deep sky object

Field of View

The term (acronym: "FOV") given to describe the amount of sky as seen through any particular telescope eyepiece or pair of binoculars

Filament

A long strand; descriptive of nebulae; a prominence in solar terminology

Finderscope

A small, low-power telescope mounted to, and aligned with, a larger telescope to help locate specific stars or areas of the sky

Fireball

Slang term given to any meteor brighter than the visual magnitude of Venus

Flamsteed Number

A catalog number assigned to specific stars in order of right ascension

Galactic

Pertaining to our galaxy; the Milky Way

Galactic Cluster

Term given to an irregular group of stars; an open cluster

Galactic Plane

The apparent angle on the sky described by the Milky Way's disc

Galaxy

Expansive star systems containing innumerable individual stars, dust, gases, planets, etc. which are held together by gravity; class types are given by description; e.g., spiral, barred or elliptical

Galilean Moons

Common term for Jupiter's four largest satellites: Io, Europa, Ganymede and Callisto.

Geosynchronous Orbit

Also referred to as geostationary orbit; the orbit of a satellite whose velocity is matched to Earth's rotation

Giant

A star larger and brighter than its similarly aged group members

Gibbous

A phase of the Moon or a planet in which the visible surface is more than half illuminated

Globular Cluster

A spherical cluster of older stars of common origin

Granulation

Term given to describe the appearance of the Sun's surface; caused by convection

Half Moon

Slang term given to either the first or third quarter Moon; disc is half illuminated

Heliocentric

Sun-centered

Hour

Sidereal time elapsed since a given object was on the meridian; expressed as a number in right ascension

Hubble's Law

velocity and distance relationship; describes the expansion of the Universe

IC

A catalog reference as assigned by J.L.E. Dreyer; "Index Catalog"

Inferior Conjunction

When an inner planet's orbit carries it between the Earth and Sun

Infrared

A portion of the electromagnetic spectrum with wavelengths longer than visible to the human eye

Interacting Galaxies

Two or more galaxies sharing common material; tidal forces

Interferometry

The act of combining more than one wavelength to further study an object

Intergalactic

The area between two separate galaxies

Jove

Alternate word for Jupiter

Jovian

Belonging to, or pertaining to Jupiter

Julian day

The number of days which have elapsed since January 1, 4713 BC; used to calculate intervals, avoiding leap years and traditional modern calendar details

Kirkwood Gap

Area between clusters of asteroids

Kuiper Belt

An orbiting group of rocky and/or icy bodies which reside at the outer reaches of our solar system

Libration

The slight changes in the visible face of the moon allowing you to see more along the lunar limb at certain times

Limb

The outer edge of the visible surface of an astronomical object bordered by background space

Light Curve

The amount of light generated by a variable star and plotted on a graph over a period of time

Light Pollution

The overpowering glow of a terrestrial light source; reduces visibility of faint objects

Light-Year

The distance light travels in the period of a year; 9,460,000,000,000 kilometers

Luminosity

The total brightness of a star or galaxy

Lunar

Pertaining to the Moon

Lunar Eclipse

When the Earth passes between the Sun and Moon; Earth's shadow on the Moon

Lyrae-Type

A kind of variable star

Magnitude

A logarithmic scale used to denote the optical brightness of a star or celestial object; smaller numbers are brighter while larger numbers are fainter; 6th magnitude is roughly 100 times fainter than first magnitude

Main Sequence

Stars undergoing nuclear fusion; represented on a graph of stellar temperatures versus stellar brightness as a band

Mare

A vast lava plain on the surface of the Moon; plural: maria

Maximum

A time when a variable star reaches its brightest point; plural: maxima

Meridian

The imaginary north and south line passing through the zenith of the visible sky

Messier Object

Any object belonging to the catalog of Charles Messier; abbreviation "M" with corresponding number

Meteor

A piece of debris entering our atmosphere; slang "shooting star"

Meteor Shower

A designated time period during which many meteors can be seen

Minimum

A time when a variable star reaches its faintest point; plural: minima

Mira-Type

A type of variable star similar to Mira

Moondog

An atmospheric phenomenon appearing like a rainbow patch near the Moon

Multiple Star

A gravitationally bound star system containing more than two members

N

Nebula

Any cloud of gas or dust in space; plural: nebulae

Nebulosity

The faint presence of an illuminated nebula

Neutron Star

A collapsed star; a remnant which only contains densely packed neutrons

NGC

Abbreviated term given to the "New General Catalog" objects; compiled by J.L.E. Dreyer

Nova

A star which can suddenly erupt to epic brightness

Observation

The act of observing an astronomical object; notes on a visual object

Occultation

When one celestial body passes in front of another, obscuring it from view

Open Cluster

A group of stars with a finite number which are gravitationally bound and move at the same speed; also known as a "galactic" cluster

Opposition

The point where a planet appears opposite the Sun in the visible sky

Orbit

The path any celestial object takes around another due to the influence of gravity

P

Parallax

The apparent change in a nearer object's position when measured against a more distant object

Parhelia

Atmospheric phenomenon resulting in a rainbow-like appearance around the Sun or Moon; a full or partial ring

Parsec

The distance at which one Astronomical Unit appears to cover one second of arc; about 3.26 light-years

Penumbra

The part of an eclipse (solar or lunar) where a portion of the Sun's limb is still visible; solar term used to denote outer region of a sunspot

Perihelion

The point in which an orbiting body is closest to the Sun

Perturbation

The change in orbital pattern of an object when influenced by another gravitational source

Phase

The variances in illumination of the Moon or planets

Photon

A particle (quantum) of light; slang for observing; e.g., "collecting photons"

Planetary Nebula

A shell of gas expelled from a dying star equal to its mass

Population I Stars

Younger stars with high chemical and metal abundances

Population II Stars

Older stars which have shed their heavy elements

Pre-Main Sequence

A star which has not yet began the process of nuclear fusion

Primary Star

The brightest of two or more components in a multiple system; often referred to as the "A" star

Proper Motion

The apparent motion of an object across the sky in relation to the background stars

Pulsar

A rotating neutron star which emits radio waves

Quantum

A fundamental unit of energy

Quark

One of the constituent particles of electrons, protons, etc.

Quasar

A compact radio source which displays a stellar appearance, but has an extremely large redshift

R

Radiant

Area of the sky from which meteors appear to originate

Radiation

Energy transmitted through space as waves

Radio Astronomy

The study of celestial bodies via the means of the radio waves they emit and absorb

Red Giant

A massive dying star; occurring at the end of a star's main sequence period

Redshift

The apparent motion of the spectral lines towards the red end of the spectrum; caused by relative velocity lengthening the light's wavelengths as the object and viewer recede from each other

Reflection Nebula

An area of gas or dust which reflects a nearby star's light

Resolution

The act of resolving an object; e.g. globular clusters, open clusters and double stars

Resolving Power

The ability of an optical aid to show two or more close objects as separate entities.

Retrograde

The apparent backwards or contrary motion of a planet across the visible sky as the Earth overtakes its orbit

Revolution

The orbit of a planet around a star; or of one body around another

Right Ascension

The angle of an object around the celestial equator as measured from the vernal equinox; expressed in hours, minutes and seconds; abbreviated RA

Rille

A long, narrow valley or ridge on the lunar surface

Rotation

The spinning of a body on its own axis

Satellite

Any small object orbiting a larger one; a moon of a planet; an artificial orbiting body

Secondary Star

The less bright companion of a double star; also referred to as the "B" star

Seeing

Slang term applied to steadiness and clarity of the atmosphere

Separation

The angular distance between components of a double star; usually expressed in terms of arcminutes

Shooting Star

Slang for meteor; e.g. "shooter"

Sidereal Time

Time required for a moon or planet to complete one revolution around its host body relative to the background stars

Skyglow

Slang term referring to light pollution, usually from major cities

Solar Flare

A violent release of energy from the Sun which releases radiation and charged particles; e.g. a "coronal mass ejection" (CME)

Solar System

Our Sun and everything that orbits it; e.g. planets, moons, comets, asteroids, etc.

Solar Wind

A high speed stream of energized particles flowing from the Sun

Solstice

Position or time of year when the Sun reaches its northernmost or southernmost declination

Spectrum

Wavelengths of light expressed in colors; plural: spectra

Spectral line

A wavelength where the intensity is greater than (emission line) or less than (absorption line) neighboring values

Spectroscopy

The analysis of light; using absorption lines to determine chemical properties; determining an object's motion and velocity

Starhop

Slang term used to describe a method of moving from star to star to locate a faint object

Star Cloud

An apparent group of stars which are independent of each other

Star Cluster

A group of stars which share the same relative motion

Star Field

Stars which are apparent in any given field of view

Sun Dog

An atmospheric phenomena which resembles a rainbow patch near the Sun

Sunspot

A darker, cooler area of the visible solar surface; highly magnetized

Supergiant Star

The largest known

Superior Conjunction

The point in an inner planet's orbit when the Sun is between it and the Earth

Supernova

The massive and violent explosion of a star; the expulsion of nearly a star's entire mass; may briefly outshine an entire galaxy in magnitude

Supernova Remnant

The shell of debris from an exploded star; rich in heavy elements

Surface Brightness

The total value of light spread out over a large area; the magnitude of an extensive object as compared to the background sky

Terminator

The edge of a lighted area; such as appears on the lunar or planetary surface

Trail

The visible path left in the sky; wake of a meteor; also known as a train

Transient Lunar Phenomenon

An unexpected brightening, outgassing, or other change on the lunar surface

Transit

The passing of an orbiting body over its parent disc; e.g., Mercury or Venus crossing the Sun as seen from Earth; a Galilean moon passing in front of Jupiter from our viewpoint; the moment a celestial body crosses the meridian

Transparency

Slang term to denote presence or absence of thin cloud cover

Ultraviolet

Frequency of the electromagnetic spectrum where the wavelengths are shorter than the bluest light visible to the human eye

Umbra

The inner part of an eclipse shadow; the inner dark portion of a sunspot

Universal Time

(UT)—astronomical time that coordinates observers around the world; based on Greenwich Mean Time (GMT)

Variable Nebula

A nebula which periodically changes in brightness

Variable Star

A star which periodically changes in brightness or magnitude

Wavelength

The peaks and troughs of any emission across the electromagnetic spectrum

White Dwarf

A small, hot star; the core remnant of a red giant

Wilson Effect

The visual magnetic warping of a sunspot's appearance as it approaches the solar limb

X-ray

A band of the electromagnetic spectrum invisible to the human eye

X-ray Binary

A binary star system whose members orbits so closely they can only be distinguished by the changes in their X-ray spectra

Yellow Star

An ordinary star (such as our Sun) at a stable point in its evolution

Young

Slang term used to denoted new features, such as on a planet; a relatively newly evolved star; a nebula or star cluster less aged than its neighbors

Zenith

The area directly overhead in the visible dome of the sky

Zodiac

The constellations which are situated on the ecliptic plane

Zodiacal Light

Slang term given to the visible cone of light which sometimes appears before or after sunrise or sunset; possible solar accretion disc remnants

Zones

Bright bands in cloud layers visible on the planets

Resources

If I were to pray for a taste which would stand by me under every variety of circumstances, and be a source of happiness and cheerfulness to me through life, and a shield against its ills, however things might go amiss, and the world frown upon me, it would be a taste for reading.

—Sir William Herschel

While there are many great resources out there to help you along your way to enjoying the hobby of astronomy, here are a few I think you'll find very useful!

www.lunar-occultations.com/iota
This is the International Occultation and Timing Association (IOTA) site. The accurate information they provide for viewers around the globe will prove invaluable.

www.heavens-above.com
This site is easy to use, concise, and offers perfect information and charts for viewing satellites and bright comets.

www.spaceweather.com
Spaceweather will keep you up to date on solar and auroral events and many other things. This is definitely a good tool!

www.astroleague.org
The Astronomical League website offers many fine observing programs for the amateur astronomer, and offers a wealth of resources, including links to current information on comets.

www.fourmilab.ch/yoursky
Yoursky is a wonderful interactive planetarium tool that allows you to create customized maps and sky views specific to your location.

www.skyandtelescope.com
Sky and Telescope magazine offers terrific online tips, charts, and information, as well as many articles for the amateur astronomer.

www.astronomy.com
Astronomy magazine also has online resources, observing tips, and more.

www.lpl.arizona.edu/alpo
The Association of Lunar and Planetary Observers (ALPO) has a comprehensive, educational, and very useful website.

www.aavso.org
The American Association of Variable Star Observers (AAVSO) will offer you the kind of information you need to study variable stars.

www.lunarrepublic.com
Lunar Republic offers up some of the very finest online reference materials available.

www.cfa.harvard.edu/iau/Ephemerides/
The IAU's Minor Planet Center offers excellent ephemerides for locating currently observable comets, asteroids, and a variety of other objects.

www.soho.com
The definitive site for studying solar phenomena!

Although there are many more terrific websites out there, I think you'll find the ones listed here some of your most often used "tools." Be sure to visit your local library as well, where you'll find other great observing books and star charts.

Clear Skies!

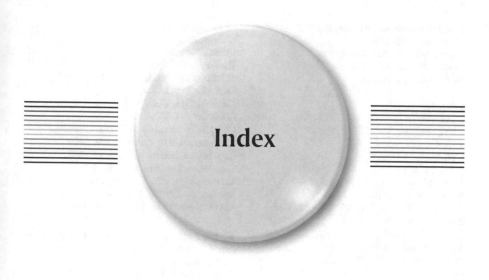

Index

Other Titles in this Series

(Continued from page ii)

Printed in the United States
By Bookmasters